Solutions Manual
Introduction to Algebra

Richard Rusczyk
Art of Problem Solving

Art of Problem Solving®

Books • Online Classes • Videos • Interactive Resources

www.aops.com

Published by: AoPS Incorporated
                  10865 Rancho Bernardo Ste 100
                  San Diego, CA 92127
                  (858) 675-4555
                  books@artofproblemsolving.com

ISBN: 978-1-934124-15-4

Art of Problem Solving® is a registered trademark of AoPS Incorporated.

Visit the Art of Problem Solving website at http://www.artofproblemsolving.com

Cover image designed by Vanessa Rusczyk using KaleidoTile software.

Cover includes the photograph "Manzanita Bark" by Vanessa Rusczyk.

Printed in the United States of America.

Second Edition. Printed in 2019.

Foreword

This book contains the full solution to every Exercise, Review Problem, and Challenge Problem in the text *Introduction to Algebra*.

In most problems, the final answer is contained in a box, $\boxed{\text{like this}}$. However, we strongly recommend against just looking up the final answer and moving on to the next problem. Instead, even if you got the right answer, read the solution in this book. It might show you a different way of solving the problem that you might not have thought of.

If you don't understand a solution, or you think you have a better way of solving the problem, or (gasp!) find an error in one of our solutions, we invite you to come to our message board at

www.artofproblemsolving.com

and discuss it. Our message board is free to use and includes thousands of the world's most eager mathematical problem-solvers.

Also, if you find any errors or typos in the text or solutions, please email them to us at

books@artofproblemsolving.com

NOTE: In many solutions in this solution manual, we use the symbol "$\Rightarrow$" to connect a string of equations in which each step follows in a clear manner from the previous equation. For example, a full solution to the equation $2x - 7 = 9$ might look like this:

$$2x - 7 = 9 \quad \Rightarrow \quad 2x = 16 \quad \Rightarrow \quad x = 8.$$

Contents

CHAPTER 1

Follow the Rules

Exercises for Section 1.2

1.2.1 In each part we follow the order of operations.

(a) Exponentiation, then multiplication, then addition:

$$3^2 + 4 \times 2 = 9 + 4 \times 2 = 9 + 8 = \boxed{17}.$$

(b) We evaluate the expressions within the parentheses first.

$$(5 - 8) \times (2 + 7) = (-3) \times (9) = \boxed{-27}.$$

(c) We must follow the order of operations within the parentheses:

$$(3^3 - 5^2) \times 5 - 8 = (27 - 25) \times 5 - 8 = 2 \times 5 - 8 = 10 - 8 = \boxed{2}.$$

(d) Parentheses, then division, then addition:

$$8/(6 - 2) + 5 = 8/4 + 5 = 2 + 5 = \boxed{7}.$$

(e) Exponentiation, then division and multiplication, then addition:

$$8^2/4^2 + 3 \times 4 = 64/16 + 3 \times 4 = 4 + 12 = \boxed{16}.$$

(f) We first take care of the expression in the exponent, then perform the exponentiation, then the multiplication:

$$11 \times 6^{(2^2-3)} = 11 \times 6^{(4-3)} = 11 \times 6^1 = 11 \times 6 = \boxed{66}.$$

Exercises for Section 1.3

1.3.1 We first write $63 - 27$ as $63 + (-27)$. Now we can use the commutative property of addition to write

$$63 + (-27) = (-27) + 63 = -27 + 63.$$

This is choice $\boxed{(D)}$.

1.3.2 In both parts, we use the commutative and associative properties of addition to reorder the terms in a way that makes computation easier.

(a) $83 - 27 - 81 = 83 - 81 - 27 = 2 - 27 = \boxed{-25}$.

(b) $273 - 8198 - 274 + 8200 = 273 - 274 - 8198 + 8200 = (273 - 274) + (-8198 + 8200) = -1 + 2 = \boxed{1}$.

1.3.3 We use the commutative and associative properties of multiplication to reorder the terms in a way that makes computation easier.

(a) $63 \times \dfrac{2}{7} \times \dfrac{2}{63} = 63 \times \dfrac{2}{63} \times \dfrac{2}{7} = \left(63 \times \dfrac{2}{63}\right) \times \dfrac{2}{7} = 2 \times \dfrac{2}{7} = \boxed{\dfrac{4}{7}}$.

(b) $\dfrac{1}{4} \times 48 \times 97 \times \dfrac{1}{12} = \dfrac{48}{4} \times 97 \times \dfrac{1}{12} = 12 \times 97 \times \dfrac{1}{12} = \left(12 \times \dfrac{1}{12}\right) \times 97 = 1 \times 97 = \boxed{97}$.

1.3.4 $\boxed{\text{Yes!}}$ If the two numbers are the same, we can reverse them, since the difference in both cases is 0. For example, $5 - 5 = 5 - 5$.

Exercises for Section 1.4

1.4.1

(a)
$$
\begin{aligned}
6 \times (3 + 5) &= & 6 \times 8 &= & 48 \\
6 \times (3 + 5) &= & 6 \times 3 + 6 \times 5 &= & 18 + 30 &= & 48
\end{aligned}
$$

(b)
$$
\begin{aligned}
(4 + 8) \times (-2) &= & 12 \times (-2) &= & -24 \\
(4 + 8) \times (-2) &= & 4 \times (-2) + 8 \times (-2) &= & -8 - 16 &= & -24
\end{aligned}
$$

(c)
$$
\begin{aligned}
7 \times (5 - 2) &= & 7 \times 3 &= & 21 \\
7 \times (5 - 2) &= & 7 \times 5 - 7 \times 2 &= & 35 - 14 &= & 21
\end{aligned}
$$

(d)
$$
\begin{aligned}
(8 - 13) \times (-3) &= & (-5) \times (-3) &= & 15 \\
(8 - 13) \times (-3) &= & 8 \times (-3) - (13) \times (-3) &= & -24 + 39 &= & 15
\end{aligned}
$$

1.4.2 Richard forgot about the negative sign before the 3. He should have written

$$(-2) \times (5 - 3) = (-2) \times 5 + (-2) \times (-3) = -10 + 6 = -4,$$

or $(-2) \times (5 - 3) = (-2) \times 5 - (-2) \times (3) = -10 + 6 = -4$.

1.4.3

(a) $88 + 16 = \boxed{4 \times (22 + 4)}$.

(b) $400 - 32 = \boxed{4 \times (100 - 8)}$.

(c) $-24 + 16 + 72 = \boxed{4 \times (-6 + 4 + 18)}$.

(d) $92 - 160 + 36 = \boxed{4 \times (23 - 40 + 9)}$.

1.4.4 We see that each term in the numerator is divisible by 11. So, we factor out the 11, and this factor cancels with the 11 in the denominator:

$$\frac{99 + 88 - 77 + 66}{11} = \frac{11 \times 9 + 11 \times 8 - 11 \times 7 + 11 \times 6}{11} = \frac{\cancel{11} \times (9 + 8 - 7 + 6)}{\cancel{11}} = 9 + 8 - 7 + 6 = \boxed{16}.$$

1.4.5

(a) We have factors of 4 and 7 in both the numerator and denominator, so we can cancel them:

$$\frac{4 \times 6 \times 7}{7 \times 4} = \frac{\cancel{4} \times 6 \times \cancel{7}}{\cancel{7} \times \cancel{4}} = \boxed{6}.$$

(b) At first it doesn't look like we have any common factors, but we can write 27 as 3×9. Then we have a 3 in both the numerator and denominator:

$$\frac{3 \times 8}{27} = \frac{3 \times 8}{3 \times 9} = \frac{\cancel{3} \times 8}{\cancel{3} \times 9} = \boxed{\frac{8}{9}}.$$

Exercises for Section 1.5

1.5.1 The left side of the equation is just $16 - 3 = 13$. The right side equals

$$2 + 4 \times 3 + 2 - 4 - 1 = 2 + 12 + 2 - 4 - 1 = 11.$$

Uh-oh. Stanley did something wrong! The left side is fine. The right side is a problem. When we subtract $4 - 1$ from $2 + 4 \times 3 + 2$, we have

$$2 + 4 \times 3 + 2 - (4 - 1) = 2 + 4 \times 3 + 2 + (-1) \times (4 - 1) = 2 + 4 \times 3 + 2 - 4 + 1.$$

Notice the '+1' at the end; Stanley had '−1' there in his expression. He wasn't careful about keeping track of his signs!

1.5.2 As long as the two sides of the equation don't both equal zero, then their reciprocals are indeed equal. Suppose we have the equation $a = b$. We can divide both sides of this equation by a to get $1 = \frac{b}{a}$. Now we can divide both sides by b to find that

$$\frac{1}{b} = \frac{b}{ab} = \frac{1}{a}.$$

So, if $a = b$, then $\dfrac{1}{a} = \dfrac{1}{b}$. (Again, this only is true if a and b are not equal to 0.)

1.5.3 Because we can take the reciprocals of both sides of an equation to get another equation, we have $1/c = 1/d$. We can then multiply this equation by $a = b$ to get $a/c = b/d$. Therefore, the quotient of the left sides of two equations equals the quotient of the right sides of the equations (as long as we aren't dividing by 0!)

Exercises for Section 1.6

1.6.1

(a) $7^3 = 7 \times 7 \times 7 = 49 \times 7 = \boxed{343}$.

(b) Since 0 times anything is 0, when we multiply 9584 0s, we get $\boxed{0}$.

(c) As discussed in the text, any nonzero number raised to the 0 power is $\boxed{1}$.

(d) $(-10)^5 = (-10) \times (-10) \times (-10) \times (-10) \times (-10) = \boxed{-100,000}$.

(e) $5^{-3} = \dfrac{1}{5^3} = \boxed{\dfrac{1}{125}}$.

(f) $\left(-\dfrac{1}{3}\right)^{-4} = \dfrac{1}{\left(-\frac{1}{3}\right)^4} = \dfrac{1}{\frac{1}{81}} = \boxed{81}$.

1.6.2

(a) We have $2^5 \times 2^9 = 2^{5+9} = 2^{14}$, so $x = \boxed{14}$.

(b) We have $\dfrac{5^9}{5^4} = 5^{9-4} = 5^5$, so $x = \boxed{5}$.

(c) We have $(4^3)^7 = 4^{3 \times 7} = 4^{21}$, so $x = \boxed{21}$.

(d) Since $4 = 2^2$, we have $\dfrac{1}{4} = \dfrac{1}{2^2} = 2^{-2}$, so $x = \boxed{-2}$.

(e) Since $9 = 3^2$ and $27 = 3^3$, we have

$$\frac{9^2 \times 3}{27^2} = \frac{(3^2)^2 \times 3^1}{(3^3)^2} = \frac{3^{2\times 2} \times 3^1}{3^{3\times 2}} = \frac{3^4 \times 3^1}{3^6} = \frac{3^{4+1}}{3^6} = \frac{3^5}{3^6} = 3^{5-6} = 3^{-1},$$

so $x = \boxed{-1}$.

(f) Since $25 = 5^2$, we have $25^3 = (5^2)^3 = 5^{2\times 3} = 5^6$. Also, we have $125 = 5^3$, so we can write the equation as $5^3 = \dfrac{5^6}{5^x}$. In order to leave 5^3 on the right, we must cancel 3 of the 6 5's in the numerator. So, we must have $x = \boxed{3}$.

1.6.3 We can apply the laws of exponents to $(3^a \times 8^2)^5$ to find that $(3^a \times 8^2)^5 = (3^a)^5 \times (8^2)^5 = 3^{a \times 5} \times 8^{2 \times 5} = 3^{5a} \times 8^{10}$. Therefore, our given equation can be written $3^{5a} \times 8^{10} = 3^{15} \times 8^b$. The exponents of the 3s are the same if $5a = 15$, so $a = \boxed{3}$, and the exponents of the 8s are the same if $b = \boxed{10}$.

1.6.4 $\boxed{\text{Yes}}$. We have $(-3)^5 = (-3)\times(-3)\times(-3)\times(-3)\times(-3) = -243$, and $-3^5 = -(3\times 3\times 3\times 3\times 3) = -243$. Recall that in the text, we found that $(-2)^4$ and -2^4 *are not* the same, but here, we see that $(-3)^5$ and -3^5 *are* the same. The difference is that in $(-2)^4$, the exponent is even, while in $(-3)^5$, the exponent is odd. Multiplying an even number of negative numbers gives a positive number, but multiplying an odd number of negative numbers gives a negative number.

1.6.5 When n is a positive odd number, we have a product of an odd number of negative numbers, which gives a negative result. If n is a negative odd number, then the result is the reciprocal of a product of an odd number of negative numbers (since $2^{-p} = \frac{1}{2^p}$), which also gives a negative result.

When n is a positive even number, we have a product of an even number of negative numbers, which gives a positive result. If n is a negative even number, then the result is the reciprocal of a product of an even number of negative numbers (since $2^{-p} = \frac{1}{2^p}$), which also gives a positive result.

Finally, if $n = 0$, then the result is 1, a positive number. So, the result is positive if and only if $\boxed{n \text{ is an even number}}$.

1.6.6

(a) We note that $32 = 2^5$ and $25 = 5^2$, and then we apply the negative exponents:

$$32 \times 25 \times 2^{-2} \times \left(\frac{5}{2}\right)^{-3} = 2^5 \times 5^2 \times \frac{1}{2^2} \times \frac{1}{(5/2)^3} = 2^5 \times 5^2 \times \frac{1}{2^2} \times \frac{1}{5^3/2^3} = 2^5 \times 5^2 \times \frac{1}{2^2} \times \frac{2^3}{5^3}.$$

Grouping the expressions with base 2 and those with base 5, we have

$$2^5 \times 5^2 \times \frac{1}{2^2} \times \frac{2^3}{5^3} = \frac{2^5 \times 2^3}{2^2} \times \frac{5^2}{5^3} = 2^{5+3-2} \times 5^{2-3} = 2^6 \times 5^{-1} = \boxed{\frac{64}{5}}.$$

(b) We note that $8 = 2^3$ and $27 = 3^3$, and then we apply our laws of exponents:

$$(-2^4 \times 3^3)^3 \times \frac{8 \times 27}{(2^3 \times 3^3)^4} = (-2^4)^3 \times (3^3)^3 \times \frac{2^3 \times 3^3}{(2^3)^4 \times (3^3)^4} = (-1 \times 2^4)^3 \times 3^{3\times 3} \times \frac{2^3 \times 3^3}{2^{3\times 4} \times 3^{3\times 4}}$$

$$= (-1)^3 \times (2^4)^3 \times 3^9 \times \frac{2^3 \times 3^3}{2^{12} \times 3^{12}} = -1 \times 2^{4\times 3} \times 3^9 \times 2^{3-12} \times 3^{3-12}$$

$$= -1 \times 2^{12} \times 3^9 \times 2^{-9} \times 3^{-9} = -1 \times 2^{12-9} \times 3^{9-9} = -1 \times 2^3 \times 3^0 = \boxed{-8}.$$

(c) First, we note that $(-40)^7 = (-1 \times 40)^7 = (-1)^7 \times 40^7 = -40^7$. Next, we note that $40 = 2^3 \times 5$, so $40^7 = (2^3 \times 5)^7 = (2^3)^7 \times 5^7 = 2^{21} \times 5^7$. Finally, we have $25 = 5^2$, so

$$\frac{(-40)^7}{2^{30} \times 25^3} = \frac{-40^7}{2^{30} \times (5^2)^3} = \frac{-2^{21} \times 5^7}{2^{30} \times 5^6} = -2^{21-30} \times 5^{7-6} = -2^{-9} \times 5^1 = -\frac{5}{2^9} = \boxed{-\frac{5}{512}}.$$

1.6.7 Since $8 = 2^3$, we can write 8^x as $(2^3)^x$. Applying the laws of exponents gives $(2^3)^x = 2^{3x}$. If this equals 2^1, then it appears that we must have $3x = 1$, which means that $x = \boxed{1/3}$. But what does a fractional exponent mean!?!? Read the next section to find out!

Exercises for Section 1.7

1.7.1

(a) Since $64 = 4^3$, we have $64^{1/3} = (4^3)^{1/3} = 4^{3\times(1/3)} = 4^1 = \boxed{4}$.

(b) Since $100^2 = 10000$, we have $10000^{1/2} = \boxed{100}$.

(c) Since $6^3 = 216$, we know that $(-6)^3 = -216$, so $(-216)^{1/3} = \boxed{-6}$.

(d) We first find the prime factorization of 7056 as $2^4 \times 3^2 \times 7^2$. We then have

$$7056^{1/2} = (2^4 \times 3^2 \times 7^2)^{1/2} = (2^4)^{1/2} \times (3^2)^{1/2} \times (7^2)^{1/2} = 2^{4 \times (1/2)} \times 3^{2 \times (1/2)} \times 7^{2 \times (1/2)} = 2^2 \times 3 \times 7 = \boxed{84}.$$

1.7.2 0 is the number whose square is 0, so $0^{1/2} = \boxed{0}$.

1.7.3

(a) Since $4 = 2^2$, we have $4^{5/2} = (2^2)^{5/2} = 2^{2 \times (5/2)} = 2^5 = \boxed{32}$.

(b) We have $(-1)^{36/5} = [(-1)^{36}]^{1/5}$. Multiplying an even number of -1s gives 1, so $[(-1)^{36}]^{1/5} = 1^{1/5} = \boxed{1}$.

(c) Since $27 = 3^3$, we have $27^{5/3} = (3^3)^{5/3} = 3^{3 \times (5/3)} = 3^5 = \boxed{243}$.

(d) First, we deal with the negative exponent by noting that

$$(-8)^{-4/3} = \frac{1}{(-8)^{4/3}}.$$

Next, since $(-2)^3 = -8$, we have $(-8)^{1/3} = -2$. Therefore, we have

$$(-8)^{-4/3} = \frac{1}{(-8)^{4/3}} = \frac{1}{[(-8)^{1/3}]^4} = \frac{1}{(-2)^4} = \boxed{\frac{1}{16}}.$$

1.7.4

(a) $\boxed{\text{Yes}}$. There is a nonnegative number whose fourth power is 4, but it is not a whole number. Since $1.4^4 \approx 3.84$ and $1.5^4 \approx 5.06$, it appears that the number equal to $4^{1/4}$ is between 1.4 and 1.5, and probably is closer to 1.4. Using a calculator, we can find that the number whose fourth power is 4 is approximately 1.414.

(b) $\boxed{\text{No}}$. The fourth power of a positive number is positive. The fourth power of a negative number is also positive. So, there is not a positive or negative number whose fourth power is -4.

1.7.5

(a) $(5^{1/3})^2 \times 5^{4/3} = 5^{(1/3) \times 2} \times 5^{4/3} = 5^{2/3} \times 5^{4/3} = 5^{(2/3)+(4/3)} = 5^2 = \boxed{25}$.

(b) $(-2^{1/3})^9 = (-1 \times 2^{1/3})^9 = (-1)^9 \times (2^{1/3})^9 = -1 \times 2^{(1/3) \times 9} = -2^3 = \boxed{-8}$.

(c) Since $12 = 2^2 \times 3$, we have

$$(12^{1/2})^{3/2} \times 3^{1/4} \times 2^{1/2} = 12^{(1/2) \times (3/2)} \times 3^{1/4} \times 2^{1/2} = 12^{3/4} \times 3^{1/4} \times 2^{1/2} = (2^2 \times 3)^{3/4} \times 3^{1/4} \times 2^{1/2}$$

$$= (2^2)^{3/4} \times 3^{3/4} \times 3^{1/4} \times 2^{1/2} = 2^{2 \times (3/4)} \times 3^{(3/4)+(1/4)} \times 2^{1/2} = 2^{3/2} \times 3^1 \times 2^{1/2}$$

$$= 2^{(3/2)+(1/2)} \times 3^1 = 2^2 \times 3^1 = \boxed{12}.$$

(d) Since $8 = 2^3$, we have $8^2 = (2^3)^2 = 2^{3 \times 2} = 2^6$, so

$$\frac{(8^2 \times 5^3)^{3/5}}{2^{3/5} \times 5^{-1/5}} = \frac{(2^6)^{3/5} \times (5^3)^{3/5}}{2^{3/5} \times 5^{-1/5}} = \frac{2^{6 \times (3/5)} \times 5^{3 \times (3/5)}}{2^{3/5} \times 5^{-1/5}} = \frac{2^{18/5} \times 5^{9/5}}{2^{3/5} \times 5^{-1/5}}$$

$$= 2^{(18/5)-(3/5)} \times 5^{(9/5)-(-1/5)} = 2^3 \times 5^2 = 8 \times 25 = \boxed{200}.$$

Exercises for Section 1.8

1.8.1

(a) Since $11^2 = 121$, we have $\sqrt{121} = \boxed{11}$.

(b) We have $\sqrt{36000} = \sqrt{36} \times \sqrt{100} \times \sqrt{10} = 6 \times 10 \times \sqrt{10} = \boxed{60\sqrt{10}}$.

(c) First, we find the prime factorization of 1323 as $3^3 \times 7^2$, so we have $\sqrt{1323} = \sqrt{3^3 \times 7^2} = \sqrt{3^3} \times \sqrt{7^2} = \sqrt{3^2} \times \sqrt{3} \times 7 = 3 \times \sqrt{3} \times 7 = \boxed{21\sqrt{3}}$.

(d) If we don't notice immediately that $6.76 = 2.6^2$, we can also write 6.76 as a fraction: $6.76 = \frac{676}{100} = \frac{169}{25}$. Now, finding the square root is much easier: $\sqrt{6.76} = \sqrt{\frac{169}{25}} = \frac{\sqrt{169}}{\sqrt{25}} = \frac{13}{5} = \boxed{2.6}$.

(e) Since $\sqrt{2} = 2^{1/2}$, we have $\left(\sqrt{2}\right)^8 = (2^{1/2})^8 = 2^{(1/2)\times 8} = 2^4 = \boxed{16}$.

(f) $(-\sqrt{27})^3 = (-1 \times \sqrt{3^3})^3 = (-1)^3 \times (3^{3/2})^3 = -3^{9/2} = -3^4 \times 3^{1/2} = \boxed{-81\sqrt{3}}$.

1.8.2

(a) $\sqrt[3]{80} = \sqrt[3]{8 \times 10} = \sqrt[3]{8} \times \sqrt[3]{10} = \boxed{2\sqrt[3]{10}}$.

(b) Since $3^4 = 81$, we have $\sqrt[4]{81} = \sqrt[4]{3^4} = \boxed{3}$.

(c) $\sqrt[4]{\dfrac{32}{625}} = \dfrac{\sqrt[4]{32}}{\sqrt[4]{625}} = \dfrac{\sqrt[4]{16} \times \sqrt[4]{2}}{\sqrt[4]{5^4}} = \boxed{\dfrac{2\sqrt[4]{2}}{5}}$.

(d) Since $(-0.1)^3 = -0.001$, we have $\sqrt[3]{-0.001} = \boxed{-0.1}$.

(e) $\left(\sqrt[3]{7}\right)^9 = \left(7^{1/3}\right)^9 = 7^{(1/3)\times 9} = 7^3 = \boxed{343}$.

(f) $\sqrt[4]{2^{4/3}} = (2^{4/3})^{1/4} = 2^{(4/3)\times(1/4)} = 2^{1/3} = \boxed{\sqrt[3]{2}}$.

1.8.3 If the prime factorization of n has any primes that are to the third power or higher, we can factor that cubed prime out to simplify $\sqrt[3]{n}$. For example, if $n = 11250 = 2 \times 3^2 \times 5^4$, then $\sqrt[3]{n} = \sqrt[3]{5^3} \times \sqrt[3]{2 \times 3^2 \times 5} = 5\sqrt[3]{90}$. If there are no primes to the third power or higher in the prime factorization of n, then we cannot simplify $\sqrt[3]{n}$.

1.8.4

(a) Since $\sqrt{28} = \sqrt{4} \times \sqrt{7} = 2\sqrt{7}$, $\sqrt{63} = \sqrt{9} \times \sqrt{7} = 3\sqrt{7}$, and $\sqrt{175} = \sqrt{25} \times \sqrt{7} = 5\sqrt{7}$, we have $\sqrt{28} + \sqrt{63} - \sqrt{175} = 2\sqrt{7} + 3\sqrt{7} - 5\sqrt{7} = (2 + 3 - 5)\sqrt{7} = \boxed{0}$.

(b) We have $\sqrt{135} = \sqrt{9} \times \sqrt{15} = 3\sqrt{15}$, $3\sqrt{1500} = 3\sqrt{100} \times \sqrt{15} = 30\sqrt{15}$, and $\sqrt{960} = \sqrt{4} \times \sqrt{240} = 2 \times \sqrt{16} \times \sqrt{15} = 8\sqrt{15}$, so

$$\sqrt{135} - 3\sqrt{1500} + \sqrt{960} = 3\sqrt{15} - 30\sqrt{15} + 8\sqrt{15} = (3 - 30 + 8)\sqrt{15} = \boxed{-19\sqrt{15}}.$$

(c) Since $5^3 = 125$ and $4^3 = 64$, we have $\sqrt[3]{375} = \sqrt[3]{125} \times \sqrt[3]{3} = 5\sqrt[3]{3}$ and $\sqrt[3]{192} = \sqrt[3]{64} \times \sqrt[3]{3} = 4\sqrt[3]{3}$. Therefore, $\sqrt[3]{375} - \sqrt[3]{192} = 5\sqrt[3]{3} - 4\sqrt[3]{3} = \boxed{\sqrt[3]{3}}$.

(d) First, we simplify all three cube roots (and simplify 12/81 to 4/27):

$$\sqrt[3]{\frac{256}{27}} = \frac{\sqrt[3]{256}}{\sqrt[3]{27}} = \frac{\sqrt[3]{2^8}}{3} = \frac{\sqrt[3]{2^3} \times \sqrt[3]{2^3} \times \sqrt[3]{2^2}}{3} = \frac{4\sqrt[3]{4}}{3},$$

$$\sqrt[3]{32} = \sqrt[3]{2^3} \times \sqrt[3]{2^2} = 2\sqrt[3]{4},$$

$$\sqrt[3]{\frac{12}{81}} = \sqrt[3]{\frac{4}{27}} = \frac{\sqrt[3]{4}}{\sqrt[3]{27}} = \frac{\sqrt[3]{4}}{3}.$$

We then have $\sqrt[3]{\frac{256}{27}} + \sqrt[3]{32} - \sqrt[3]{\frac{12}{81}} = \frac{4\sqrt[3]{4}}{3} + 2\sqrt[3]{4} - \frac{\sqrt[3]{4}}{3} = \left(\frac{4}{3} + 2 - \frac{1}{3}\right)\sqrt[3]{4} = \boxed{3\sqrt[3]{4}}$.

1.8.5

(a) We have $\sqrt{3} \times \sqrt[3]{3} = 3^{1/2} \times 3^{1/3} = 3^{(1/2)+(1/3)}$. Since $\frac{1}{2} + \frac{1}{3} = \frac{3}{6} + \frac{2}{6} = \frac{5}{6}$, we have $x = \boxed{5/6}$.

(b) We have $\sqrt{8} \times \sqrt[5]{4} = (2^3)^{1/2} \times (2^2)^{1/5} = 2^{3/2} \times 2^{2/5} = 2^{(3/2)+(2/5)}$. Since $\frac{3}{2} + \frac{2}{5} = \frac{15}{10} + \frac{4}{10} = \frac{19}{10}$, we have $x = \boxed{19/10}$.

1.8.6 We simplify the radicals, and then combine the results:

$$\frac{\sqrt{600} - \sqrt{150} + 3\sqrt{54}}{6\sqrt{32} - 2\sqrt{50} - \sqrt{288}} = \frac{\sqrt{100} \times \sqrt{6} - \sqrt{25} \times \sqrt{6} + 3\sqrt{9} \times \sqrt{6}}{6\sqrt{16} \times \sqrt{2} - 2\sqrt{25} \times \sqrt{2} - \sqrt{144} \times \sqrt{2}}$$

$$= \frac{10\sqrt{6} - 5\sqrt{6} + 9\sqrt{6}}{24\sqrt{2} - 10\sqrt{2} - 12\sqrt{2}} = \frac{(10 - 5 + 9)\sqrt{6}}{(24 - 10 - 12)\sqrt{2}}$$

$$= \frac{14\sqrt{6}}{2\sqrt{2}} = \frac{14}{2} \times \frac{\sqrt{6}}{\sqrt{2}} = 7\sqrt{\frac{6}{2}} = 7\sqrt{3}.$$

Therefore, $\boxed{a = 7 \text{ and } b = 3}$.

1.8.7 Simplifying the radicals gives

$$\frac{-\sqrt{20} + 2\sqrt{245}}{\sqrt{270} - \sqrt{120}} = \frac{-\sqrt{4} \times \sqrt{5} + 2\sqrt{49} \times \sqrt{5}}{\sqrt{9} \times \sqrt{30} - \sqrt{4} \times \sqrt{30}} = \frac{-2\sqrt{5} + 14\sqrt{5}}{3\sqrt{30} - 2\sqrt{30}} = \frac{12\sqrt{5}}{\sqrt{30}}.$$

Uh-oh, that's not in the form $a\sqrt{b}$, where a and b are integers. Even if we combine the radicals into the square root of a fraction, we'll end up with $12\sqrt{5/30} = 12\sqrt{1/6} = 12/\sqrt{6}$, which isn't in the right form, either. However, we can be a little sneaky here, and write $12\sqrt{5}$ as $2 \times \sqrt{36} \times \sqrt{5}$, which equals $2\sqrt{180}$. Aha! Now we can simplify:

$$\frac{12\sqrt{5}}{\sqrt{30}} = \frac{2\sqrt{180}}{\sqrt{30}} = 2\sqrt{\frac{180}{30}} = 2\sqrt{6}.$$

Therefore, $\boxed{a = 2 \text{ and } b = 6}$.

As another solution, consider what happens when you multiply both the numerator and denominator of $12/\sqrt{6}$ by $\sqrt{6}$. Why doesn't this change the value of the fraction? Why did we choose to multiply by $\sqrt{6}$?

CHAPTER **2**

_x_ **Marks the Spot**

Exercises for Section 2.1

2.1.1

(a) $r - 7 = 3 - 7 = \boxed{-4}$.

(b) $-3r = -3(3) = \boxed{-9}$.

(c) $\sqrt{r^2 + (r+1)^2} = \sqrt{3^2 + (3+1)^2} = \sqrt{9 + 4^2} = \sqrt{9 + 16} = \sqrt{25} = \boxed{5}$.

(d) $5r/3 - 9/r = 5 \cdot 3/3 - 9/3 = 15/3 - 3 = 5 - 3 = \boxed{2}$.

2.1.2

(a) $13 - s = 13 - (-4) = 13 + 4 = \boxed{17}$.

(b) $\sqrt{-9s} = \sqrt{-9 \cdot (-4)} = \sqrt{36} = \boxed{6}$.

(c) $-s^2 + 4s - 12 = -(-4)^2 + 4(-4) - 12 = -(16) - 16 - 12 = \boxed{-44}$.

(d) $(s - 8)/3 = (-4 - 8)/3 = (-12)/3 = \boxed{-4}$.

Exercises for Section 2.2

In most of the parts below, we've added considerably more detail than you'll likely have to do. Many of the steps shown can be done in your head once you have more experience.

2.2.1

(a) $(3x - 8) + (5x + 7) = 3x - 8 + 5x + 7 = 3x + 5x - 8 + 7 = \boxed{8x - 1}$.

(b) $(3 - 3x) + (-19x + 27) = 3 - 3x - 19x + 27 = -3x - 19x + 3 + 27 = \boxed{-22x + 30}$.

2.2.2

(a) $t^3 \cdot t^4 = t^{3+4} = \boxed{t^7}$.

(b) $(16x^2)(4x^5) = 16 \cdot x^2 \cdot 4 \cdot x^5 = (16 \cdot 4) \cdot (x^2 \cdot x^5) = \boxed{64x^7}$.

(c) $(y^5)^9 = y^{5 \cdot 9} = \boxed{y^{45}}$.

(d) $(3x^2)^6 = 3^6 \cdot (x^2)^6 = 3^6 \cdot x^{2 \cdot 6} = \boxed{729x^{12}}$.

2.2.3

(a) $\dfrac{p^7}{p^2} = p^{7-2} = \boxed{p^5}$.

(b) $\dfrac{25z^3}{30z^7} = \dfrac{25}{30} \cdot \dfrac{z^3}{z^7} = \dfrac{5}{6} \cdot z^{3-7} = \dfrac{5}{6} \cdot z^{-4} = \boxed{\dfrac{5}{6z^4}}$.

(c) $\dfrac{(4x^3)(2x^5)}{6x^4} = \dfrac{8x^{3+5}}{6x^4} = \dfrac{8x^8}{6x^4} = \dfrac{8}{6} \cdot \dfrac{x^8}{x^4} = \dfrac{4}{3} \cdot x^{8-4} = \dfrac{4}{3} \cdot x^4 = \boxed{\dfrac{4x^4}{3}}$.

(d) $\dfrac{24t^3}{15t^4} \cdot \dfrac{5t^8}{3t^6} = \dfrac{24}{15} \cdot \dfrac{t^3}{t^4} \cdot \dfrac{5}{3} \cdot \dfrac{t^8}{t^6} = \dfrac{8}{5} \cdot \dfrac{1}{t} \cdot \dfrac{5}{3} \cdot t^2 = \left(\dfrac{8}{5} \cdot \dfrac{5}{3}\right) \cdot \left(\dfrac{1}{t} \cdot t^2\right) = \dfrac{8}{3} \cdot t = \boxed{\dfrac{8t}{3}}$.

2.2.4

(a) $3^{-1} = \dfrac{1}{3^1} = \boxed{\dfrac{1}{3}}$.

(b) $5^{-3} = \dfrac{1}{5^3} = \boxed{\dfrac{1}{125}}$.

(c) $\dfrac{2x^{-3}}{x^5} = \dfrac{2 \cdot \frac{1}{x^3}}{x^5} = \dfrac{\frac{2}{x^3}}{x^5} = \dfrac{2}{x^3 \cdot x^5} = \boxed{\dfrac{2}{x^8}}$.

(d) $\dfrac{4}{x^{-4}} = \dfrac{4}{\frac{1}{x^4}} = \boxed{4x^4}$.

2.2.5

(a) $(2t - 7) + (2t - 7) = 2t + 2t - 7 - 7 = \boxed{4t - 14}$.

(b) $(2t - 7) + (2t - 7) + (2t - 7) = 2t + 2t + 2t - 7 - 7 - 7 = \boxed{6t - 21}$.

(c) $(2t - 7) + (2t - 7) + (2t - 7) + (2t - 7) = 2t + 2t + 2t + 2t - 7 - 7 - 7 - 7 = \boxed{8t - 28}$.

Did you notice anything interesting?

2.2.6 We could just successively multiply together r^4's until we have r^{20}, or we could notice that multiplying a bunch of r^4 terms together is the same as raising r^4 to some power. Specifically, if we multiply p copies of r^4 together, we have $(r^4)^p$, which equals r^{4p}. We want this to equal r^{20}, so we need a p that makes $4p$ equal to 20. Since $4 \cdot 5 = 20$, we need $p = 5$. Therefore, if we multiply $\boxed{5}$ r^4's together, we will have r^{20}.

2.2.7 No, it is not correct! As we mentioned in the text, we cannot cancel terms in the numerator and denominator when either of these is a sum. Specifically, we cannot cancel the x's in $\frac{5+3x}{x}$. If you're not convinced, take a look at what happens when we let $x = 5$. Then, we have $\frac{5+3x}{x} = \frac{5+3 \cdot 5}{5} = \frac{20}{5} = 4$, which clearly isn't equal to 8!

2.2.8

(a) We have $\sqrt{96t^6} = \sqrt{96} \cdot \sqrt{t^6} = \sqrt{16} \cdot \sqrt{6} \cdot \sqrt{t^6} = 4\sqrt{6} \cdot \sqrt{t^6}$. Since t is positive, we have $\sqrt{t^6} = (t^6)^{1/2} = t^{6(1/2)} = t^3$, so $4\sqrt{6} \cdot \sqrt{t^6} = \boxed{4t^3\sqrt{6}}$.

(b) $\sqrt[3]{125x^{12}} = \sqrt[3]{125} \cdot \sqrt[3]{x^{12}} = 5(x^{12})^{1/3} = 5x^{(12)(1/3)} = \boxed{5x^4}$.

(c) We have $\sqrt[3]{27p^5} = \sqrt[3]{27} \cdot \sqrt[3]{p^3} \cdot \sqrt[3]{p^2} = 3p\sqrt[3]{p^2}$, and we have

$$\frac{2\sqrt[3]{p^8}}{p} = \frac{2\sqrt[3]{p^6} \cdot \sqrt[3]{p^2}}{p} = \frac{2p^2\sqrt[3]{p^2}}{p} = 2p\sqrt[3]{p^2}.$$

Therefore, we have $\sqrt[3]{27p^5} - \dfrac{2\sqrt[3]{p^8}}{p} = 3p\sqrt[3]{p^2} - 2p\sqrt[3]{p^2} = (3p - 2p)\sqrt[3]{p^2} = \boxed{p\sqrt[3]{p^2}}$, which we can also write as $p \cdot p^{2/3} = p^{1+(2/3)} = \boxed{p^{5/3}}$.

Exercises for Section 2.3

2.3.1

(a) $2(2t - 7) = 2 \cdot (2t) + 2 \cdot (-7) = \boxed{4t - 14}$.

(b) $x(x + 9) = x \cdot x + x \cdot 9 = \boxed{x^2 + 9x}$.

(c)

$$(x^3 - 2x^2 + x + 1) \cdot (3x^2) = (x^3) \cdot (3x^2) + (-2x^2) \cdot (3x^2) + (x) \cdot (3x^2) + (1) \cdot (3x^2)$$
$$= \boxed{3x^5 - 6x^4 + 3x^3 + 3x^2}$$

2.3.2

(a) $(3x + 7) - (4x + 9) = 3x + 7 - 4x - 9 = 3x - 4x + 7 - 9 = \boxed{-x - 2}$.

(b) $(r^2 + 3r - 2) - (r^2 + 7r - 5) = r^2 + 3r - 2 - r^2 - 7r + 5 = r^2 - r^2 + 3r - 7r - 2 + 5 = \boxed{-4r + 3}$.

(c) $3(t + 7) - t(t + 9) = 3t + 3 \cdot 7 - t^2 - 9t = 3t + 21 - t^2 - 9t = -t^2 + 3t - 9t + 21 = \boxed{-t^2 - 6t + 21}$.

2.3.3

(a) Both terms have a factor of 6, so $12a - 18 = 6 \cdot (2a) - 6 \cdot (3) = \boxed{6(2a - 3)}$.

(b) Both terms have a factor of x, so $7x^2 - 30x = x \cdot (7x) - x \cdot (30) = \boxed{x(7x - 30)}$.

(c) Every term has a factor of 4, so $-8t^2 + 4t + 16 = 4 \cdot (-2t^2) + 4 \cdot t + 4 \cdot 4 = \boxed{4(-2t^2 + t + 4)}$.

(d) Both 9 and z are factors of each term, so we can factor out $9z$:

$$9z^3 - 27z^2 + 27z = (9z) \cdot (z^2) - (9z) \cdot (3z) + (9z) \cdot 3 = \boxed{9z(z^2 - 3z + 3)}.$$

Notice that in each part we are careful to keep track of the positive and negative signs.

2.3.4

(a) If you put 4 into the machine, the machine adds 8 to get 12, then multiplies this by 6 to get 72. It then subtracts 12 from this to get 60, and divides this by 2 to get 30. Then the machine adds 18 to your number to get 22 and subtracts this from the 30 it had earlier, to get $30 - 22 = 8$.

If you put 6 into the machine, it adds 8 to get 14. Then it multiplies by 6 to get 84, then subtracts 12 to get 72. It then divides this by 2 to get 36 and subtracts $6 + 18 = 24$ from 36 to get 12. Hmmm. Maybe the machine will always double your initial number.

(b) If you put x into the machine, the machine will add 8 to get $x + 8$, then multiply by 6 to get $6(x + 8)$. The machine will subtract 12 from this product to get $6(x + 8) - 12$. We can simplify this expression by expanding the product $6(x + 8)$:

$$6(x + 8) - 12 = 6x + 48 - 12 = 6x + 36.$$

The machine will then divide this by 2 to get $(6x + 36)/2$. We can simplify this fraction because we can factor the numerator:

$$\frac{6x + 36}{2} = \frac{6(x + 6)}{2} = 3(x + 6) = 3x + 18.$$

From this, the machine will subtract 18 more than x:

$$3x + 18 - (x + 18) = 3x + 18 - x - 18 = 3x - x + 18 - 18 = 2x.$$

Indeed, the machine is just a very fancy way to double the number put into it.

2.3.5 Dividing by 2 is the same as multiplying by 1/2, so this is just the distributive property:

$$\frac{2x + 48}{2} = (2x + 48) \cdot \left(\frac{1}{2}\right) = (2x) \cdot \left(\frac{1}{2}\right) + (48) \cdot \left(\frac{1}{2}\right) = \frac{2x}{2} + \frac{48}{2}.$$

2.3.6 We can factor the expression $x - 3$ out of each term:

$$2x(x - 3) + 3(x - 3) = 2x \cdot (x - 3) + 3 \cdot (x - 3) = \boxed{(2x + 3)(x - 3)}.$$

If you don't quite see how this works, suppose we put A in place of $x - 3$ everywhere in the original expression. Then we can see the factoring more clearly:

$$2xA + 3A = 2x \cdot A + 3 \cdot A = (2x + 3)A.$$

Putting $x - 3$ back in for A, we have our factorization: $(2x + 3)(x - 3)$.

Exercises for Section 2.4

2.4.1

(a) $\dfrac{-5x^2 + 25}{5x} = \dfrac{5(-x^2 + 5)}{5x} = \dfrac{\cancel{5}(-x^2 + 5)}{\cancel{5}x} = \boxed{\dfrac{-x^2 + 5}{x}}.$

(b) $\dfrac{3r^3 - 21r}{9r^2} = \dfrac{3r(r^2 - 7)}{9r^2} = \dfrac{3}{9} \cdot \dfrac{r}{r^2} \cdot (r^2 - 7) = \dfrac{1}{3} \cdot \dfrac{1}{r} \cdot (r^2 - 7) = \boxed{\dfrac{r^2 - 7}{3r}}.$

(c) $\dfrac{7t}{3t^2 - 8t} = \dfrac{7t}{t(3t - 8)} = \dfrac{7\cancel{t}}{\cancel{t}(3t - 8)} = \boxed{\dfrac{7}{3t - 8}}.$

(d) $\dfrac{3x^2 + 9x}{4x^3 + 12x^2} = \dfrac{3x(x + 3)}{4x^2(x + 3)} = \dfrac{3\cancel{x}\cancel{(x + 3)}}{4\cancel{x} \cdot x \cdot \cancel{(x + 3)}} = \boxed{\dfrac{3}{4x}}.$

2.4.2 Our common denominator is $7x$, so we have

$$\frac{3}{7x} - \frac{6x}{7} = \frac{3}{7x} - \frac{6x}{7} \cdot \frac{x}{x} = \frac{3}{7x} - \frac{6x^2}{7x} = \boxed{\frac{3 - 6x^2}{7x}}.$$

2.4.3 Our denominators are t and 7, so the simplest common denominator is $7t$:

$$\frac{2t}{7} + \frac{9 - 2t}{t} = \frac{2t}{7} \cdot \frac{t}{t} + \frac{9 - 2t}{t} \cdot \frac{7}{7} = \frac{2t^2}{7t} + \frac{63 - 14t}{7t} = \boxed{\frac{2t^2 - 14t + 63}{7t}}.$$

2.4.4 Our denominators are $3t^3$ and $6t^2$, so the simplest common denominator is $6t^3$. We need to multiply $3t^3$ by 2 to get $6t^3$, and we need to multiply $6t^2$ by t to get $6t^3$:

$$\frac{3t - 1}{3t^3} + \frac{5}{6t^2} = \frac{3t - 1}{3t^3} \cdot \frac{2}{2} + \frac{5}{6t^2} \cdot \frac{t}{t} = \frac{6t - 2}{6t^3} + \frac{5t}{6t^3} = \boxed{\frac{11t - 2}{6t^3}}.$$

2.4.5 We can make our denominators the same by multiplying the numerator and denominator of the second fraction by $x - 1$:

$$\frac{3x}{x(x - 1)} + \frac{2}{x} = \frac{3x}{x(x - 1)} + \frac{2}{x} \cdot \frac{(x - 1)}{(x - 1)} = \frac{3x}{x(x - 1)} + \frac{2(x - 1)}{x(x - 1)} = \frac{3x + 2(x - 1)}{x(x - 1)} = \frac{3x + 2x - 2}{x(x - 1)} = \boxed{\frac{5x - 2}{x(x - 1)}}.$$

2.4.6 The denominators of the fractions are z and $z - 1$. We'll need one of each of these factors in our denominator, so our simplest common denominator is $z(z - 1)$. We therefore have:

$$\begin{aligned}
2 + \frac{3}{z} - \frac{z - 2}{z - 1} &= 2 \cdot \frac{z(z - 1)}{z(z - 1)} + \frac{3}{z} \cdot \frac{(z - 1)}{(z - 1)} - \frac{z - 2}{z - 1} \cdot \frac{z}{z} \\
&= \frac{2z(z - 1)}{z(z - 1)} + \frac{3(z - 1)}{z(z - 1)} - \frac{(z - 2)(z)}{z(z - 1)} \\
&= \frac{2z^2 - 2z}{z(z - 1)} + \frac{3z - 3}{z(z - 1)} - \frac{z^2 - 2z}{z(z - 1)} \\
&= \frac{(2z^2 - 2z) + (3z - 3) - (z^2 - 2z)}{z(z - 1)} \\
&= \frac{2z^2 - 2z + 3z - 3 - z^2 + 2z}{z(z - 1)} \\
&= \boxed{\frac{z^2 + 3z - 3}{z(z - 1)}}.
\end{aligned}$$

Review Problems

2.17

(a) $x^2 - 3 = 3^2 - 3 = 9 - 3 = \boxed{6}$.

(b) $3x/4 + 7/4 = 3 \cdot 3/4 + 7/4 = 9/4 + 7/4 = (9 + 7)/4 = 16/4 = \boxed{4}$.

(c) $(2x - 3)(2x + 3) = (2 \cdot 3 - 3)(2 \cdot 3 + 3) = (6 - 3)(6 + 3) = (3)(9) = \boxed{27}$.

(d) $2^{2x} = 2^{2 \cdot 3} = 2^6 = \boxed{64}$.

2.18

(a) $-t + 4 = -(-7) + 4 = 7 + 4 = \boxed{11}$.

(b) $(5 - t)^2/(t + 6) = [5 - (-7)]^2/(-7 + 6) = (5 + 7)^2/(-1) = 12^2/(-1) = 144/(-1) = \boxed{-144}$.

(c) $t^5/t^3 = t^{5-3} = t^2 = (-7)^2 = \boxed{49}$.

(d) $2t^2 - 3t/7 + 8 = 2(-7)^2 - 3(-7)/7 + 8 = 2(49) - (-21)/7 + 8 = 98 - (-3) + 8 = \boxed{109}$.

2.19

(a) $p^8 \cdot p^3 \cdot p^4 = p^{8+3+4} = \boxed{p^{15}}$.

(b) $(4x^4)(6x^6) = 4 \cdot x^4 \cdot 6 \cdot x^6 = (4 \cdot 6) \cdot (x^4 \cdot x^6) = 24 \cdot (x^{4+6}) = \boxed{24x^{10}}$.

2.20   There are several ways to approach this problem. First, we might note that we need to multiply 3 by 12 in order to get 36, and we need to multiply y^5 by y^3 to get $y^{5+3} = y^8$. So, we must multiply $3y^5$ by $\boxed{12y^3}$ to get $36y^8$:

$$(3y^5)(12y^3) = (3 \cdot 12) \cdot (y^5 \cdot y^3) = 36y^8.$$

We also could use division. For example, because 15 divided by 3 equals 5, we know that 3 times 5 equals 15. Similarly, we divide $36y^8$ by $3y^5$ to find what we must multiply $3y^5$ by to get $36y^8$:

$$\frac{36y^8}{3y^5} = \frac{36}{3} \cdot \frac{y^8}{y^5} = 12y^3.$$

If we multiply both sides of $\dfrac{36y^8}{3y^5} = 12y^3$ by $3y^5$, we have

$$3y^5 \cdot \frac{36y^8}{3y^5} = (3y^5)(12y^3).$$

This tells us that $36y^8 = (3y^5)(12y^3)$, so we must multiply $3y^5$ by $12y^3$ to get $36y^8$.

2.21

(a) $(x^5)^4 = x^{5 \cdot 4} = \boxed{x^{20}}$.

(b) $(2k)^4(3k^2)^3 = (2^4 k^4)[3^3 \cdot (k^2)^3] = (16k^4)(27k^6) = (16 \cdot 27) \cdot (k^4 \cdot k^6) = \boxed{432k^{10}}$.

(c) We have $\sqrt{484t^9} = \sqrt{484} \cdot \sqrt{t^9}$. Since $484 = 2^2 \cdot 11^2$, we have $\sqrt{484} = 2 \cdot 11 = 22$. We also have $\sqrt{t^9} = \sqrt{t^8} \cdot \sqrt{t} = t^4 \sqrt{t}$, so $\sqrt{484t^9} = \sqrt{484} \cdot \sqrt{t^9} = \boxed{22t^4 \sqrt{t}}$.

(d) We have $\sqrt[3]{243y^{12}} = \sqrt[3]{243} \cdot \sqrt[3]{y^{12}}$. Finding the prime factorization of 243 gives $243 = 3^5$, so we can factor out 3^3 to simplify $\sqrt[3]{243}$. This gives us $\sqrt[3]{243} = \sqrt[3]{3^5} = \sqrt[3]{3^3} \cdot \sqrt[3]{3^2} = 3\sqrt[3]{9}$. We also have $\sqrt[3]{y^{12}} = (y^{12})^{1/3} = y^{12(1/3)} = y^4$, so $\sqrt[3]{243y^{12}} = \sqrt[3]{243} \cdot \sqrt[3]{y^{12}} = \left(3\sqrt[3]{9}\right) \cdot y^4 = \boxed{3y^4 \sqrt[3]{9}}$.

(e) Applying the laws of exponents gives $r^{4/3} \cdot (r^{1/3})^2 = r^{4/3} \cdot r^{(1/3)(2)} = r^{4/3} \cdot r^{2/3} = r^{(4/3)+(2/3)} = \boxed{r^2}$.

(f) We apply the laws of exponents, and then group the constants together and group the variables together to find

$$(4r^5)^{1/2}(81r^7)^{-1/2} = (4^{1/2})(r^5)^{1/2}(81^{-1/2})(r^7)^{-1/2} = 2r^{5/2} \cdot \frac{1}{81^{1/2}} \cdot r^{-7/2}$$

$$= 2 \cdot \frac{1}{9} \cdot r^{(5/2)+(-7/2)} = \frac{2}{9} \cdot r^{-1} = \boxed{\frac{2}{9r}}.$$

2.22

(a) $\dfrac{r^8}{r^{12}} = r^{8-12} = r^{-4} = \boxed{\dfrac{1}{r^4}}$.

(b) $\dfrac{16t^3}{14t^3} = \dfrac{2 \cdot 8t^3}{2 \cdot 7 \cdot t^3} = \boxed{\dfrac{8}{7}}$.

(c) $\dfrac{(-8u^4)(2u^3)}{(4u^2)(6u^3)} = \dfrac{(-8 \cdot 2)(u^4 \cdot u^3)}{(4 \cdot 6)(u^2 \cdot u^3)} = \dfrac{-16u^7}{24u^5} = \dfrac{-16}{24} \cdot \dfrac{u^7}{u^5} = -\dfrac{2}{3} \cdot u^{7-5} = \boxed{-\dfrac{2u^2}{3}}$.

(d) $\dfrac{3r^2}{2r^8} \cdot \dfrac{6r^4}{5r^2} \cdot \dfrac{r^8}{3r^2} = \dfrac{3}{2r^6} \cdot \dfrac{6r^2}{5} \cdot \dfrac{r^6}{3} = \dfrac{3 \cdot 6r^2 \cdot r^6}{2r^6 \cdot 5 \cdot 3} = \dfrac{6r^2}{10} = \boxed{\dfrac{3r^2}{5}}$.

2.23

(a) $(-2)^{-3} = \dfrac{1}{(-2)^3} = \boxed{-\dfrac{1}{8}}$.

(b) $\dfrac{1}{4^{-2}} = 4^2 = \boxed{16}$.

(c) $\dfrac{6^{-1}r^{-3}r^2}{r^5} = \dfrac{\frac{r^2}{6^1r^3}}{r^5} = \dfrac{r^2}{6^1r^3r^5} = \dfrac{r^2}{6r^8} = \boxed{\dfrac{1}{6r^6}}$. Note that we could have immediately gone from $\dfrac{6^{-1}r^{-3}r^2}{r^5}$ to $\dfrac{r^2}{6^1r^3r^5}$ by moving the expressions with negative exponents to the denominator and changing the sign of their exponents. Make sure you see why!

(d) $\dfrac{3^{-2}}{x^{-7}} = \dfrac{\frac{1}{3^2}}{\frac{1}{x^7}} = \dfrac{x^7}{3^2} = \boxed{\dfrac{x^7}{9}}$. Once again, we could have gone straight from $\dfrac{3^{-2}}{x^{-7}}$ to $\dfrac{x^7}{3^2}$ by noting that we can change the 3^{-2} in the numerator to a 3^2 in the denominator (because $3^{-2} = 1/3^2$) and we change the x^{-7} in the denominator into an x^7 in the numerator (because $1/x^{-7} = x^7$). Make sure you see why!

2.24

(a) $16\left(\dfrac{x}{2} - \dfrac{3}{4}\right) = 16 \cdot \left(\dfrac{x}{2}\right) - 16\left(\dfrac{3}{4}\right) = \dfrac{16x}{2} - \dfrac{48}{4} = \boxed{8x - 12}$

(b) $x^2(2 - 3x) = x^2 \cdot 2 - x^2 \cdot (3x) = 2x^2 - 3x^3 = \boxed{-3x^3 + 2x^2}$. Make sure you see how we were careful to keep our signs correct.

2.25 We multiply each term in the parentheses by $3x$ to expand:

$$3x\left(\dfrac{x}{3} + \dfrac{3}{x} + \dfrac{1}{3x}\right) = (3x) \cdot \left(\dfrac{x}{3}\right) + (3x) \cdot \left(\dfrac{3}{x}\right) + (3x)\left(\dfrac{1}{3x}\right)$$
$$= x^2 + 9 + 1$$
$$= \boxed{x^2 + 10}.$$

2.26

(a) $(7 - 3x) + (5x - 8) = 7 - 3x + 5x - 8 = -3x + 5x + 7 - 8 = \boxed{2x - 1}$.

(b) $(y^3 - 2y + 1) - (y^2 - 2y + 8) = y^3 - 2y + 1 - y^2 + 2y - 8 = y^3 - y^2 - 2y + 2y + 1 - 8 = \boxed{y^3 - y^2 - 7}$.

2.27 We first expand each product with the distributive property, then we combine like terms:

$$2(t^2 - 4t + 1) - t(t + 7) = 2t^2 - 8t + 2 - t^2 - 7t = 2t^2 - t^2 - 8t - 7t + 2 = \boxed{t^2 - 15t + 2}.$$

2.28 Jerri's 7 cans together have $7x$ balls. Including her 8 extras, she has $7x + 8$ tennis balls. If she doubles this, she has $2(7x + 8) = 2(7x) + 2(8) = 14x + 16$ tennis balls. Sam has $14x$ balls in his cans, plus 3 extras, for a total of $14x + 3$. Therefore, Jerri has

$$(14x + 16) - (14x + 3) = 14x + 16 - 14x - 3 = 14x - 14x + 16 - 3 = \boxed{13}$$

more tennis balls than Sam. We also could have noticed that if Jerri doubles her tennis balls by doubling the cans to 14 and the extras to 16, she then has the same number of cans as Sam does. However, she has $16 - 3 = 13$ more extras, so she has 13 more balls.

2.29

(a) Each term has a factor of x, so $x^4 - 6x = x \cdot (x^3) + x \cdot (-6) = \boxed{x(x^3 - 6)}$.

(b) Each term has a factor of 4, so $16r^3 - 4 = 4 \cdot (4r^3) + 4 \cdot (-1) = \boxed{4(4r^3 - 1)}$.

(c) Each term has a factor of 8 and a factor of x^2, so we can factor out $8x^2$:

$$-24x^2 + 8x^5 = (8x^2) \cdot (-3) + (8x^2) \cdot (x^3) = \boxed{(8x^2)(-3 + x^3)}.$$

(d) Each term has a factor of 6 and a factor of u, so we can factor out $6u$:

$$42u^3 + 36u^2 - 72u = (6u) \cdot (7u^2) + (6u) \cdot (6u) + (6u) \cdot (-12) = \boxed{6u(7u^2 + 6u - 12)}.$$

2.30

(a) If I start with 6, I first subtract from 7 to get 1. I multiply this by 3 to get 3. When I subtract 36 from 8 times 6, I get 12. Half of this is 6. When I add this to my 3 from earlier, I get $\boxed{9}$.

(b) Let's see what happens if we play this game with x. First, we subtract x from 7: $7 - x$. Then, we multiply by 3 to get $3(7 - x) = 3 \cdot 7 - 3x = 21 - 3x$. Next, we must find half the difference when 36 is subtracted from 8 times x. This expression is

$$\frac{1}{2}(8x - 36) = \frac{1}{2}(8x) - \frac{1}{2}(36) = 4x - 18.$$

We must add this to our $21 - 3x$ from before:

$$(21 - 3x) + (4x - 18) = 21 - 3x + 4x - 18 = -3x + 4x + 21 - 18 = x + 3.$$

Therefore, MathWizard's trick is just a fancy way of adding 3 to a number. So, to figure out a friend's number from the result of her trick, all MathWizard has to do is subtract 3 from the result.

2.31

(a) The factor $x^2 - 3$ is common to both terms, so we can factor it out:

$$2x(x^2 - 3) + 5(x^2 - 3) = (2x) \cdot (x^2 - 3) + (5) \cdot (x^2 - 3) = \boxed{(2x + 5)(x^2 - 3)}.$$

(b) The factor $2d + 7$ is common to both terms, so we can factor it out:

$$3(2d + 7) - 5d(2d + 7) = (3) \cdot (2d + 7) + (-5d) \cdot (2d + 7) = \boxed{(3 - 5d)(2d + 7)}.$$

2.32

(a) $\dfrac{3x^2 - 6}{9} = \dfrac{3 \cdot (x^2 - 2)}{3 \cdot 3} = \boxed{\dfrac{x^2 - 2}{3}}.$

(b) $\dfrac{18 - 36x}{2 - 4x} = \dfrac{18(1 - 2x)}{2(1 - 2x)} = \dfrac{18}{2} = \boxed{9}.$

(c) We can factor an a out of the numerator and an a^2 out of the denominator:

$$\frac{a^3 - a}{a^4 - a^2} = \frac{a(a^2 - 1)}{a^2(a^2 - 1)} = \frac{a}{a^2} = \boxed{\frac{1}{a}}.$$

(d) We can factor $15z^3$ out of the numerator and $12z$ out of the denominator:

$$\frac{15z^5 + 15z^4 + 15z^3}{12z^3 + 12z^2 + 12z} = \frac{15z^3(z^2 + z + 1)}{12z(z^2 + z + 1)} = \frac{15z^3}{12z} = \frac{15}{12} \cdot z^{3-1} = \boxed{\frac{5z^2}{4}}.$$

2.33 To divide one fraction by a second fraction, we multiply the first fraction by the reciprocal of the second. Therefore,

$$\frac{\dfrac{3x}{4x - 4}}{\dfrac{9x^2}{x - 1}} = \frac{3x}{4x - 4} \cdot \frac{x - 1}{9x^2} = \frac{3x(x - 1)}{9x^2(4x - 4)}.$$

Now we use factoring and canceling:

$$\frac{3x(x-1)}{9x^2(4x-4)} = \frac{\cancel{3}\cancel{x}(\cancel{x-1})}{3 \cdot \cancel{3} \cdot x \cdot \cancel{x} \cdot 4 \cdot \cancel{(x-1)}} = \boxed{\frac{1}{12x}}.$$

2.34

(a) We multiply our first denominator, 5, by $8x$ to make it equal to the second denominator:

$$\frac{3x}{5} - \frac{11}{40x} = \frac{3x}{5} \cdot \frac{8x}{8x} - \frac{11}{40x} = \frac{24x^2}{40x} - \frac{11}{40x} = \boxed{\frac{24x^2 - 11}{40x}}.$$

(b) Our denominators are $5r$ and 2, so the simplest common denominator is $10r$:

$$\frac{2-r^2}{5r} - \frac{r}{2} = \frac{2-r^2}{5r} \cdot \frac{2}{2} - \frac{r}{2} \cdot \frac{5r}{5r} = \frac{(2-r^2)(2)}{10r} - \frac{5r^2}{10r} = \frac{4-2r^2-5r^2}{10r} = \boxed{\frac{-7r^2+4}{10r}}.$$

(c) Our denominators are $8z^3$ and $2z^4$. We need a z^4 in the common denominator to take care of the z's, and we need an 8 to take care of the constants. So, our simplest common denominator is $8z^4$:

$$\begin{aligned}
\frac{2}{8z^3} - \frac{3-z}{2z^4} &= \frac{2}{8z^3} \cdot \frac{z}{z} - \frac{3-z}{2z^4} \cdot \frac{4}{4} \\
&= \frac{2z}{8z^4} - \frac{(3-z)(4)}{8z^4} \\
&= \frac{2z - (3-z)(4)}{8z^4} \\
&= \frac{2z - 12 + 4z}{8z^4} \\
&= \frac{6z - 12}{8z^4}
\end{aligned}$$

We're not quite finished! We can factor a 2 out of the numerator and simplify further:

$$\frac{6z-12}{8z^4} = \frac{2(3z-6)}{2 \cdot (4z^4)} = \boxed{\frac{3z-6}{4z^4}}.$$

We also could have simplified the first fraction in our original expression before finding the common denominator.

(d) We can clearly simplify our second fraction before trying to combine the two. Maybe we can simplify the first fraction, too. We can factor an a out of the numerator of the first fraction to get

$$\frac{a^3-a}{a^2-1} = \frac{a\cancel{(a^2-1)}}{\cancel{a^2-1}} = a.$$

Similarly, we can simplify our second fraction: $\dfrac{7a^3}{49a^4} = \dfrac{1}{7a}$.

So, we now have $\dfrac{a^3-a}{a^2-1} + \dfrac{7a^3}{49a^4} = a + \dfrac{1}{7a}$. Clearly, $7a$ is the common denominator:

$$a + \frac{1}{7a} = a \cdot \frac{7a}{7a} + \frac{1}{7a} = \frac{7a^2}{7a} + \frac{1}{7a} = \boxed{\frac{7a^2+1}{7a}}.$$

2.35 Our denominators are z^2 and $z^2 + 1$, so the best we can do for a common denominator is $z^2(z^2 + 1)$:

$$\frac{1}{z^2 + 1} - \frac{1}{z^2} = \frac{1}{z^2 + 1} \cdot \frac{z^2}{z^2} - \frac{1}{z^2} \cdot \frac{z^2 + 1}{z^2 + 1} = \frac{z^2}{z^2(z^2 + 1)} - \frac{z^2 + 1}{z^2(z^2 + 1)} = \frac{z^2 - (z^2 + 1)}{z^2(z^2 + 1)} = \boxed{-\frac{1}{z^2(z^2 + 1)}}.$$

Challenge Problems

2.36 First we factor out the expression $r^2 + 1$ that is common to both terms:

$$2r^2(r^2 + 1) - 8r(r^2 + 1) = (2r^2) \cdot (r^2 + 1) + (-8r) \cdot (r^2 + 1) = (2r^2 - 8r)(r^2 + 1).$$

We're not finished! We can factor a 2 and an r out of $2r^2 - 8r$:

$$2r^2 - 8r = (2r)(r) - (2r)(4) = 2r(r - 4).$$

Therefore, we have

$$(2r^2 - 8r)(r^2 + 1) = \boxed{2r(r - 4)(r^2 + 1)}.$$

2.37 The three denominators we have are $2z - 1$, z, and $2z^2 - z$. We might then use the product of these as our common denominator. However, before we do that we see that we can factor a z out of the last denominator, $2z^2 - z$:

$$2z^2 - z = z(2z - 1).$$

So, our three denominators are $2z - 1$, z, and $z(2z - 1)$. Therefore, our simplest common denominator is $z(2z - 1)$. Now, we're ready to write all the terms with this denominator:

$$\begin{aligned}
2 + \frac{4}{2z - 1} - \frac{3}{z} + \frac{z}{2z^2 - z} &= 2 \cdot \frac{z(2z - 1)}{z(2z - 1)} + \frac{4}{2z - 1} \cdot \frac{z}{z} - \frac{3}{z} \cdot \frac{(2z - 1)}{(2z - 1)} + \frac{z}{z(2z - 1)} \\
&= \frac{2z(2z - 1)}{z(2z - 1)} + \frac{4z}{z(2z - 1)} - \frac{3(2z - 1)}{z(2z - 1)} + \frac{z}{z(2z - 1)} \\
&= \frac{2z(2z - 1) + 4z - 3(2z - 1) + z}{z(2z - 1)} \\
&= \frac{4z^2 - 2z + 4z - 6z + 3 + z}{z(2z - 1)} \\
&= \boxed{\frac{4z^2 - 3z + 3}{z(2z - 1)}}.
\end{aligned}$$

2.38 If the expression is the same for all values of x, then the x's in $3(x + 7)$ must cancel out with those in $\_\_(2x + 9)$. When we expand $3(x + 7)$, we have $3x + 21$, so when we expand $\_\_(2x + 9)$, we must have a $3x$ term, so that when we subtract:

$$3(x + 7) - \_\_(2x + 9),$$

the x terms cancel. Because the expansion of $\_\_(2x + 9)$ must have $3x$, the number in the blank must be $\boxed{\frac{3}{2}}$. Checking our answer, we find

$$3(x + 7) - \frac{3}{2}(2x + 9) = 3x + 21 - \frac{3}{2}(2x) - \frac{3}{2}(9) = 3x + 21 - 3x - \frac{27}{2} = 21 - \frac{27}{2} = \frac{15}{2}.$$

The x's do indeed cancel out when we put 3/2 in the blank, so the expression is the same for all values of x.

2.39   At first we seem stuck, but we notice that the last two terms in our expression have 8 as a factor. So, we try factoring the 8 out of those two terms:

$$2r(r-7) + 8r - 56 = 2r(r-7) + 8(r) - (8)(7) = 2r(r-7) + 8(r-7).$$

Now we have a common factor of $r-7$ in both terms! We can factor that out, as well as a common factor of 2:

$$2r(r-7) + 8(r-7) = (2r+8)(r-7) = \boxed{2(r+4)(r-7)}.$$

2.40

(a) $x(x+2) = x \cdot x + x \cdot (2) = \boxed{x^2 + 2x}$.

(b)   The distributive property tells us that

$$a(b+c) = ab + ac.$$

The variables a, b, and c can stand for whole expressions. For example, if we let a be $x+1$, then we have

$$(x+1)(b+c) = (x+1)(b) + (x+1)(c).$$

Now, we let $b = x$ and $c = 2$, so we have

$$(x+1)(x+2) = (x+1)(x) + (x+1)(2).$$

We continue by expanding the products on the right:

$$(x+1)(x+2) = (x^2 + x) + (2x + 2) = x^2 + x + 2x + 2 = \boxed{x^2 + 3x + 2}.$$

(c)   If we have

$$x^2 + 5x + 4 = (x + \underline{})(x + \underline{}),$$

then when we multiply out the right side, we must get $x^2 + 5x + 4$. When we multiply out the right side, the only way we will get a term without an x is when we multiply the two numbers that we put in the blanks. So, the numbers we put in the blanks must multiply to 4. So, we'll try 2 and 2, then try 4 and 1. We have

$$(x+2)(x+2) = (x+2)(x) + (x+2)(2) = x^2 + 2x + 2x + 4 = x^2 + 4x + 4.$$

As expected, the constant is 4, but the x term is $4x$, not $5x$. So, we'll have to try 4 and 1:

$$(x+4)(x+1) = (x+4)(x) + (x+4)(1) = x^2 + 4x + x + 4 = x^2 + 5x + 4.$$

Success! Our factorization is $x^2 + 5x + 4 = \boxed{(x+4)(x+1)}$. Of course, due to the commutative property of multiplication, we could also have written this as $x^2 + 5x + 4 = (x+1)(x+4)$.

2.41 We'll focus on the constant and the x term separately. Since the simplest common denominator of Alice's and Bob's expressions is $4x^2$, we know that neither of their expressions has a power of x higher than x^2. Also, at least one of their expressions has a factor of 4 in the denominator (not just 2), but neither of them has any factors besides 2 or 4 (or else we'd need more factors in the common denominator). We also know that it's possible that one of the denominators has 1 as its constant (such as $\frac{1}{x^2}$). Finally, because the simplest common denominator of Bob's and Carol's expressions has a different constant term than the simplest common denominator of Alice's and Carol's expressions, we know that Bob and Alice must have different constants in their denominator.

Turning to "the simplest common denominator of Bob's and Carol's expressions is $12x^3$," we see that Carol's fraction must have x^3 in the denominator, since we know the power of x in Bob's denominator cannot be higher than x^2. Moreover, since there is a factor of 3 in the simplest common denominator of Bob's and Carol's fractions and we know that Bob's constant can only be 1, 2, or 4, we know that Carol's fraction has a factor of 3 in its denominator.

So, we know that Carol's denominator must have a factor of 3 and a factor of x^3. What about a factor of 2? Because "the simplest common denominator of Alice and Carol's expressions is $6x^3$," we know that Carol cannot have a factor of 4 in her denominator. She can, however have a factor of 2. But she doesn't have to. Here are two possibilities that show that the two possible expressions Carol could have are $\boxed{\dfrac{1}{3x^3} \text{ and } \dfrac{1}{6x^3}}$:

Alice	:	$\dfrac{1}{2x^2}$		Alice	:	$\dfrac{1}{2x^2}$
Bob	:	$\dfrac{1}{4x^2}$		Bob	:	$\dfrac{1}{4x^2}$
Carol	:	$\dfrac{1}{3x^3}$		Carol	:	$\dfrac{1}{6x^3}$

2.42

(a) $1 + 2 + 3 = \boxed{6}$.

(b) $1 + 2 + 3 + 4 = (1 + 2 + 3) + 4 = 6 + 4 = \boxed{10}$.

(c) $1 + 2 + 3 + 4 + 5 = (1 + 2 + 3 + 4) + 5 = 10 + 5 = \boxed{15}$.

(d) We see that $1 + 2 + 3 = (3 \cdot 4)/2$, $1 + 2 + 3 + 4 = (4 \cdot 5)/2$, and $1 + 2 + 3 + 4 + 5 = (5 \cdot 6)/2 = 15$. We guess that

$$1 + 2 + 3 + 4 + 5 + 6 + 7 + 8 + 9 + 10 = \frac{10 \cdot 11}{2} = 55.$$

When we add the 10 numbers, we see that we are right.

(e) We guess that $1 + 2 + 3 + \cdots + (n - 1) + n = \dfrac{n \cdot (n + 1)}{2}$.

(f) If we add $n + 1$ to both sides of our guess from the previous part, we have

$$1 + 2 + 3 + \cdots + (n - 1) + n + (n + 1) = \frac{n \cdot (n + 1)}{2} + n + 1.$$

We can write the right side with a common denominator, 2:

$$\frac{n \cdot (n+1)}{2} + n + 1 = \frac{n \cdot (n+1)}{2} + \frac{2(n+1)}{2} = \frac{n \cdot (n+1) + 2(n+1)}{2}.$$

We can factor $(n+1)$ out of both terms in the numerator:

$$\frac{n \cdot (n+1) + 2 \cdot (n+1)}{2} = \frac{(n+2)(n+1)}{2}.$$

Hmmm... This looks a lot like the $\frac{n(n+1)}{2}$ that we started with. In fact, what we did was start with

$$1 + 2 + 3 + \cdots + (n-1) + n = \frac{n \cdot (n+1)}{2},$$

then add $n+1$ to both sides. The result we got was:

$$1 + 2 + 3 + \cdots + (n-1) + n + (n+1) = \frac{(n+1)(n+2)}{2}.$$

This means that if

$$1 + 2 + 3 + \cdots + (n-1) + n = \frac{n \cdot (n+1)}{2} \tag{2.1}$$

is true for some n, then when we add $n+1$, we get

$$1 + 2 + 3 + \cdots + (n-1) + n + (n+1) = \frac{(n+1)(n+2)}{2},$$

which is the exact same equation as Equation (2.1), but with $n+1$ in place of n. In other words, we showed that if our formula works for n, then it will work from $n+1$. We have already seen that it works when $n = 3$. Because we know it works for $n = 3$, it must work for $n = 4$. Since it works for $n = 4$, it must work for $n = 5$. Since it works for $n = 5$, it works for $n = 6$. And we can keep going like this forever! So, we have shown that the formula

$$1 + 2 + 3 + \cdots + (n-1) + n = \frac{n \cdot (n+1)}{2}$$

does indeed work for all positive integers n.

The process we used to prove this is called mathematical induction. If you didn't quite follow it, don't worry! You won't need induction for this book, and you'll get a much more thorough introduction to it in later math books.

3

Exercises for Section 3.1

3.1.1

(a) Adding 7 to both sides gives $x = 14 + 7 = \boxed{21}$.

(b) First we simplify the left side, which gives $16 = 2 + r$. Subtracting 2 from both sides gives us $r = \boxed{14}$.

(c) We simplify the right side to get $-3 + y = 2.5$. Adding 3 to both sides gives $y = 2.5 + 3 = \boxed{5.5}$.

(d) First we perform the arithmetic on the left side: $\frac{1}{3} - 3 = \frac{1}{3} - \frac{9}{3} = -\frac{8}{3}$. Now our equation is $-\frac{8}{3} = \frac{2}{3} + x$. Subtracting $\frac{2}{3}$ from both sides gives $x = -\dfrac{8}{3} - \dfrac{2}{3} = \boxed{-\dfrac{10}{3}}$.

3.1.2

(a) Dividing both sides by 3 gives us $x = 24/3 = \boxed{8}$.

(b) Dividing both sides by 2 gives us $\dfrac{-1.2}{2} = r$, so $r = \boxed{-0.6}$.

(c) Multiplying both sides by 3 gives us $y = 3 \cdot \dfrac{2}{9} = \dfrac{6}{9} = \boxed{\dfrac{2}{3}}$.

(d) The coefficient of s is $-3/8$, so we multiply both sides by its reciprocal, $-8/3$:

$$\left(-\frac{8}{3}\right)\left(-\frac{3s}{8}\right) = \left(-\frac{8}{3}\right)(-6).$$

Since $\left(-\frac{8}{3}\right)\left(-\frac{3}{8}\right) = 1$, we just have s on the left:

$$s = \frac{(-8)(-6)}{3} = \boxed{16}.$$

3.1.3 Just as we can get rid of the 3 in the denominator of the equation $\frac{z}{3} = 5$ by multiplying both sides of the equation by 3, we can get rid of the 3 in the denominator of this equation by multiplying by 3:

$$3\left(\frac{x-1}{3}\right) = 3 \cdot 5.$$

On the left we have $\dfrac{\cancel{3}(x-1)}{\cancel{3}} = x - 1$. On the right we have 15, so our equation is $x - 1 = 15$. Adding 1 to both sides gives $x = \boxed{16}$.

We can also reason our way to the answer. The equation tells us that $x - 1$ divided by 3 equals 5. We know that 15 is the number that gives us 5 when divided by 3. So, $x - 1$ must equal 15. Therefore, x is 16, as before.

3.1.4 Just as we got rid of the 3 in the previous problem by multiplying both sides by 3, we get rid of it in this problem by dividing both sides by 3:

$$\frac{3(r-5)}{3} = \frac{24}{3}.$$

The 3's on the left cancel, leaving $r - 5$. On the right we have $24/3 = 8$, so our equation is $r - 5 = 8$. Adding 5 to both sides of this equation gives $r = \boxed{13}$.

Again, we could reason our way to the answer. Since 3 times $r - 5$ equals 24, we know that $r - 5$ must be 8, since we must multiply 3 by 8 to get 24. Since $r - 5 = 8$, we have $r = 13$, as before.

3.1.5 Since $x = 3$ is a solution to the equation, we can let $x = 3$ in the equation. This gives us the equation

$$\frac{3}{a} = 7.$$

We wish to find a. We don't know a whole lot about dealing with equations when the variable is in the denominator. However, we do know that if an equation is true, we can take the reciprocal of both sides to get another true equation. This gives us

$$\frac{a}{3} = \frac{1}{7}.$$

Multiplying both sides by 3 gives $a = \boxed{3/7}$. We could also have solved the equation $\frac{3}{a} = 7$ by cross-multiplying to get $7a = 3$ (the denominator of the 7 on the right side is just 1). We then divide this equation by 7 to get $a = 3/7$, as before.

Exercises for Section 3.2

3.2.1

(a) We add 4 to both sides to get $3x = 21$, then divide by 3 to find $x = \boxed{7}$.

(b) We first add $2r$ to both sides to get all the variable terms on one side. This gives us $4 = 17 + 7r$. We then subtract 17 from both sides to find $7r = 4 - 17 = -13$. Dividing both sides by 7 gives $r = \boxed{-13/7}$.

(c) First we get all the terms with y on one side by subtracting $1.7y$ from both sides. This gives us $4 + 0.6y = -20$. Subtracting 4 from both sides gives $0.6y = -24$. Dividing both sides by 0.6 isolates y and gives us $y = -24/(0.6) = -240/6 = \boxed{-40}$.

(d) We first get rid of all the fractions by multiplying both sides by 4. On the left we have

$$4\left(-2t + \frac{3}{2}\right) = 4(-2t) + 4\left(\frac{3}{2}\right) = -8t + \frac{12}{2} = -8t + 6.$$

On the right we have

$$4\left(\frac{t}{4} - 12\right) = 4\left(\frac{t}{4}\right) - 4(12) = t - 48,$$

so our equation is now $-8t + 6 = t - 48$. Subtracting t from both sides to get all the variable terms on one side gives $-9t + 6 = -48$. Subtracting 6 from both sides gives $-9t = -54$. Finally, we divide both sides by -9 to find $t = \boxed{6}$.

We could also have isolated t starting from $-8t + 6 = t - 48$ by adding $8t$ to both sides to get $6 = 9t - 48$, adding 48 to both sides of this to get $54 = 9t$, then dividing by 9 to find $t = 6$, as before.

(e) First we simplify the left side by combining the u terms, to get $-14u - 5 = 3 - 14u$. Uh-oh. Adding $14u$ to both sides gives us $-5 = 3$, which is never true! Therefore, there are $\boxed{\text{no solutions}}$ to this equation.

(f) First we simplify the left side to get $-2y + 8 = 8 - 17y$. Adding $17y$ to both sides gives $15y + 8 = 8$, and subtracting 8 from both sides gives $15y = 0$. Dividing both sides by 15 gives us $y = \boxed{0}$.

(g) Expanding the products on both sides gives $-3r - 21 = 15 - 5r$. Adding $5r$ to collect the terms with r on one side gives $2r - 21 = 15$. Adding 21 to both sides gives us $2r = 36$, and dividing both sides by 2 finishes our solution: $r = 36/2 = \boxed{18}$.

(h) We start by cross-multiplying, which gives us $3(x - 7) = 5(x - 2)$. (This is the same as multiplying both sides of the equation by 5, then multiplying both sides by 3.) Expanding both sides of this equation gives $3x - 21 = 5x - 10$. Subtracting $5x$ from both sides gives $-2x - 21 = -10$. Adding 21 to both sides gives $-2x = 11$, and dividing both sides by -2 gives $x = \boxed{-11/2}$.

3.2.2 He knew he made a mistake because when $x = 5$, the left side of the equation is a whole number, but the right side is not a whole number. Therefore, the two sides can't be equal when $x = 5$.

3.2.3 We first simplify the right side to get $3y + 2a = 3y + 10$. Subtracting $3y$ from both sides leaves $2a = 10$. Dividing by 2 gives us $a = \boxed{5}$. When $a = 5$, the equation is $3y + 10 = 4y + 7 - y + 3$, which simplifies to $3y + 10 = 3y + 10$. Since the two sides of this equation are identical, every value of y satisfies the equation.

3.2.4 When Joan solves the equation she wrote down, she adds 7 to both sides to get $3x = 45$. Then she divides both sides by 3 to find $x = 15$. The teacher says the correct answer is 6 lower than this, or $x = 9$. The teacher then says the coefficient of x was wrong but that everything else was right. In other words, the correct equation was $\_\_ \cdot (x) - 7 = 38$. We know that $x = 9$ solves this equation, so we have $\_\_ \cdot (9) - 7 = 38$. Adding 7 to both sides gives $\_\_ \cdot (9) = 45$. So, the coefficient should have been $\boxed{5}$, not 3.

Exercises for Section 3.3

3.3.1 Let n be the number. Three times the number is $3n$ and "seven more than double the number" is $7 + 2n$. These two must be equal, so we have $3n = 7 + 2n$. Subtracting $2n$ from both sides gives $n = \boxed{7}$.

3.3.2 Suppose I am y years old now. My brother is 4 times as old as I am, so my brother's age is $4y$. Six years from now, my brother will be $4y + 6$ years old and I will be $y + 6$ years old. Since my brother will be twice as old as I am at that time, we must have

$$4y + 6 = 2(y + 6).$$

Expanding the right side gives $4y + 6 = 2y + 12$. Subtracting $2y$ from both sides gives $2y + 6 = 12$, and subtracting 6 from both sides gives $2y = 6$. Dividing both sides by 2 gives $y = 3$. The question asks for the age of my brother now, which is $4y = \boxed{12}$. (Always make sure you answer the question.)

3.3.3 Let x be the number of paces. First we consider the phrase:

Think ye of the number that is seventeen more than the quotient when six times the number of paces to the treasure is divided by two

The quotient when 6 times the number is divided by 2 is $\frac{6x}{2}$. Seventeen more than this is $17 + \frac{6x}{2}$. We then turn to the phrase:

The sum when twelve more than nine minus fifteen is added to four times the number of paces to the treasure

Turning these words into math gives $12 + 9 - 15 + 4x$. So, our equation is

$$17 + \frac{6x}{2} = 12 + 9 - 15 + 4x.$$

Simplifying both sides gives us $17 + 3x = 6 + 4x$. Subtracting $3x$ from both sides gives $17 = 6 + x$, and subtracting 6 from both sides gives $x = \boxed{11}$. So, we should take 11 paces to get the treasure.

3.3.4 Let there be n nickels. Since there are three times as many dimes as nickels, there are $3n$ dimes. The value of the nickels is $0.05n$ dollars and the value of the dimes is $0.1(3n) = 0.3n$. Since together these add up to \$36.05, we have the equation $0.05n + 0.3n = 36.05$. Simplifying the left side gives $0.35n = 36.05$. Dividing both sides by 0.35 gives $n = (36.05)/(0.35) = 3605/35 = \boxed{103}$ nickels.

3.3.5 Let Cindy's number be c. She subtracted 9 then divided the result by 3, so she computed $(c - 9)/3$. She got an answer of 43 when she did this, so $(c - 9)/3 = 43$. We multiply both sides by 3 to get $c - 9 = 129$. We then add 9 to both sides to find $c = 138$. Now, we compute the answer Cindy should have found. She was supposed to subtract 3 from this number, which would have given her $138 - 3 = 135$. Then, she was supposed to divide by 9, which would give her $135/9 = \boxed{15}$.

3.3.6 Let Walter be y years old in 1994. So, he was born in the year $1994 - y$. Since Walter's grandmother is twice as old as Walter in 1994, she is $2y$ years old in 1994. So, she was born in the year $1994 - 2y$. The sum of their birth years is 3838, so we have

$$(1994 - y) + (1994 - 2y) = 3838.$$

Simplifying the left side gives us $3988 - 3y = 3838$. Subtracting 3988 from both sides gives $-3y = -150$ and dividing both sides of this by -3 gives $y = 50$. Therefore, in 1999, Walter will be $y + 5 = \boxed{55}$ years old.

Exercises for Section 3.4

3.4.1 We first combine the two fractions with x in the denominator to give $\frac{3}{x} - \frac{3}{5} = \frac{1}{5}$. Adding $\frac{3}{5}$ to both sides gives us $\frac{3}{x} = \frac{4}{5}$. Cross-multiplying this equation (or multiplying both sides by $5x$) gives $4x = 15$. Dividing both sides of this equation by 4 gives us $x = \boxed{15/4}$.

3.4.2 We first get all the terms with r in them on one side by adding $\sqrt{2r}$ to both sides, which gives $4 = \sqrt{2r} - 6 + \sqrt{2r}$. We can combine the two $\sqrt{2r}$ terms since $\sqrt{2r} + \sqrt{2r} = 2\sqrt{2r}$, so our equation now is $4 = 2\sqrt{2r} - 6$. We add 6 to both sides to get $10 = 2\sqrt{2r}$. Dividing both sides by 2 gives us $5 = \sqrt{2r}$. We now get rid of the square root sign by squaring both sides, which gives us $25 = 2r$. Dividing both sides of this equation by 2 gives us $r = \boxed{25/2}$.

Because we squared the equation as a step, we have to go back and check our solution to make sure it isn't extraneous. When $r = 25/2$, we have $\sqrt{2r} = \sqrt{25} = 5$, and our equation then reads $4 - 5 = 5 - 6$, which is indeed a valid equation. Therefore, $r = 25/2$ is not extraneous.

3.4.3 We let the numerator be x, so the denominator is $3x - 7$. Because the fraction equals 2/5, we have $x/(3x - 7) = 2/5$. Multiplying both sides by $5(3x - 7)$ (or cross-multiplying) gives $5x = 2(3x - 7)$. Expanding the right side gives $5x = 6x - 14$. Subtracting $6x$ from both sides gives $-x = -14$, so we find $x = \boxed{14}$.

3.4.4 We first simplify the left side, which gives $3 + 2\sqrt[4]{2 - z} = \sqrt[4]{2 - z}$. We then group all the terms with z in them by subtracting $2\sqrt[4]{2 - z}$ from both sides, which gives $3 = -\sqrt[4]{2 - z}$. Multiplying both sides by -1 then gives us $-3 = \sqrt[4]{2 - z}$. At this point we might already realize there are $\boxed{\text{no solutions}}$, since the fourth root of a number must be positive.

However, suppose we didn't realize there are no solutions at this point. We can raise both sides to the fourth power to get rid of the radical:

$$(-3)^4 = (\sqrt[4]{2 - z})^4.$$

On the left we have $(-3)^4 = 81$ and on the right we just have $2 - z$, so our equation is $81 = 2 - z$. Solving this equation gives us $z = -79$. We substitute this back into the original equation to check if it is extraneous. When $z = -79$, the left side of our original equation is $12 + 2\sqrt[4]{81} - 9 = 12 + 2(3) - 9 = 12 + 6 - 9 = 9$. The right side of the equation is $\sqrt[4]{81} = 3$. These two are not equal! So, the solution $z = -79$ is extraneous and is therefore not a valid solution. There are no solutions to the original equation.

3.4.5

(a) We take the reciprocal of both sides of the equation $r = \frac{1}{x-1}$ to give us $\frac{1}{r} = x - 1$. Adding 1 to both sides gives us $x = \boxed{\dfrac{1}{r} + 1}$.

(b) Substituting our expression from part (a) into $xr + \frac{2}{3} = 2r$ gives us

$$\left(\frac{1}{r} + 1\right)r + \frac{2}{3} = 2r.$$

Expanding the left side gives us $\frac{1}{r} \cdot r + 1 \cdot r + \frac{2}{3} = 2r$, so $1 + r + \frac{2}{3} = 2r$. This is just a linear equation! Simplifying the left side gives $\frac{5}{3} + r = 2r$. Subtracting r from both sides gives $r = \boxed{\dfrac{5}{3}}$. Substituting our value for r into our expression for x in part (a) gives us

$$x = \frac{1}{r} + 1 = \frac{1}{\frac{5}{3}} + 1 = \frac{3}{5} + 1 = \boxed{\frac{8}{5}}.$$

Review Problems

3.19

(a) Subtracting 2 from both sides gives $r = \boxed{4}$.

(b) We combine the terms with t on the left and simplify the right to get $t - 7 = -7$. Adding 7 to both sides gives $t = \boxed{0}$.

(c) First, we simplify both sides to get $10 + t = 4$. Subtracting 10 from both sides gives $t = \boxed{-6}$.

(d) First, we simplify the left side. Because $4x - 3x + 2x - x = 2x$ and $4 - 3 + 2 - 1 = 2$, our equation is $\frac{2x}{2} - \frac{2}{4} = -2$. Simplifying the fractions on the left side gives $x - \frac{1}{2} = -2$. Adding $\frac{1}{2}$ to both sides gives $x = \boxed{-\dfrac{3}{2}}$.

3.20

(a) Dividing both sides by 5 gives $t = \boxed{7}$.

(b) Dividing both sides by -2.5 gives $x = 24/(-2.5) = -24/(5/2) = \boxed{-48/5}$. We can also write this as $x = -9.6$.

(c) First we combine the expressions on the left to give $-3y/2 = -21$. Multiplying both sides by $-2/3$ isolates y on the left and gives us $y = (-21)(-2/3) = (7)(2) = \boxed{14}$.

(d) We get rid of the fraction by multiplying both sides by 7, which gives $3x - 5 = 4 \cdot 7 = 28$. Adding 5 to both sides gives $3x = 33$. Dividing by 3 gives $x = \boxed{11}$.

3.21

(a) We collect the z terms on one side by adding $2z$ to both sides, which gives $5z - 5 = 15$. Adding 5 to both sides gives $5z = 20$. Dividing both sides by 5 gives $z = \boxed{4}$.

(b) First, we simplify both sides to get $2.1x - 3.8 = 2.1x - 3.8$. If we subtract $2.1x$ from both sides of this equation, we have $-3.8 = -3.8$, which is always true. Therefore, our equation is true for $\boxed{\text{all values of } x}$.

(c) We first simplify the left side. We combine the y terms and we combine the constants:

$$\frac{9-2y}{4} + \frac{y+2}{2} = \frac{9}{4} - \frac{2y}{4} + \frac{y}{2} + \frac{2}{2} = \frac{9}{4} - \frac{y}{2} + \frac{y}{2} + 1 = \frac{9}{4} + 1 + \frac{y}{2} - \frac{y}{2} = \frac{13}{4}.$$

The y terms canceled each other out! The remaining equation is just $13/4 = 6$, which is never true. Therefore, there are $\boxed{\text{no solutions}}$ to this equation.

(d) We first simplify the left side and expand the right side. On the left, we have:

$$3 - \frac{8-2y}{5} = 3 - \left(\frac{8}{5} - \frac{2y}{5}\right) = 3 - \frac{8}{5} + \frac{2y}{5} = \frac{15}{5} - \frac{8}{5} + \frac{2y}{5} = \frac{7}{5} + \frac{2y}{5}.$$

On the right, we have $2(y - 9) = 2y - 18$. So, our equation is now

$$\frac{7}{5} + \frac{2y}{5} = 2y - 18.$$

We combine y terms on the right by subtracting $2y/5$ from both sides. Since $2y - \frac{2y}{5} = \frac{10y}{5} - \frac{2y}{5} = \frac{8y}{5}$, this makes our equation

$$\frac{7}{5} = \frac{8y}{5} - 18.$$

Adding 18 to both sides gives us $\frac{7}{5} + 18 = \frac{7}{5} + \frac{90}{5} = \frac{97}{5}$ on the left, so we have $\frac{97}{5} = \frac{8y}{5}$. Multiplying this equation by 5 gives us $97 = 8y$, and dividing both sides by 8 gives $y = \boxed{97/8}$.

Notice that we could have saved ourselves some of the headache of dealing with fractions by multiplying the original equation by 5, to get $15 - (8 - 2y) = 10(y - 9)$. Simplifying the left and expanding the right gives $7 + 2y = 10y - 90$. Subtracting $2y$ from both sides and adding 90 to both sides gives $97 = 8y$, from which we find $97/8 = y$, as before.

3.22 Let my number be x, so 5 more than one-half my number is $5 + \frac{x}{2}$. Since this must equal my number, we have $5 + \frac{x}{2} = x$. Subtracting $\frac{x}{2}$ from both sides gives $\frac{x}{2} = 5$, and multiplying both sides by 2 gives $x = \boxed{10}$.

3.23   All five choices make the left side a whole number. Only choice (C) makes the right side an integer, so only choice (C) can possibly be correct. (Check it, and you'll see that it is!)

3.24   Let w be the weight of one widget. Since the weight of the bag of widgets drops 8 kg when 2 widgets are removed, we have $2w = 8$. Dividing by 2 gives $w = 4$, so each widget weighs 4 kg. Since the bag weighs 1 kg, the total weight of the widgets in the final bag is $73 - 1 = 72$ kg. Suppose there are n widgets in the bag. Each widget weighs 4 kg and all of them together weigh 72 kg, so we must have $4n = 72$. Dividing both sides of this equation by 4 gives $n = \boxed{18}$ widgets left. (You might have done a lot of this figuring in your head without even making linear equations. That's great! As you see, even if you've never seen linear equations before this chapter, you already know how to solve many of them.)

3.25   Let my age now be x. Six years ago, I was $x - 6$ years old. Four years from now I will be $x + 4$ years old. Since my age six years ago is one-half my age four years from now, we must have:

$$x - 6 = \frac{x + 4}{2}.$$

We get rid of the fractions by multiplying both sides by 2, to get $2(x - 6) = x + 4$. Expanding the left gives $2x - 12 = x + 4$. Subtracting x from both sides gives $x - 12 = 4$ and adding 12 to both sides gives $x = 16$. Since I am 16 years old now, I will be $16 + 5 = \boxed{21}$ years old five years from now.

3.26   *Solution 1: Collect like terms first.* We get all the terms with z on the right by subtracting $\frac{2}{z}$ from both sides. On the right side, we'll now have:

$$\frac{5}{z} - 4 - \frac{2}{z} = \frac{5}{z} - \frac{2}{z} - 4 = \frac{5 - 2}{z} - 4 = \frac{3}{z} - 4.$$

So, our equation now is $5 = \frac{3}{z} - 4$, and adding 4 to both sides makes the equation $9 = \frac{3}{z}$. We multiply both sides by z to get z out of the denominator. This gives us $9z = 3$. (We could also view this step as cross-multiplying the equation $\frac{9}{1} = \frac{3}{z}$.) Dividing both sides of $9z = 3$ by 9 gives us $z = 3/9 = \boxed{1/3}$. This solution doesn't make any denominators equal to 0, so it is a valid solution.

Solution 2: Get rid of the fractions first. We first multiply both sides of the equation by z to get rid of the denominators. This gives us

$$z\left(\frac{2}{z} + 5\right) = z\left(\frac{5}{z} - 4\right).$$

Expanding both sides gives us $2 + 5z = 5 - 4z$. Adding $4z$ to both sides and subtracting 2 from both sides gives $9z = 3$. Dividing both sides of this equation by 9 leaves $z = \boxed{1/3}$.

3.27 We can combine the two terms on the left side to get $\dfrac{1 + 2x}{x - 1} = 5$. We then multiply both sides of this equation by $x - 1$ to get rid of the fractions. This gives us $1 + 2x = 5(x - 1)$. Expanding the right side gives $1 + 2x = 5x - 5$. Subtracting $5x$ from both sides gives $1 - 3x = -5$, and subtracting 1 from both sides of this equation yields $-3x = -6$. Dividing both sides of this equation by -3 gives us our answer, $x = \boxed{2}$.

3.28 Since each T-shirt costs \$5 more than a pair of socks and a pair of socks costs \$4, each T-shirt costs \$4 + \$5 = \$9. Each league member needs 2 pairs of socks and 2 T-shirts. Together, these cost $2(\$4) + 2(\$9) = \$8 + \$18 = \$26$. So, the cost is \$26 per league member. Suppose there are m members in the league. Since the cost is \$26 per member, the total cost is \$26m. We are told that the total cost is \$2366, so we must have \$26m = \$2366. Dividing both sides of this equation by \$26, we find that there are $m = \boxed{91}$ league members.

3.29 Let x be the number. Converting the words in the problem into an equation gives us $3 + \dfrac{1}{x} = \dfrac{7}{x}$. Subtracting $\dfrac{1}{x}$ from both sides gives $3 = \dfrac{6}{x}$. Multiplying both sides of this equation by x gives $3x = 6$, and dividing both sides of this equation by 3 gives $x = \boxed{2}$.

3.30 Let x be my age now. So, five years ago I was $x - 5$ years old and my grandfather's age was $5(x - 5)$. Therefore, my grandfather is now $5(x - 5) + 5 = 5x - 25 + 5 = 5x - 20$ years old. Three years from now, I will be $x + 3$ years old and my grandfather will be $5x - 20 + 3 = 5x - 17$ years old. Since my grandfather will then be three times as old as I am then, we have $3(x + 3) = 5x - 17$. Therefore, we have $3x + 9 = 5x - 17$, so $26 = 2x$, which means that $x = 13$. Therefore, I am $\boxed{13}$ years old now.

3.31 Let x be the smallest of the five integers, so that the other four are $x + 1$, $x + 2$, $x + 3$, and $x + 4$. The sum of these five is $x + (x + 1) + (x + 2) + (x + 3) + (x + 4) = 5x + 10$. We are told that this sum is 6 greater than the greatest of the five integers. The greatest of the five integers is $x + 4$, so we have $5x + 10 = x + 4 + 6$. Simplifying the right side gives $5x + 10 = x + 10$. Subtracting x from both sides and subtracting 10 from both sides gives $4x = 0$. Dividing by 4 gives us $x = 0$, so the smallest of the five integers is $\boxed{0}$.

3.32 If $x = 3$ is a solution to the equation, then the equation is true when we set $x = 3$. Letting $x = 3$ in the given equation makes the equation

$$b(3^2) + 3 \cdot 3 - 2b = 0.$$

Simplifying the left side gives us $9b + 9 - 2b = 0$. Combining the terms with b in them gives $7b + 9 = 0$. Subtracting 9 from both sides gives $7b = -9$, and dividing both sides by 7 gives $b = \boxed{-9/7}$.

3.33 We combine the $\sqrt[3]{z}$ terms by adding $\sqrt[3]{z}$ to both sides. On the left, we then have

$$\sqrt[3]{z} - 13 + \sqrt[3]{z} = (\sqrt[3]{z} + \sqrt[3]{z}) - 13 = 2\sqrt[3]{z} - 13.$$

So, our equation now is $2\sqrt[3]{z} - 13 = 5$. Adding 13 to both sides gives us $2\sqrt[3]{z} = 18$. Dividing by 2 gives us $\sqrt[3]{z} = 9$. We isolate z now by cubing both sides, since $(\sqrt[3]{z})^3 = z$. Therefore, our solution is $z = 9^3 = \boxed{729}$.

3.34   We start by simplifying the fraction on the right side:

$$\frac{z}{z+z+z+z} = \frac{\cancel{z}}{4\cancel{z}} = \frac{1}{4}.$$

Our equation is now $\dfrac{4-z}{4+z} = \dfrac{1}{4}$. Cross-multiplying gives $4(4-z) = 1(4+z)$. Expanding both sides gives $16 - 4z = 4 + z$. Adding $4z$ to both sides gives $16 = 4 + 5z$. Subtracting 4 from both sides results in $12 = 5z$. Finally, we divide both sides by 5 to find $z = \boxed{12/5}$. This value of z does not make any of the denominators 0 in the original equation, so it is a valid solution.

Challenge Problems

3.35   We first simplify the right side by combining the two r terms to get $3r^2 + r = 27 + r$. We then get all the r terms on one side by subtracting r from both sides. This gives us $3r^2 = 27$. Dividing both sides by 3 to isolate r^2 gives us $r^2 = 9$. We have to be a little careful here: there are two values of r that have 9 as their square. Both $\boxed{3 \text{ and } -3}$ have 9 as their square, so they are both solutions to the equation.

3.36   At first, it doesn't look like we can combine the two terms with y in them. However, we notice that 16 is the fourth power of 2, so $\sqrt[4]{16} = 2$. Therefore, we have

$$\sqrt[4]{16y} = \sqrt[4]{16} \cdot \sqrt[4]{y} = 2\sqrt[4]{y}.$$

So, our equation is $\sqrt[4]{y} + 2\sqrt[4]{y} - 2 = 4$. Combining the two $\sqrt[4]{y}$ terms on the left, we have $3\sqrt[4]{y} - 2 = 4$. Adding 2 to both sides gives $3\sqrt[4]{y} = 6$, and dividing both sides of this equation by 3 gives $\sqrt[4]{y} = 2$. We can isolate y by raising both sides of this equation to the fourth power, because $(\sqrt[4]{y})^4 = y$. So, we find $y = 2^4 = \boxed{16}$.

Because we raised the equation to an even power as a step in our solution, we must check to make sure our answer is not extraneous. Letting $y = 16$ in the original equation makes the left side

$$\sqrt[4]{16} + \sqrt[4]{16 \cdot 16} - 2 = 2 + \sqrt[4]{256} - 2 = 2 + 4 - 2 = 4.$$

Therefore, $y = 16$ does indeed satisfy the original equation.

3.37   Since the denominators of both fractions on the left side are the same, we can combine the two fractions by adding the numerators. This gives us

$$\frac{7}{2 + \sqrt{y}} = 1.$$

Multiplying both sides by $2 + \sqrt{y}$ gets rid of the fractions and gives us the equation $7 = 2 + \sqrt{y}$. Subtracting 2 from both sides gives $5 = \sqrt{y}$. Squaring both sides of this equation isolates y because $(\sqrt{y})^2 = y$. This gives us the solution $y = \boxed{25}$. Once again, we raised an equation to an even power as one step in our solution. So, we have to make sure our solution is not extraneous. Letting $y = 25$ in the left side of the original equation gives

$$\frac{3}{2 + \sqrt{25}} + \frac{4}{2 + \sqrt{25}} = \frac{3}{2 + 5} + \frac{4}{2 + 5} = \frac{3}{7} + \frac{4}{7} = 1,$$

so $y = 25$ is indeed a valid solution.

3.38 Let x be the number of each coin Stan has in his bag. Therefore, the quarters are together worth $0.25x$, the dimes are worth $0.10x$, and the nickels are worth $0.05x$. Combining these, the total value of the money in the bag is $0.25x + 0.10x + 0.05x = 0.40x$. In order for this to be a whole number of dollars, the expression $0.40x$ must equal a whole number. Trying $x = 1, 2, 3$, and 4 fails to make this expression a whole number, but for $x = 5$, we have $0.40(5) = 2$. So, the smallest amount of money Stan could have is $\boxed{\$2}$.

3.39

(a) The smallest integer is $x-1$ and the largest is $x+1$, so the sum of the three integers is $(x-1)+x+(x+1) = \boxed{3x}$.

(b) We could let x be the smallest integer, then write the other 22 integers in terms of x, but the previous part gives us an idea. We instead let x be the middle integer. Then, when we write all the integers in terms of x, our sum is

$$(x - 11) + (x - 10) + \cdots + (x - 1) + x + (x + 1) + \cdots + (x + 10) + (x + 11).$$

When we add all of these up, the -11 will cancel with the $+11$, the -10 will cancel with the $+10$, and so on, leaving only the 23 x terms left. Therefore,

$$(x - 11) + (x - 10) + \cdots + (x - 1) + x + (x + 1) + \cdots + (x + 10) + (x + 11) = 23x.$$

Since this sum equals 2323, we have $23x = 2323$. Dividing by 23 gives $x = 101$. So, the largest of the integers is $101 + 11 = \boxed{112}$.

3.40 We start by multiplying both sides by $1 - \frac{2}{z}$ in order to get rid of the complicated fraction. On the left we have

$$\frac{3}{1 - \frac{2}{z}} \cdot \left(1 - \frac{2}{z}\right) = \frac{3\left(1 - \frac{2}{z}\right)}{1 - \frac{2}{z}} = 3.$$

On the right, we have

$$3z \left(1 - \frac{2}{z}\right) = 3z - 3z \cdot \frac{2}{z} = 3z - \frac{6z}{z} = 3z - 6.$$

So, our equation is $3 = 3z - 6$. Adding 6 to both sides gives $9 = 3z$, and dividing by 3 gives $z = \boxed{3}$. Since $z = 3$ doesn't make any denominator equal to 0, it is a valid solution.

3.41 We start by simplifying both sides of the equation, which gives $x + 7 = x - b$. We could solve this equation for b to find $b = -7$, but that's not what we're asked to do! Lois picked b. However, she picked a b such that the left side, $x + 7$, is always 3 greater than the right side, $x - b$. So, we must have

$$(x + 7) - (x - b) = 3.$$

Simplifying the left side gives $(x + 7) - (x - b) = x + 7 - x + b = 7 + b$, so our equation is now $7 + b = 3$. Subtracting 7 from both sides gives $b = \boxed{-4}$. (Indeed, the original equation has no solution: Lois is being mean to Clark.)

3.42 At first it might look like we cannot easily combine the two terms on the left side because they have different denominators. However, the second denominator is just the negative of the first:

$$-(t-2) = -t - (-2) = -t + 2 = 2 - t.$$

Therefore, we can make the two denominators the same by multiplying the numerator and denominator of the second fraction by -1. This allows us to add the fractions easily:

$$\frac{3}{t-2} + \frac{9}{2-t} = \frac{3}{t-2} + \frac{9}{2-t} \cdot \frac{-1}{-1} = \frac{3}{t-2} + \frac{-9}{t-2} = \frac{3-9}{t-2} = \frac{-6}{t-2}.$$

Now our equation is $\dfrac{-6}{t-2} = 12$. Multiplying both sides by $t - 2$ gets rid of the fraction and leaves us $-6 = 12(t - 2)$. We might expand the right side, or first divide by 6 to give $-1 = 2(t - 2)$. Expanding the right side gives $-1 = 2t - 4$. Adding 4 to both sides of this gives $3 = 2t$, and dividing both sides of this equation by 2 gives us the solution $t = \boxed{3/2}$. This value of t does not make any of our denominators equal to 0, so it is a valid solution.

3.43 We first multiply both sides by $\sqrt{3 - \sqrt[3]{t}}$ to get rid of the nasty denominator. On the left we have

$$\cancel{\sqrt{3 - \sqrt[3]{t}}} \cdot \frac{\sqrt{3 + \sqrt[3]{t}}}{\cancel{\sqrt{3 - \sqrt[3]{t}}}} = \sqrt{3 + \sqrt[3]{t}},$$

and on the right we have $3\sqrt{3 - \sqrt[3]{t}}$, so our equation is

$$\sqrt{3 + \sqrt[3]{t}} = 3\sqrt{3 - \sqrt[3]{t}}.$$

Now, we have our next hurdle: the square root signs. Squaring both sides will take care of the big square root signs, because

$$\left(\sqrt{3 + \sqrt[3]{t}}\right)^2 = 3 + \sqrt[3]{t} \qquad \text{and} \qquad \left(3\sqrt{3 - \sqrt[3]{t}}\right)^2 = 3^2\left(\sqrt{3 - \sqrt[3]{t}}\right)^2 = 9(3 - \sqrt[3]{t}) = 27 - 9\sqrt[3]{t}.$$

So, we now have the equation $3 + \sqrt[3]{t} = 27 - 9\sqrt[3]{t}$. Now we're in familiar territory! Adding $9\sqrt[3]{t}$ to both sides gives $3 + 10\sqrt[3]{t} = 27$. Subtracting 3 from both sides gives $10\sqrt[3]{t} = 24$. Dividing both sides by 10 gives us $\sqrt[3]{t} = 12/5$. Finally, we can raise both sides of this equation to the third power to get $t = (12/5)^3 = \boxed{1728/125}$.

We can check this solution by substituting into the original left side of the equation. We know that when $t = 1728/125$, we have $\sqrt[3]{t} = 12/5 = 2.4$, so our left side is

$$\frac{\sqrt{3 + \sqrt[3]{t}}}{\sqrt{3 - \sqrt[3]{t}}} = \frac{\sqrt{3 + 2.4}}{\sqrt{3 - 2.4}} = \frac{\sqrt{5.4}}{\sqrt{0.6}} = \sqrt{\frac{5.4}{0.6}} = \sqrt{9} = 3.$$

3.44 We start by getting rid of the nasty denominator. We can do so by multiplying both sides by $\sqrt{x+1} - \sqrt{x-1}$. On the left, we have

$$(\sqrt{x+1} - \sqrt{x-1}) \cdot \frac{\sqrt{x+1} + \sqrt{x-1}}{\sqrt{x+1} - \sqrt{x-1}} = \frac{(\cancel{\sqrt{x+1} - \sqrt{x-1}})(\sqrt{x+1} + \sqrt{x-1})}{\cancel{\sqrt{x+1} - \sqrt{x-1}}} = \sqrt{x+1} + \sqrt{x-1}.$$

On the right, we have $3(\sqrt{x+1} - \sqrt{x-1}) = 3\sqrt{x+1} - 3\sqrt{x-1}$, so our equation is

$$\sqrt{x+1} + \sqrt{x-1} = 3\sqrt{x+1} - 3\sqrt{x-1}.$$

We have two $\sqrt{x+1}$ terms, so we can combine them by subtracting $\sqrt{x+1}$ from both sides. This will leave $\sqrt{x-1}$ on the left, and on the right we'll have

$$3\sqrt{x+1} - 3\sqrt{x-1} - \sqrt{x+1} = (3\sqrt{x+1} - \sqrt{x+1}) - 3\sqrt{x-1} = (3-1)\sqrt{x+1} - 3\sqrt{x-1} = 2\sqrt{x+1} - 3\sqrt{x-1}.$$

Now we have the equation $\sqrt{x-1} = 2\sqrt{x+1} - 3\sqrt{x-1}$. We can combine the two $\sqrt{x-1}$ terms by adding $3\sqrt{x-1}$ to both sides, which gives us $4\sqrt{x-1} = 2\sqrt{x+1}$. We can simplify this equation a little by dividing both sides by 2, which gives $2\sqrt{x-1} = \sqrt{x+1}$. Now our main obstacles are those square root signs. We can get rid of both of them by squaring both sides of the equation. On the left we have $(2\sqrt{x-1})^2 = 2^2(\sqrt{x-1})^2 = 4(x-1) = 4x-4$. On the right, squaring $\sqrt{x+1}$ gives us $x+1$, so our equation now is just $4x - 4 = x + 1$. Solving this equation, we find $x = \boxed{5/3}$. Make sure you plug this solution back in and see that it isn't extraneous!

4 {.chapter-number}

Exercises for Section 4.1

4.1.1

(a) $y - 2x = 6 - 2(-2) = 6 + 4 = \boxed{10}$.

(b) $3xy = 3(-2)(6) = \boxed{-36}$.

(c) $2x^2y + xy^2 = 2(-2)^2(6) + (-2)(6)^2 = 2(4)(6) + (-2)(36) = 48 - 72 = \boxed{-24}$.

(d) $\dfrac{x^2}{y+6} = \dfrac{(-2)^2}{6+6} = \dfrac{4}{12} = \boxed{\dfrac{1}{3}}$.

(e) $x^y = (-2)^6 = \boxed{64}$.

(f) $(2x - y)(2x + y) = [2(-2) - 6][2(-2) + 6] = (-4 - 6)(-4 + 6) = (-10)(2) = \boxed{-20}$.

4.1.2

(a) $ab + bc + ca = (3/2)(-1) + (-1)(6) + (6)(3/2) = -3/2 - 6 + 9 = \boxed{3/2}$.

(b) $ab^2c = (3/2)(-1)^2(6) = (3/2)(1)(6) = \boxed{9}$.

(c) $\dfrac{(2a + b)(c - 2)}{abc} = \dfrac{[2(3/2) + (-1)][6 - 2]}{(3/2)(-1)(6)} = \dfrac{(3 - 1)(4)}{-9} = \boxed{-\dfrac{8}{9}}$.

(d) $ca^b = (6)(3/2)^{-1} = (6)(2/3) = \boxed{4}$.

Exercises for Section 4.2

4.2.1

(a) $(2a - 3b) + (4a + 7b) = 2a - 3b + 4a + 7b = (2a + 4a) + (-3b + 7b) = \boxed{6a + 4b}$.

(b) $(6x - 9y + 2z) + (3y - 2z + 9x) = 6x - 9y + 2z + 3y - 2z + 9x = (6x + 9x) + (-9y + 3y) + (2z - 2z) = \boxed{15x - 6y}$.

(c) $\dfrac{d^2}{2} + 3c^2 - 7d^2 + 5c^2 = (3c^2 + 5c^2) + \left(\dfrac{d^2}{2} - 7d^2\right) = 8c^2 + \left(\dfrac{d^2}{2} - \dfrac{14d^2}{2}\right) = \boxed{8c^2 - \dfrac{13d^2}{2}}$.

(d) $\dfrac{a}{d} + \dfrac{3a}{d} + \dfrac{2}{d} + \dfrac{2a-2}{d} = \dfrac{a+3a+2+(2a-2)}{d} = \dfrac{(a+3a+2a)+(2-2)}{d} = \boxed{\dfrac{6a}{d}}.$

4.2.2 In each part we handle each variable and the constants separately.

(a) $x \cdot y \cdot y \cdot z \cdot y \cdot z \cdot x = (x \cdot x) \cdot (y \cdot y \cdot y) \cdot (z \cdot z) = \boxed{x^2 y^3 z^2}.$

(b) $a \cdot b \cdot a^3 \cdot a^2 \cdot b^2 = (a \cdot a^3 \cdot a^2) \cdot (b \cdot b^2) = (a^{1+3+2}) \cdot (b^{1+2}) = \boxed{a^6 b^3}.$

(c) $(3x^3 y^2 z)(2xy^5 z^5) = (3 \cdot 2) \cdot (x^3 \cdot x) \cdot (y^2 \cdot y^5) \cdot (z \cdot z^5) = 6x^{3+1} y^{2+5} z^{1+5} = \boxed{6x^4 y^7 z^6}.$

(d) $(3r^3)(2s^5)(2rs)(4r^2 s^3) = (3 \cdot 2 \cdot 2 \cdot 4) \cdot (r^3 \cdot r \cdot r^2) \cdot (s^5 \cdot s \cdot s^3) = 48r^{3+1+2} s^{5+1+3} = \boxed{48r^6 s^9}.$

4.2.3 In each part we handle each variable and the constants separately.

(a) $(x^7 y^3)^4 = (x^7)^4 (y^3)^4 = x^{7 \cdot 4} y^{3 \cdot 4} = \boxed{x^{28} y^{12}}.$

(b) $(-3v^3 z^4)^5 = (-3)^5 (v^3)^5 (z^4)^5 = -243 v^{3 \cdot 5} z^{4 \cdot 5} = \boxed{-243 v^{15} z^{20}}.$

4.2.4 Again, in each part we handle each variable and the constants separately.

(a) $\dfrac{14a^2 b^3}{21a^3 b^7} = \dfrac{14}{21} \cdot \dfrac{a^2}{a^3} \cdot \dfrac{b^3}{b^7} = \dfrac{2}{3} a^{2-3} b^{3-7} = \dfrac{2}{3} a^{-1} b^{-4} = \boxed{\dfrac{2}{3ab^4}}.$

(b) $\dfrac{-2x^3 y^5 z}{-8x^3 y^2 z^3} = \dfrac{-2}{-8} \cdot \dfrac{x^3}{x^3} \cdot \dfrac{y^5}{y^2} \cdot \dfrac{z}{z^3} = \dfrac{1}{4} \cdot 1 \cdot y^{5-2} z^{1-3} = \dfrac{1}{4} y^3 z^{-2} = \boxed{\dfrac{y^3}{4z^2}}.$

4.2.5 We could either note that $(32x^3 t^8)/(2xt^3) = 16x^2 t^5$, or we could reason our way to the answer as follows. To make 2 into 32, we must multiply by 16. We must multiply x by x^2 to get x^3, and we must multiply t^3 by t^5 to get t^8. So, we must multiply $2xt^3$ by $\boxed{16x^2 t^5}$ to get $32x^3 t^8$:

$$(2xt^3)(16x^2 t^5) = (2 \cdot 16)(x \cdot x^2)(t^3 \cdot t^5) = 32x^3 t^8.$$

Exercises for Section 4.3

4.3.1

(a) $3(2r - 8s) = 3(2r) - 3(8s) = \boxed{6r - 24s}.$

(b) $(x + y - 3z) \cdot (2x) = x(2x) + y(2x) - 3z(2x) = \boxed{2x^2 + 2xy - 6xz}.$

4.3.2 We have to keep careful track of our signs!

(a) $(x + 2y) - (3x - 2y) = x + 2y - 3x + 2y = (x - 3x) + (2y + 2y) = \boxed{-2x + 4y}.$

(b) $2(t^2 - 2ts + s^2) - 4(t^2 + 2ts + s^2) = 2t^2 - 4ts + 2s^2 - 4t^2 - 8ts - 4s^2 = (2t^2 - 4t^2) + (-4ts - 8ts) + (2s^2 - 4s^2) = \boxed{-2t^2 - 12ts - 2s^2}.$

4.3.3

(a) Both terms have a factor of 8, so $-8x + 24y = 8(-x) + 8(3y) = \boxed{8(-x + 3y)}$.

(b) Both terms have a factor of 5, a factor of x, and a factor of y:

$$20x^2y - 5xy = (5xy)(4x) + (5xy)(-1) = (5xy)(4x - 1) = \boxed{5xy(4x - 1)}.$$

(c) All three terms have a factor of r:

$$3r^3t^2 - 3r^2t + 7r = r(3r^2t^2) + r(-3rt) + r(7) = \boxed{r(3r^2t^2 - 3rt + 7)}.$$

(d) Each term has a factor of 3, a factor of a, and a factor of c:

$$-9a^3c^2 + 18a^2c^3 - 3abc = 3ac(-3a^2c) + 3ac(6ac^2) + 3ac(-b) = \boxed{3ac(-3a^2c + 6ac^2 - b)}.$$

4.3.4 We first factor the denominator of the first expression and the numerator of the second expression. We find

$$3a^2 - 6b = 3(a^2 - 2b) \qquad \text{and} \qquad 9a^3 - 18ab = 9a(a^2 - 2b).$$

So, our expression equals

$$\frac{2}{3(a^2 - 2b)} \cdot \frac{9a(a^2 - 2b)}{10a^2} = \frac{2 \cdot 9a(a^2 - 2b)}{3 \cdot 10a^2(a^2 - 2b)}.$$

Now we can do some canceling:

$$\frac{2 \cdot 9a(a^2 - 2b)}{3 \cdot 10a^2(a^2 - 2b)} = \frac{2 \cdot 3 \cdot 3\cancel{a}(\cancel{a^2 - 2b})}{3 \cdot 2 \cdot 5a \cdot \cancel{a}(\cancel{a^2 - 2b})} = \boxed{\frac{3}{5a}}.$$

4.3.5 Adding 2 copies of $-4x + 3y$ is the same as multiplying $-4x + 3y$ by 2. Similarly, adding 4 copies of $-4x + 3y$ is the same as multiplying $-4x + 3y$ by 4. So, finding the number of copies of $-4x + 3y$ that we must add together to get $-24x + 18y$ is the same as finding the number we must multiply $-4x + 3y$ by to get $-24x + 18y$. Since $(-4) \cdot 6 = -24$ and $3 \cdot 6 = 18$, we see that the desired number is $\boxed{6}$:

$$6(-4x + 3y) = 6(-4x) + 6(3y) = -24x + 18y.$$

4.3.6 Both of the terms $2x(y + 1)$ and $-6x^2(y + 1)$ have a factor of $y + 1$, so we can factor it out:

$$2x \cdot (y + 1) - 6x^2 \cdot (y + 1) = (2x - 6x^2)(y + 1).$$

We're not finished! We can factor a 2 and an x out of both terms of $2x - 6x^2$:

$$(2x - 6x^2)(y + 1) = [2x(1) + 2x(-3x)](y + 1) = [2x(1 - 3x)](y + 1) = \boxed{2x(1 - 3x)(y + 1)}.$$

4.3.7 We apply the distributive property, just as we have in many problems before:

$$a(b + c) = ab + ac.$$

Here, we let $a = x + 7$, $b = y$, and $c = -4$, so we have

$$(x + 7)(y - 4) = (x + 7)(y) + (x + 7)(-4).$$

We have multiplied $(x + 7)$ by each term in $(y - 4)$ and added the results. On the right we now have expressions we know how to expand:

$$(x + 7)(y) + (x + 7)(-4) = x \cdot y + 7 \cdot y + x \cdot (-4) + 7 \cdot (-4) = \boxed{xy + 7y - 4x - 28}.$$

Exercises for Section 4.4

4.4.1 Our denominators are x and y; the simplest common denominator is xy:

$$\frac{-4}{x} + \frac{7}{y} = \frac{-4}{x} \cdot \frac{y}{y} + \frac{7}{y} \cdot \frac{x}{x} = \frac{-4y}{xy} + \frac{7x}{xy} = \boxed{\frac{-4y + 7x}{xy}}.$$

4.4.2 Our denominators are $9x^2$ and $3x$. We can multiply $3x$ by $3x$ to get $9x^2$, so the simplest common denominator is $9x^2$:

$$\frac{2y}{9x^2} - \frac{6-y}{3x} = \frac{2y}{9x^2} - \frac{6-y}{3x}\cdot\frac{3x}{3x} = \frac{2y}{9x^2} - \frac{(6-y)(3x)}{9x^2} = \frac{2y}{9x^2} - \frac{18x-3xy}{9x^2} = \frac{2y-(18x-3xy)}{9x^2} = \boxed{\frac{-18x+2y+3xy}{9x^2}}.$$

4.4.3 Our denominators are $6ab^2$ and $9a^2b$. We tackle the constants and each variable separately. First, the least common multiple of 6 and 9 is 18, so the constant in our simplest common denominator is 18. One denominator has a and the other has a^2, so the simplest common denominator has a factor of a^2. Similarly, one denominator has b^2 and the other has b, so the simplest common denominator has a factor of b^2. So, our simplest common denominator is $18a^2b^2$.

We must multiply $6ab^2$ by $3a$ to get $18a^2b^2$, and we must multiply $9a^2b$ by $2b$ to get $18a^2b^2$. We're ready to add our fractions now:

$$\begin{aligned}
\frac{2+a}{6ab^2} + \frac{9-b}{9a^2b} &= \frac{2+a}{6ab^2}\cdot\frac{3a}{3a} + \frac{9-b}{9a^2b}\cdot\frac{2b}{2b}\\
&= \frac{(2+a)(3a)}{18a^2b^2} + \frac{(9-b)(2b)}{18a^2b^2}\\
&= \frac{(2+a)(3a) + (9-b)(2b)}{18a^2b^2}\\
&= \frac{6a + 3a^2 + 18b - 2b^2}{18a^2b^2}\\
&= \boxed{\frac{3a^2 + 6a - 2b^2 + 18b}{18a^2b^2}}.
\end{aligned}$$

4.4.4 We start by simplifying each fraction by first factoring the numerator and denominator of each as much as possible. We find

$$\frac{8r-8s}{2r^2-2rs} = \frac{8(r-s)}{2r(r-s)} = \frac{2\cdot 4\cancel{(r-s)}}{\cancel{2}r\cancel{(r-s)}} = \frac{4}{r}$$

and

$$\frac{3r^2}{rs-r} = \frac{3r\cdot\cancel{r}}{\cancel{r}(s-1)} = \frac{3r}{s-1}.$$

So, now our addition is simply $\dfrac{4}{r} + \dfrac{3r}{s-1}$. Since our denominators, r and $s-1$, have no common factors, the simplest common denominator is their product $r(s-1)$:

$$\frac{4}{r} + \frac{3r}{s-1} = \frac{4}{r}\cdot\frac{s-1}{s-1} + \frac{3r}{s-1}\cdot\frac{r}{r} = \frac{4(s-1)}{r(s-1)} + \frac{3r^2}{r(s-1)} = \frac{4(s-1)+3r^2}{r(s-1)} = \boxed{\frac{3r^2+4s-4}{r(s-1)}}.$$

Exercises for Section 4.5

4.5.1

(a) We isolate x by first isolating $\frac{x}{a}$ by subtracting b from both sides to get $\frac{x}{a} = c - b$. We then isolate x by multiplying both sides by a to find $x = \boxed{a(c - b)}$. (We can also write this final expression as $x = ac - ab$.)

(b) Isolating b is easier than isolating x; all we have to do is subtract $\dfrac{x}{a}$ from both sides to get

$$b = \boxed{c - \frac{x}{a}}.$$

(c) Isolating a is a little trickier. However, looking back at the first part, we see our opportunity to isolate a. There, we found that $x = a(c - b)$. We isolate a by dividing by $c - b$, to get $a = \boxed{\dfrac{x}{c - b}}$.

4.5.2 First, we move all terms with x to the left by adding $2x$ to both sides. This gives us $3xy + 2x + 4 = 8y$. We isolate the terms with x on the left side by subtracting 4 from both sides, which gives $3xy + 2x = 8y - 4$. We factor x out of both terms on the left to give $x(3y + 2) = 8y - 4$. Dividing both sides by $3y + 2$ isolates x and gives us

$$x = \boxed{\frac{8y - 4}{3y + 2}}.$$

Review Problems

4.18

(a) $ab + 2b + 3a = (-8)(1/2) + 2(1/2) + 3(-8) = -4 + 1 - 24 = \boxed{-27}$.

(b) $\dfrac{a}{b} = \dfrac{-8}{1/2} = \boxed{-16}$.

(c) $4(a^2b + b^2a) = 4[(-8)^2(1/2) + (1/2)^2(-8)] = 4[(64)(1/2) + (1/4)(-8)] = 4[32 - 2] = 4(30) = \boxed{120}$.

(d) $(a - 2)\sqrt{-ab} = (-8 - 2)\sqrt{-(-8)(1/2)} = -10\sqrt{4} = \boxed{-20}$.

4.19

(a) $t = 3r - 2s = 3(-2) - 2(6) = -6 - 12 = \boxed{-18}$. We can now use $t = -18$ for the remaining parts.

(b) $(2t - r)(s - 1) = [2(-18) - (-2)](6 - 1) = (-36 + 2)(5) = (-34)(5) = \boxed{-170}$.

(c) $3r - 2s - t = 3(-2) - 2(6) - (-18) = -6 - 12 + 18 = \boxed{0}$. Note that because $t = 3r - 2s$, we could have tackled this part by noting that $3r - 2s - t = (3r - 2s) - (3r - 2s) = 0$.

(d) $\dfrac{t}{r} + \dfrac{t}{s} = \dfrac{-18}{-2} + \dfrac{-18}{6} = 9 - 3 = \boxed{6}$.

4.20   When $a = -5b$ we have

$$2a+3b+7-a+2b+5 = 2(-5b)+3b+7-(-5b)+2b+5 = -10b+3b+7+5b+2b+5 = (-10b+3b+5b+2b)+(7+5) = \boxed{12}.$$

4.21

(a) $(6a - 7b) + (3a - 2b + 9) = 6a - 7b + 3a - 2b + 9 = (6a + 3a) + (-7b - 2b) + 9 = \boxed{9a - 9b + 9}$.

(b) $(6ab+2ac+3bc)+(2ab-3ac-4bc) = 6ab+2ac+3bc+2ab-3ac-4bc = (6ab+2ab)+(2ac-3ac)+(3bc-4bc) = \boxed{8ab - ac - bc}$.

4.22

(a) $r^2 \cdot t \cdot r^3 \cdot t^3 = (r^2 \cdot r^3) \cdot (t \cdot t^3) = r^{2+3}t^{1+3} = \boxed{r^5 t^4}$

(b) $(3x^3yz^2)(2x^3y^3) = (3 \cdot 2)(x^3 \cdot x^3)(y \cdot y^3)(z^2) = (6)(x^{3+3})(y^{1+3})(z^2) = \boxed{6x^6y^4z^2}$.

(c) $(x^2y^5)^7 = (x^2)^7(y^5)^7 = x^{2 \cdot 7}y^{5 \cdot 7} = \boxed{x^{14}y^{35}}$.

(d) $(2a^3)^3(3ab^2)^2 = [2^3(a^3)^3][3^2a^2(b^2)^2] = (8a^9)(9a^2b^4) = (8 \cdot 9)(a^9 \cdot a^2)(b^4) = \boxed{72a^{11}b^4}$.

(e) $(-2c^3d)^2 + (4cd^2)(-3c^5) = [(-2)^2(c^3)^2d^2] + [4 \cdot (-3)](c \cdot c^5)(d^2) = 4c^6d^2 - 12c^6d^2 = \boxed{-8c^6d^2}$.

(f) First we simplify $\sqrt{4a^4b^8}$:

$$\sqrt{4a^4b^8} = (4a^4b^8)^{1/2} = 4^{1/2}(a^4)^{1/2}(b^8)^{1/2} = 2a^2b^4.$$

So, we have

$$a^3b^2\sqrt{4a^4b^8} = a^3b^2(2a^2b^4) = 2(a^3a^2)(b^2b^4) = \boxed{2a^5b^6}.$$

4.23   As with our other products raised to powers, we handle the constant and each variable separately:

$$(16x^8y^4z^{16})^{1/4} = (16)^{1/4}(x^8)^{1/4}(y^4)^{1/4}(z^{16})^{1/4} = \boxed{2x^2yz^4}.$$

4.24

(a) $\dfrac{15a^2b^3}{5a^3b} = \dfrac{15}{5} \cdot \dfrac{a^2}{a^3} \cdot \dfrac{b^3}{b} = 3a^{2-3}b^{3-1} = 3a^{-1}b^2 = \boxed{\dfrac{3b^2}{a}}$.

(b) $\dfrac{2x^3z^5}{(2xz)^4} = \dfrac{2x^3z^5}{16x^4z^4} = \dfrac{2}{16} \cdot \dfrac{x^3}{x^4} \cdot \dfrac{z^5}{z^4} = \dfrac{1}{8}x^{3-4}z^{5-4} = \dfrac{1}{8}x^{-1}z^1 = \boxed{\dfrac{z}{8x}}$.   Once you're more comfortable with

algebra, you won't have to show all these steps. You'll see $\dfrac{2x^3z^5}{16x^4z^4}$, and you'll think, "The 2 in the numerator cancels with one of the 2's in 16 to leave 8 in the denominator. There's one more x in the denominator than in the numerator, so after we cancel x's, we'll just have an x in the denominator. Similarly, there's one more z in the numerator than in the denominator, so there will be one z left in the numerator after canceling."

4.25   In each part, we just use the distributive property:

(a) $2(a + b - 3c - 5) = 2a + 2b + 2(-3c) + 2(-5) = \boxed{2a + 2b - 6c - 10}$.

(b) $2x^2y\left(x^3y + \dfrac{4}{xy}\right) = 2x^2y(x^3y) + 2x^2y\left(\dfrac{4}{xy}\right) = 2x^{2+3}y^{1+1} + \dfrac{8x^2y}{xy} = \boxed{2x^5y^2 + 8x}$.

4.26 We need an expression such that

$$(2x - 3y + z) - (\text{Expression}) = 3x + 2y.$$

We tackle each variable separately. We must subtract $-x$ from $2x$ to get $3x$: $2x - (-x) = 2x + x = 3x$. We must subtract $-5y$ from $-3y$ to get $+2y$: $-3y - (-5y) = -3y + 5y = 2y$. We have no z in our final result, so to eliminate the z in $(2x - 3y + z)$, we must subtract one z. Putting these together, we need to subtract $\boxed{-x - 5y + z}$ from $2x - 3y + z$ to get $3x + 2y$:

$$(2x - 3y + z) - (-x - 5y + z) = 2x - 3y + z - (-x) - (-5y) - z = (2x + x) + (-3y + 5y) + (z - z) = 3x + 2y.$$

Notice that $(2x - 3y + z) - (3x + 2y) = -x - 5y + z$, which is our answer. Is this a coincidence?

4.27

(a) $-3w - 2x + 5 - (2w - 3x - 4) = -3w - 2x + 5 - 2w - (-3x) - (-4) = -3w - 2x + 5 - 2w + 3x + 4 = (-3w - 2w) + (-2x + 3x) + (5 + 4) = \boxed{-5w + x + 9}$.

(b)

$$2(r^2 - 3s) - 3(2r^2 + 2r - s) = 2r^2 + 2(-3s) + (-3)(2r^2) + (-3)(2r) + (-3)(-s)$$
$$= 2r^2 - 6s - 6r^2 - 6r + 3s$$
$$= \boxed{-4r^2 - 6r - 3s}.$$

4.28 When $x = -y$, we have

$$4(x^2 - 2y + 7) - 2(2y^2 + 4x + 2) = 4[(-y)^2 - 2y + 7] - 2[2y^2 + 4(-y) + 2]$$
$$= 4(y^2 - 2y + 7) - 2(2y^2 - 4y + 2)$$
$$= 4y^2 - 8y + 28 - 4y^2 + 8y - 4$$
$$= (4y^2 - 4y^2) + (-8y + 8y) + (28 - 4)$$
$$= \boxed{24}.$$

4.29

(a) Both terms have a factor of 7: $7x - 35y^2 = 7(x) + 7(-5y^2) = \boxed{7(x - 5y^2)}$.

(b) Both terms have a factor of 3 and a factor of a:

$$21ab^2 - 24a = 3a(7b^2) + 3a(-8) = \boxed{3a(7b^2 - 8)}.$$

(c) Both terms have a factor of x^2: $-20x^3 - 13x^2yz = x^2(-20x) + x^2(-13yz) = \boxed{x^2(-20x - 13yz)}$.

(d) All three terms have a factor of 6, a factor of r, a factor of s, and a factor of t. So, we can factor out $6rst$ from each term:

$$6rst(2t) + 6rst(4r) + 6rst(-3s) = \boxed{6rst(2t + 4r - 3s)}.$$

4.30   Our denominators are a, b, and c. These denominators have no factors in common, so the simplest common denominator is abc:

$$\frac{1}{a} + \frac{1}{b} + \frac{1}{c} = \frac{1}{a} \cdot \frac{bc}{bc} + \frac{1}{b} \cdot \frac{ac}{ac} + \frac{1}{c} \cdot \frac{ab}{ab} = \frac{bc}{abc} + \frac{ac}{abc} + \frac{ab}{abc} = \boxed{\frac{bc + ac + ab}{abc}}.$$

4.31   Our denominators are $14y^2z^4$ and $18x^3z^2$. We handle each variable and the constant separately.

First, the least common multiple of 14 and 18 is 126. (We can see this by noting that $14 = 2 \cdot 7$ and $18 = 2 \cdot 9$, so the denominator needs a factor of 2, a factor of 7, and a factor of 9, which gives us $2 \cdot 7 \cdot 9 = 126$ as the smallest possible constant for the common denominator.)

Turning to the variables, one denominator has x^3 and the other has no x at all, so we need x^3 in the common denominator. Similarly, one denominator has y^2 and the other has no y, so we need y^2 in the common denominator. Finally, one denominator has z^4 and the other has z^2. We need the higher of these powers of z in the denominator (make sure you see why), so the denominator needs a z^4. Putting these together with our constant from earlier, our simplest common denominator is $126x^3y^2z^4$.

We must multiply $14y^2z^4$ by $9x^3$ to get $126x^3y^2z^4$, and we must multiply $18x^3z^2$ by $7y^2z^2$ to get $126x^3y^2z^4$, so we have:

$$\frac{3x}{14y^2z^4} - \frac{5y}{18x^3z^2} = \frac{3x}{14y^2z^4} \cdot \frac{9x^3}{9x^3} - \frac{5y}{18x^3z^2} \cdot \frac{7y^2z^2}{7y^2z^2}$$

$$= \frac{(3x)(9x^3)}{126x^3y^2z^4} - \frac{(5y)(7y^2z^2)}{126x^3y^2z^4}$$

$$= \frac{27x^4}{126x^3y^2z^4} - \frac{35y^3z^2}{126x^3y^2z^4}$$

$$= \boxed{\frac{27x^4 - 35y^3z^2}{126x^3y^2z^4}}.$$

4.32   We start by simplifying both fractions. We do so by first factoring the numerator and denominator of each fraction. We find

$$\frac{2a^2 - 4a}{3a - 6} = \frac{2a(a - 2)}{3(a - 2)} = \frac{2a\cancel{(a - 2)}}{3\cancel{(a - 2)}} = \frac{2a}{3}$$

and

$$\frac{2b^2 - 4b}{8b - 16} = \frac{2b(b - 2)}{8(b - 2)} = \frac{2b\cancel{(b - 2)}}{8\cancel{(b - 2)}} = \frac{2b}{8} = \frac{b}{4}.$$

So, our addition is now $\dfrac{2a}{3} + \dfrac{b}{4}$. This is much simpler. Our simplest common denominator is 12, and we have

$$\frac{2a}{3} + \frac{b}{4} = \frac{8a}{12} + \frac{3b}{12} = \boxed{\frac{8a + 3b}{12}}.$$

4.33   We first isolate $3x$ by subtracting both $2y$ and z from both sides. This gives us $3x = 4 - 2y - z$. We then isolate x by dividing both sides by 3, which gives us

$$x = \boxed{\frac{4 - 2y - z}{3}}.$$

Challenge Problems

4.34 Let's first simplify the numerator and the denominator. The numerator is

$$\frac{3r^2t - 6rt^2}{6r^2t^3} = \frac{3rt(r - 2t)}{6r^2t^3} = \frac{\cancel{3}\cancel{r}t(r - 2t)}{2 \cdot \cancel{3}r \cdot \cancel{r} \cdot t \cdot t \cdot t} = \frac{r - 2t}{2rt^2}.$$

The denominator is

$$\frac{8t^3r - 4t^2r^2}{6r^3t^4} = \frac{4rt^2(2t - r)}{6r^3t^4} = \frac{2 \cdot 2 \cdot \cancel{r} \cdot \cancel{t^2}(2t - r)}{\cancel{2} \cdot 3\cancel{r} \cdot r^2 \cdot \cancel{t^2} \cdot t^2} = \frac{2(2t - r)}{3r^2t^2}.$$

So, our giant fraction is a little simpler now: $\dfrac{\dfrac{r - 2t}{2rt^2}}{\dfrac{2(2t - r)}{3r^2t^2}}$. To divide one fraction by a second fraction, we

multiply the first fraction by the reciprocal of the second. So, our expression equals:

$$\frac{r - 2t}{2rt^2} \cdot \frac{3r^2t^2}{2(2t - r)} = \frac{3r^2t^2(r - 2t)}{4rt^2(2t - r)} = \frac{3 \cdot \cancel{rt^2} \cdot r \cdot \cancel{(r - 2t)}}{4 \cdot \cancel{rt^2} \cdot (-1) \cdot \cancel{(r - 2t)}} = \boxed{-\frac{3r}{4}}.$$

4.35

(a) We use the distributive property, $a(b + c) = ab + ac$. Here, we let the entire expression $x + 1$ be a, so that

$$(x + 1)(y + 1) = (x + 1)(y) + (x + 1)(1) = (xy + y) + (x + 1) = \boxed{xy + x + y + 1}.$$

(b) We do the same thing, but with different numbers:

$$(x + 3)(y - 7) = (x + 3)(y) + (x + 3)(-7) = (xy + 3y) + (-7x - 21) = \boxed{xy - 7x + 3y - 21}.$$

4.36 To get a better idea how to approach the problem, we think about a simpler, but similar, problem: By what fraction can we multiply A to get B? This is much simpler indeed:

$$A \cdot \frac{B}{A} = B.$$

We multiply A by the fraction $\frac{B}{A}$ to get B. Returning to our question, we see that all we have to do is view $\dfrac{3x^3}{2y^5}$ as A and $\dfrac{6y^2}{5x^2}$ as B. We have to multiply A by $\frac{B}{A}$ to get B. Using our big nasty fractions for A and B, we have

$$\frac{B}{A} = \frac{\dfrac{6y^2}{5x^2}}{\dfrac{3x^3}{2y^5}} = \frac{6y^2}{5x^2} \cdot \frac{2y^5}{3x^3} = \frac{12y^7}{15x^5} = \boxed{\frac{4y^7}{5x^5}}.$$

4.37

(a) Because the simplest common denominator of Fiona's and Henry's fractions is $4a^3b$, neither of their fractions has a power of b higher than b. However, the simplest common denominator of George's and Henry's fractions is $20a^3b^2$, so one of the denominators of these two fractions must have a b^2. We know Henry's denominator doesn't have a power of b higher than just b, so it must be $\boxed{\text{George's}}$ fraction that has a denominator with $\boxed{b^2}$.

(b) Because the simplest common denominator of Fiona's and Henry's fractions is $4a^3b$, neither of their fractions has a constant in their denominator that is larger than 4. However, the simplest common denominator of George's and Henry's fractions is $20a^3b^2$, so one of the denominators of these two fractions must have a factor of 5. We know Henry's doesn't have such a factor, since Henry's denominator's constant is at most 4. Therefore, George's denominator has a factor of 5, and therefore must be larger than 4. So, we know that the denominator of $\boxed{\text{George's}}$ fraction has the largest constant.

(c) We know at least one of the fractions has a factor of 5, and one has a factor of 4. There are no higher powers of 2 or higher powers of 5 in any of the denominators, nor other factors (otherwise, these would show up in some of the three simplest common denominators we are given). Therefore, the constant in the common denominator of all three fractions is 20. Similarly, we know that the highest power of a that appears in the denominators is a^3 (in Henry's fraction – see if you can figure out why) and the highest power of b is b^2. Therefore, the simplest common denominator is $\boxed{20a^3b^2}$.

4.38

(a) We have $(x-2)(x+2) = (x-2)(x) + (x-2)(2) = x^2 - 2x + 2x - 4 = \boxed{x^2 - 4}$.

(b) The previous part gives us a big clue! The previous part showed us that

$$(x-2)(x+2) = x^2 - 2^2.$$

This should make us wonder what happens if we replace the 2's with y's. Let's see:

$$(x-y)(x+y) = (x-y)(x) + (x-y)(y) = x^2 - xy + xy - y^2 = x^2 - y^2.$$

Aha! It worked! So, we have the factorization $x^2 - y^2 = \boxed{(x-y)(x+y)}$.

4.39 In order to get x into the expression $\dfrac{a+b}{a-b}$, we solve the equation $x = \dfrac{a}{b}$ for a and substitute it into the expression. Multiplying both sides of $x = \dfrac{a}{b}$ by b gives $bx = a$. Replacing a in our fraction with this expression gives

$$\frac{a+b}{a-b} = \frac{bx+b}{bx-b}.$$

We can factor a b out of the numerator and denominator of the fraction:

$$\frac{bx+b}{bx-b} = \frac{b(x+1)}{b(x-1)} = \boxed{\frac{x+1}{x-1}}.$$

4.40 First we note that the expression on the left can never be 0, because the only way a fraction can equal 0 is if the numerator is 0. Here, the numerator is constant at 1, so the numerator can never equal 0. Therefore, if $a = 0$, then there is no solution because the fraction on the left can never equal 0. However, are there any other illegal values of a?

We can see that $a = 1$ is impossible in two different ways. First, if $a = 1$, then we must have

$$\frac{1}{1 + \frac{1}{x}} = 1,$$

which means that $1 = 1 + \frac{1}{x}$, which is impossible because we cannot ever have $\frac{1}{x} = 0$. The other way we can see that $a = 1$ makes the equation have no solution for x is to solve the equation for x in terms of a. We start by taking the reciprocal of both sides to get

$$1 + \frac{1}{x} = \frac{1}{a}.$$

We can isolate $\frac{1}{x}$ by subtracting 1 from both sides:

$$\frac{1}{x} = \frac{1}{a} - 1.$$

We then find x in terms of a by taking the reciprocal of both sides:

$$x = \frac{1}{\frac{1}{a} - 1}.$$

Immediately, we see that $a = 0$ is a problem, since we'd then be dividing by 0. We also have to worry about the whole denominator $\frac{1}{a} - 1$. This equals 0 when $a = 1$, so if $a = 1$, there's no solution for x.

Therefore, there are no solutions for x in our equation if $a = \boxed{0 \text{ or } 1}$.

4.41 We know that x cannot be 3, since then we would be dividing by 0 on the left. However, it's not at all clear how this tells us what values of a lead to no solution for x. So, we borrow a tactic from the previous problem and solve the equation for x in terms of a. We hope that this will give us a denominator with a in it that will guide us to the answer.

We start by multiplying both sides by $x - 3$, which gives $6x - a = 3(x - 3) = 3x - 9$. We subtract $3x$ from both sides to get $3x - a = -9$. We then isolate $3x$ by adding a to both sides, which gives $3x = a - 9$. Dividing both sides by 3 gives us

$$x = \frac{a - 9}{3}.$$

Hmmm... This doesn't seem to immediately eliminate any values of a. But we do know that we cannot have $x = 3$, so whatever value of a gives us $x = 3$ is the value of a that gives us no valid solution for x. Therefore, we must solve the equation

$$3 = \frac{a - 9}{3}.$$

Solving this equation gives us $a = \boxed{18}$. We should be curious why this value of a gives us no solution for x in the original equation. Let's substitute $a = 18$ into the equation and take a look:

$$\frac{6x - 18}{x - 3} = 3.$$

We can factor the numerator of the left side, then simplify:

$$\frac{6x - 18}{x - 3} = \frac{6(x - 3)}{x - 3} = 6.$$

This makes the equation $6 = 3$, which clearly has no solution.

5

Multi-Variable Linear Equations

Exercises for Section 5.1

5.1.1

(a) We add $7y$ to both sides to get $2x = 7y + 5$. Dividing by 2 gives $\boxed{x = \dfrac{7y + 5}{2}}$.

(b) We subtract $2x$ from both sides to get $-7y = -2x + 5$. Dividing by -7 gives $\boxed{y = \dfrac{2x - 5}{7}}$.

5.1.2 We first solve the equation for q in terms of p. We add $4p$ to both sides to get $5q = 1 + 4p$. Dividing both sides by 5 gives $q = (1 + 4p)/5$. Now, we can just pick values of p and find corresponding values of q. For example, when $p = 1$, we have $q = (1 + 4 \cdot 1)/5 = 1$, so $(p, q) = (1, 1)$ satisfies the equation. When $p = 2$, we have $q = (1 + 8)/5 = 9/5$, and when $p = 3$, we have $q = (1 + 12)/5 = 13/5$. Therefore, three ordered pairs (p, q) that satisfy the equation are $(p, q) = (1, 1)$; $(2, 9/5)$; and $(3, 13/5)$.

These are not the only solutions. To test any other solutions you found, put your value for p into the equation $q = (1 + 4p)/5$. If the resulting q matches your q, then your ordered pair satisfies the original equation. If it doesn't match, then something went wrong.

5.1.3 We use the similar problem in the text as a guide. We make a list of solutions to both equations and try to find a number that is on both lists. We solve the equation $2x - 3y = -5$ for x by adding $3y$ to both sides to get $2x = -5 + 3y$, then dividing both sides by 2 to get $x = (-5 + 3y)/2$. So, when $y = 1$, we have $x = (-5 + 3)/2 = -1$. Therefore $(-1, 1)$ is a solution. Similarly, we let $y = 2, 3, 4$, and 5, and we find that $(x, y) = (1/2, 2), (2, 3), (7/2, 4)$, and $(5, 5)$ are all solutions.

Turning to the other equation, we again solve for x. Adding $2y$ to both sides, we have $5x = 4 + 2y$. Dividing by 5 gives $x = (4 + 2y)/5$. We try the same y values as we did with the previous equation, and we find that when $y = 3$, we have $x = (4 + 6)/5 = 2$. So, the pair $(x, y) = \boxed{(2, 3)}$ satisfies both equations.

5.1.4

(a) We solve $5x - 6y = 1$ for x in terms of y and find $x = (1 + 6y)/5$. Letting $y = -1$, we have $x = (1 - 6)/5 = -1$. When $y = 0$, we have $x = 1/5$. When $y = 1$, we have $x = (1 + 6)/5 = 7/5$. So, three pairs that satisfy the first equation are $(x, y) = (-1, -1), (1/5, 0)$, and $(7/5, 1)$.

Turning to the second equation, we solve $15x - 18y = 3$ for x in terms of y to find $x = (3 + 18y)/15$.

When $y = -1$, we have $x = (3 - 18)/15 = -1$. When $y = 0$, we have $x = 3/15 = 1/5$. When $y = 1$, we have $x = (3 + 18)/15 = 21/15 = 7/5$. So, the pairs $(-1, -1)$, $(1/5, 0)$, and $(7/5, 1)$ all satisfy the second equation as well.

(b) We take a closer look at our two equations to see why the three solutions we found to the first equation also worked for the second equation. We saw that solving $5x - 6y = 1$ for x in terms of y gives $x = (1 + 6y)/5$. When we solve $15x - 18y = 3$ for x in terms of y, we have $x = (3 + 18y)/15$. Looking more closely at the right side, we see that we can simplify it:

$$x = \frac{3 + 18y}{15} = \frac{\cancel{3}(1 + 6y)}{\cancel{3}(5)} = \frac{1 + 6y}{5}.$$

Aha! Now we see why the solutions to the first equation work for the second as well. In both equations, when we solve for x in terms of y, we get $x = (1 + 6y)/5$. Therefore, for any value of y, the corresponding value of x is the same in both equations. So, all pairs (x, y) that satisfy one equation will also satisfy the other.

5.1.5 We could solve the equation for x in terms of y, but we're looking for a little more clever solution, since we can't write anything down. So, using the discussion in the text as inspiration, we look for ways to change x and y without changing the value of $3x - 5y$. We see that if we increase x by 5 and increase y by 3, then $3x - 5y$ remains unchanged:

$$3(x + 5) - 5(y + 3) = 3x + 15 - 5y - 15 = 3x - 5y.$$

So, because $(1.2, 1.1)$ gives us $3x - 5y = -1.9$, we know we can increase x by 5 and y by 3 to get another solution: $(6.2, 4.1)$. We can check this solution to make sure we didn't make a mistake: $3(6.2) - 5(4.1) = 18.6 - 20.5 = -1.9$, as expected. So, we can continue increasing x by 5 and y by 3 to find more points: $(x, y) = (11.2, 7.1), (16.2, 10.1), (21.2, 13.1), \ldots$

These aren't the only solutions; you may have found many others.

Exercises for Section 5.2

5.2.1

(a) Solving the first equation for y gives $y = 10 - 2x$. Substituting this in the second equation gives

$$3x - 4(10 - 2x) = 37 \quad \Rightarrow \quad 3x - 40 + 8x = 37 \quad \Rightarrow \quad 11x - 40 = 37 \quad \Rightarrow \quad 11x = 77 \quad \Rightarrow \quad x = 7.$$

Therefore, $y = 10 - 2x = 10 - 2(7) = -4$, so our solution is $(x, y) = \boxed{(7, -4)}$.

(b) Dividing the second equation by 2 gives $y = (3x + 4)/2$, and substituting this into the first equation gives

$$5x = 6\left(\frac{3x + 4}{2}\right) - 4 \quad \Rightarrow \quad 5x = 3(3x + 4) - 4 \quad \Rightarrow \quad 5x = 9x + 8 \quad \Rightarrow \quad -4x = 8 \quad \Rightarrow \quad x = -2.$$

So, $y = [3(-2) + 4]/2 = (-6 + 4)/2 = -1$, and our solution is $(x, y) = \boxed{(-2, -1)}$.

Notice we also could have multiplied the second equation by 3 to get $6y = 9x + 12$ and substituted this for $6y$ in the first equation.

(c) We start by getting rid of the fractions. We multiply both equations by 6 to get rid of all the fractions. The first equation then becomes $4r + 5s = 33$ and the second equation becomes $4s = 14 + 3r$. Dividing the second equation by 4 gives $s = (14 + 3r)/4$. We then substitute this into the first equation:

$$4r + 5\left(\frac{14 + 3r}{4}\right) = 33 \quad \Rightarrow \quad 4r + \frac{70 + 15r}{4} = 33 \quad \Rightarrow \quad \frac{16r}{4} + \frac{70 + 15r}{4} = 33 \quad \Rightarrow \quad \frac{31r + 70}{4} = 33.$$

Cross-multiplying gives $31r + 70 = 132$. Subtracting 70 from both sides gives $31r = 62$ and dividing this by 31 gives $r = 2$. So, $s = (14 + 3 \cdot 2)/4 = 5$ and our solution is $(r, s) = \boxed{(2, 5)}$.

(d) First, we organize each equation by getting the variables on one side and the constants on the other. This makes our equations $2.2x - 3.1y = -3.2$ and $0.4x + y = 8.8$. Solving the second equation for y in terms of x gives $y = 8.8 - 0.4x$. Substituting this into our other equation gives

$$2.2x - 3.1(8.8 - 0.4x) = -3.2 \quad \Rightarrow \quad 2.2x - 27.28 + 1.24x = -3.2 \quad \Rightarrow \quad 3.44x = 24.08 \quad \Rightarrow \quad x = 7.$$

So, $y = 8.8 - 0.4x = 6$, and our solution is $(x, y) = \boxed{(7, 6)}$.

5.2.2

(a) If $x = 7$, we have $7 = 2 - t$, so $t = -5$. Therefore, $y = 4(-5) + 7 = \boxed{-13}$.

(b) If $x = -3$, then $-3 = 2 - t$, so $t = 5$. Therefore, $y = 4(5) + 7 = \boxed{27}$.

(c) In the first two parts, we found y given x by first finding t. Therefore, we try that tactic here by first solving $x = 2 - t$ for t in terms of x: $t = 2 - x$. Now, we substitute this expression into $y = 4t + 7$:

$$y = 4(2 - x) + 7 = 8 - 4x + 7 = -4x + 15.$$

So, we have $\boxed{y = -4x + 15}$.

5.2.3

(a) Adding $6y$ to both parts gives $5x = 1 + 6y$, so $x = (1 + 6y)/5$.

(b) We have

$$15\left(\frac{1 + 6y}{5}\right) - 18y = 3 \quad \Rightarrow \quad 3(1 + 6y) - 18y = 3 \quad \Rightarrow \quad 3 + 18y - 18y = 3 \quad \Rightarrow \quad 3 = 3.$$

Hmmm... This equation is always true, no matter what x and y are.

(c) In our first two parts, we solved the first equation for x in terms of y, and found that the pair $(x, y) = (\frac{1+6y}{5}, y)$ is always a solution to the first equation. We then substituted this pair into the second equation, and found that the resulting equation is always true, no matter what y is. In other words, any pair that satisfies the first equation also satisfies the second equation.

5.2.4 Instead of substituting for a variable, we notice that 273 appears on the right side of both equations. Since we know $273 = 13p - 92q$ from the first equation, we substitute this expression for 273 in the second equation to get $12p - 91q = 13p - 92q$, or $q = p$. Substituting this in either original equation gives $-79p = 273$, so $p = -273/79$. Since $q = p$, our solution is $(p, q) = \boxed{(-273/79, -273/79)}$.

Exercises for Section 5.3

5.3.1

(a) Adding the two equations gives $5x = 20$, so $x = 4$. Substituting this into the first equation gives $12 - 7y = 14$, so $y = -2/7$. Therefore, our solution is $(x, y) = \boxed{(4, -2/7)}$.

(b) Multiplying the first equation by 2 gives $10u = -14 - 4v$. Adding this to the second equation gives $13u = -39$, so $u = -3$. Substituting this into $5u = -7 - 2v$ gives $-15 = -7 - 2v$, so $v = 4$ and our solution is $(u, v) = \boxed{(-3, 4)}$.

(c) First, we simplify both equations by expanding the right sides and multiplying by 13 to get rid of the fractions. The first equation becomes

$$13\left(\frac{2x}{13} + 2y\right) = 13(-2y - 2) \quad \Rightarrow \quad 2x + 26y = -26y - 26 \quad \Rightarrow \quad x = -26y - 13.$$

The second equation becomes

$$13\left(-\frac{3x}{13}\right) = 13(-30 + 5y) \quad \Rightarrow \quad -3x = -390 + 65y.$$

Multiplying $x = -26y - 13$ by 3 gives $3x = -78y - 39$. Adding this to $-3x = -390 + 65y$ gives $0 = -13y - 429$, so $y = -429/13 = -33$. So, $x = -26y - 13 = 845$ and our solution is $(x, y) = \boxed{(845, -33)}$.

(d) We simplify the second equation by getting all the variables on one side and all the constants on the other, giving $3.75a - 6b = 27$. Multiplying the first equation by 1.5 gives $-3.75a + 7.5b = 37.5$. Adding these two equations gives $1.5b = 64.5$, so $b = (64.5)/(1.5) = 43$. Substituting this into the original first equation gives $-2.5a + 5(43) = 25$. Rearranging this equation gives $-2.5a = -190$, so $a = 76$. Therefore, our solution is $(a, b) = \boxed{(76, 43)}$.

5.3.2

(a) Putting all the variables on the left in the second equation gives $14x + 21y = 49$. Multiplying the first equation by 7 gives $14x + 21y = 49$, so now our system of equations is

$$14x + 21y = 49,$$
$$14x + 21y = 49.$$

These two equations are the same, so every solution to the first equation is a solution to the second equation. We can describe these solutions parametrically by letting $x = t$ and solving our first equation for y in terms of t:

$$2t + 3y = 7 \quad \Rightarrow \quad 3y = 7 - 2t \quad \Rightarrow \quad y = \frac{7 - 2t}{3}.$$

So, for any value of t, the pair $(x, y) = \boxed{\left(t, \dfrac{7 - 2t}{3}\right)}$ satisfies both equations.

(b) We multiply the first equation by 5 to get rid of the fractions. This gives $3x - 4y = 15$. We rewrite the second equation with the variables in alphabetical order to get $-6x + 8y = 5$. We multiply

$3x - 4y = 15$ by 2, hoping to then add the equations to eliminate x. This gives $6x - 8y = 30$, so our system of equations is now

$$6x - 8y = 30,$$
$$-6x + 8y = 5.$$

Adding these equation not only eliminates x, but it also knocks out y and leaves $0 = 35$. Obviously this equation is never true, so there are $\boxed{\text{no solutions}}$ to the original system of equations.

5.3.3 Putting the x and y terms of the second equation on the left gives $6x + 15y = 16 + a$. Multiplying the first equation by 3 gives $6x + 15y = -24$. So, our system now is

$$6x + 15y = -24,$$
$$6x + 15y = 16 + a.$$

This system has infinitely many solutions only if the two right sides are the same, which would make the two equations the same. So, we must have $-24 = 16 + a$, so $a = \boxed{-40}$.

5.3.4 We can subtract the second equation from the first to get $by - cy = d - e$. Factoring the left gives $y(b - c) = d - e$. Because we know that b and c are not equal, we know that $b - c$ is not 0. Therefore, we can divide by $b - c$ to solve for y: $y = (d - e)/(b - c)$. We can then substitute this into either of our original equations to solve for x. In other words, we are guaranteed to be able to find exactly one solution (x, y).

Exercises for Section 5.4

5.4.1 We found $e + m = 241$ in the text. Subtracting this from $2e + m = 362$ gives $e = 121$. Therefore, $m = 241 - e = 120$, so Tweedledee weighs $\boxed{121 \text{ pounds}}$ and Tweedledum weighs $\boxed{120 \text{ pounds}}$.

5.4.2 Suppose there are c chickens and p pigs. From the legs, we have $2c + 4p = 40$. Dividing this by 2 gives $c + 2p = 20$. From the heads, we have $c + p = 16$. Subtracting this from $c + 2p = 20$ gives $p = 4$. So, $c = 16 - p = \boxed{12}$ chickens.

5.4.3 Let Eric weigh e and Bob weigh b, where b is larger than e (because Eric weighs less than Bob). We are first told that $b + e = 9(b - e)$. We are also told that $b - e = b + e - 240$. Solving this equation for e gives $e = 120$, and substituting this into the first equation gives $b + 120 = 9(b - 120)$, so $b + 120 = 9b - 1080$, which gives us $b = 1200/8 = \boxed{150 \text{ pounds}}$.

5.4.4 *Solution 1: Find the weight of each ball.* Let the green balls weigh g pounds and the red balls weigh r pounds. So, we have $5g + 2r = 10$ and $g + 4r = 7$. Solving the second equation for g gives $g = 7 - 4r$. Substituting this into the first equation gives $5(7 - 4r) + 2r = 10$. Expanding and simplifying the left side gives $35 - 18r = 10$, and solving this equation gives $r = 25/18$. So, $g = 7 - 4r = 7 - \frac{50}{9} = \frac{13}{9}$. Therefore, 8 green balls and 8 red ones together weigh

$$8g + 8r = 8(g + r) = 8\left(\frac{13}{9} + \frac{25}{18}\right) = 8\left(\frac{26}{18} + \frac{25}{18}\right) = 8 \cdot \frac{51}{18} = \boxed{\frac{68}{3} \text{ pounds}}.$$

Solution 2: Find the sum of the weights of a green and a red ball. As before, we have $5g + 2r = 10$ and $g + 4r = 7$. Before solving this system of equations, we note that we seek $8g + 8r$, which equals $8(g + r)$.

So, if we can find $g + r$, we can find the total weight of the balls *without finding the weight of each ball*. Looking at our equations, we see a total of $6g$ and $6r$ on the left, so adding the two equations can get us to $g + r$. Adding the equations gives $6g + 6r = 17$, and dividing both sides by 6 gives $g + r = 17/6$. Therefore, we have $8g + 8r = 8(g + r) = 8(17/6) = \boxed{68/3 \text{ pounds}}$.

5.4.5 We are told that $a = 6b$. To convert our other information into an equation, we note that when it is b minutes after 3 o'clock, it is $b + 60$ minutes after 2 o'clock. We are told that this time is 15 minutes later than when the time was a minutes after 2 o'clock. So, we must have $b + 60 = a + 15$. Letting $a = 6b$ in this equation gives $b + 60 = 6b + 15$, from which we find $b = 9$. Therefore, it was $\boxed{3{:}09}$ when she looked at her watch for the second time.

Exercises for Section 5.5

5.5.1 Multiplying the second equation by 3 gives $\dfrac{6}{x} - \dfrac{15}{y} = 48$. Subtracting this from the first equation gives $\dfrac{22}{y} = -44$. Taking the reciprocal of this equation gives $y/22 = -1/44$, so $y = -22/44 = -1/2$. Substituting this into the first equation gives $\dfrac{6}{x} - 14 = 4$, so $6/x = 18$. Therefore, $x = 1/3$, and our solution is $(x, y) = \boxed{(1/3, -1/2)}$.

5.5.2 Let the side length of the larger square be x and the side length of the smaller square be y. We are told $x^2 + y^2 = 65$ and $x^2 - y^2 = 33$. Adding these two equations gives $2x^2 = 98$, so $x^2 = 49$. Since x must be positive, we have $x = 7$. Substituting this into either equation above gives us $y^2 = 16$. Since y must be positive, we have $y = 4$. The perimeter of the larger square is $4x$ and that of the smaller square is $4y$, so the sum of their perimeters is $4x + 4y = 4(x + y) = \boxed{44}$.

5.5.3 From our first equation, we have $\sqrt[3]{r} = 21 - 9\sqrt{s}$. Substituting this into the second equation gives
$$10(21 - 9\sqrt{s}) - \sqrt{s} = 28 \quad \Rightarrow \quad 210 - 90\sqrt{s} - \sqrt{s} = 28 \quad \Rightarrow \quad -91\sqrt{s} = -182 \quad \Rightarrow \quad \sqrt{s} = 2.$$
Squaring this equation tells us $s = 4$. So, $\sqrt[3]{r} = 21 - 9\sqrt{4} = 21 - 18 = 3$. Cubing $\sqrt[3]{r} = 3$ gives us $r = 27$. Therefore, $(r, s) = \boxed{(27, 4)}$.

Exercises for Section 5.6

5.6.1 Adding the first and third equations gives $-x + 3z = 20$. Adding 3 times the middle equation to the last equation eliminates y to give $-11x + 16z = 101$. Multiplying $-x + 3z = 20$ by -11 gives $11x - 33z = -220$, and adding this to $-11x + 16z = 101$ gives $-17z = -119$. Therefore, $z = (-119)/(-17) = 7$. Substituting this into $-x + 3z = 20$ gives $-x + 21 = 20$, from which $x = 1$. Substituting $x = 1$ and $z = 7$ into our original first equation gives $1 + 3y + 14 = 6$, from which we have $y = -3$. So, we have $(x, y, z) = \boxed{(1, -3, 7)}$.

5.6.2 We solve for x in the second equation to find $x = -y + 4z + 6$. Substituting this into the first equation gives
$$2(-y + 4z + 6) - 5y + 3z = 25 \quad \Rightarrow \quad -2y + 8z + 12 - 5y + 3z = 25 \quad \Rightarrow \quad -7y + 11z = 13.$$

Substituting for x in the last equation gives

$$3(-y + 4z + 6) + 3y - z = -4 \quad \Rightarrow \quad -3y + 12z + 18 + 3y - z = -4 \quad \Rightarrow \quad 11z = -22 \quad \Rightarrow \quad z = -2.$$

Letting $z = -2$ in $-7y + 11z = 13$ gives us $-7y - 22 = 13$, from which we get $y = -5$. So, $x = -y + 4z + 6 = 5 - 8 + 6 = 3$. So, $(x, y, z) = \boxed{(3, -5, -2)}$.

Notice that we could have eliminated x and y by adding 3 times the second equation to the third equation.

5.6.3 From the last row, we know that the constant sum equals $25 + z + 21 = z + 46$. So, this gets us focusing on z. We look at its column, which has sum $24 + x + z$. This sum equals our row sum from before, so $24 + x + z = z + 46$. This tells us that $x = 22$. By looking at v's row and column, we have $v + 18 + 25 = v + 24 + w$, so $w = 19$. Now, we have our constant sum. From the $25 - x - w$ diagonal, we see that this sum is $25 + x + w = 25 + 22 + 19 = 66$. Therefore, from z's column we have $z + 22 + 24 = 66$, which gives $z = 20$, and from y's column we have $21 + y + 19 = 66$, which gives $y = 26$. So, $y + z = \boxed{46}$.

5.6.4

(a) We can eliminate c from both equations by adding twice the second equation to the first equation. This gives $4a + b = 19$, so $b = 19 - 4a$. There are infinitely many solutions to this equation. We can pick any a to get a corresponding b. When $a = 4$, we have $b = 3$. Substituting this into $2a + 3b - 4c = 7$ gives $8 + 9 - 4c = 7$, from which we find $c = 5/2$. If we let $a = 4$ and $b = 3$ in $a - b + 2c = 6$, we have $c = 5/2$ as well, so $(a, b, c) = (4, 3, 5/2)$ satisfies both equations. There are many, many other solutions. In fact, as we will see, there are infinitely many solutions!

(b) If we let $a = 5$, we have $b = 19 - 4a = -1$. Substituting these into $2a + 3b - 4c = 7$ gives $10 - 3 - 4c = 7$, so $c = 0$. If we let $a = 5$ and $b = -1$ in $a - b + 2c = 6$, we also get $c = 0$, so $(a, b, c) = (5, -1, 0)$ also satisfies both equations.

(c) In part (a) we found $b = \boxed{19 - 4a}$. If we substitute this into our first equation, we have

$$2a + 3(19 - 4a) - 4c = 7 \quad \Rightarrow \quad 2a + 57 - 12a - 4c = 7 \quad \Rightarrow \quad -4c = 10a - 50.$$

Isolating c gives us

$$c = \frac{10a - 50}{-4} = \boxed{\frac{-5a + 25}{2}}.$$

However, what if we substitute $b = 19 - 4a$ into the second equation instead? Then, we get

$$a - (19 - 4a) + 2c = 6 \quad \Rightarrow \quad a - 19 + 4a + 2c = 6 \quad \Rightarrow \quad 2c = -5a + 25 \quad \Rightarrow \quad c = \frac{-5a + 25}{2},$$

which is exactly the same as when we substituted into the first equation. Hmmm... This suggests that the triple

$$(a, b, c) = \left(a, 19 - 4a, \frac{-5a + 25}{2} \right)$$

satisfies both equations, *no matter what a is*! Let's check:

$$2a + 3b - 4c = 2a + 3(19 - 4a) - 4\left(\frac{-5a + 25}{2} \right) = 2a + 57 - 12a - 2(-5a + 25) = -10a + 57 + 10a - 50 = 7,$$

so the triple satisfies the first equation. Now, for the second equation:

$$a - b + 2c = a - (19 - 4a) + 2\left(\frac{-5a + 25}{2}\right) = a - 19 + 4a - 5a + 25 = (a + 4a - 5a) + (-19 + 25) = 6,$$

so the triple does indeed satisfy the second equation. Therefore, for any value of a, the triple $(a, b, c) = \left(a, 19 - 4a, \frac{-5a+25}{2}\right)$ satisfies both of our equations. Therefore, this system of equations has $\boxed{\text{infinitely many solutions}}$.

Review Problems

5.20 Solving for a in terms of b gives $a = 3 + \frac{5b}{3}$. There are infinitely many solutions to this equation. To avoid fractions, we choose multiples of 3 for b. When $b = 0$, we have $a = 3$. When $b = 3$, $a = 3 + 15/3 = 3 + 5 = 8$, and when $b = 6$, we have $a = 3 + 30/3 = 13$. So, three ordered pairs that satisfy the equation are $(a, b) = (3, 0), (8, 3)$, and $(13, 6)$. There are infinitely many others.

5.21

(a) Solving the first equation for x gives $x = -5 - 4y$. Substituting this into the second equation gives

$$3(-5 - 4y) - 8y = 45 \quad \Rightarrow \quad -15 - 12y - 8y = 45 \quad \Rightarrow \quad -20y = 60 \quad \Rightarrow \quad y = -3.$$

Therefore, $x = -5 - 4y = -5 - 4(-3) = -5 + 12 = 7$, so our solution is $(x, y) = \boxed{(7, -3)}$.

(b) Solving the first equation for b gives $b = 3a - 11$. Substituting this into the second equation gives

$$6a + 4(3a - 11) = 1 \quad \Rightarrow \quad 6a + 12a - 44 = 1 \quad \Rightarrow \quad 18a = 45 \quad \Rightarrow \quad a = \frac{5}{2}.$$

Therefore, $b = 3a - 11 = \frac{15}{2} - 11 = -\frac{7}{2}$. So, we have $(a, b) = \boxed{\left(\frac{5}{2}, -\frac{7}{2}\right)}$.

5.22

(a) Adding twice the second equation to the first eliminates y to give $19x = -152$, so $x = -8$. Substituting this in our first equation gives $-40 - 6y = -64$, from which $y = 4$. So, the solution is $(x, y) = \boxed{(-8, 4)}$.

(b) Doubling the first equation gives $6x + 16y = -14$. Subtracting this from the original second equation eliminates both x and y (uh-oh), and leaves $0 = 18$. This equation is never true, so our original system has $\boxed{\text{no solutions}}$.

5.23

(a) Adding twice the first equation to the second gives $11x = 55$, so $x = 5$. Substituting this into the first equation gives $15 - 4y = 26$, from which we have $y = -11/4$. So, our solution is $(x, y) = \boxed{(5, -11/4)}$.

(b) First, we simplify both equations. We multiply the first equation by 6 to get rid of the fractions, which gives $2r - s = 3$. For the second equation, we put all the terms with variables on the left and the constants on the right to get $-5r + 5s = 10$. Dividing this equation by 5 gives us $-r + s = 2$.

Adding this equation to the $2r - s = 3$ from earlier gives $r = 5$. Substituting this into $-r + s = 2$ gives $s = 7$, so we have $(r, s) = \boxed{(5, 7)}$.

(c) We simplify both equations by bringing all the variable terms to the left and putting all the constants on the right. This makes our system

$$5a + 2b = 13,$$
$$10a + 4b = 26.$$

Multiplying the first equation by 2 makes it $10a + 4b = 26$, which is exactly the same as our second equation. Therefore, there are infinitely many solutions to this system of equations. We can use a parameter to express these solutions. We let $a = t$, and we solve $5t + 2b = 13$ for b to get $b = (13 - 5t)/2$. Therefore, all pairs $(a, b) = \boxed{\left(t, \dfrac{13 - 5t}{2}\right)}$ are solutions to this system, where t can take on any value.

Note: This is not the only way to express the solutions to this system. For example, we could have let $b = t$ and solved for a in terms of t.

(d) Putting all the variable terms on the left and the constants on the right makes our system

$$5x - 6y = 12,$$
$$2x - 7y = 14.$$

Multiplying the first equation by -2 and the second by 5 sets us up to eliminate the x terms:

$$-10x + 12y = -24,$$
$$10x - 35y = 70.$$

Adding these two equations gives $-23y = 46$, so $y = -2$. Substituting this into either of our original equations gives $x = 0$, so our solution is $(x, y) = \boxed{(0, -2)}$.

5.24 First we bring all the x and y terms to the left, and put all the other terms on the right. This makes our system

$$3x + 2y = 8,$$
$$6x + 4y = 2a - 7.$$

Multiplying the first equation by 2 makes its coefficients of x and y match those of the second equation:

$$6x + 4y = 16,$$
$$6x + 4y = 2a - 7.$$

If $2a - 7 = 16$, these equations will be the same, so the system will have infinitely many solutions. If $2a - 7$ does not equal 16, then this system will have no solutions, since $6x + 4y$ cannot equal two different numbers. Solving $2a - 7 = 16$ gives us $a = \boxed{23/2}$.

5.25 We collect all the x and y terms on the left and put the constants on the right:

$$3x - 5y = -2.3,$$
$$6x - cy = 9.3.$$

We multiply the first equation by -2 to set up eliminating the x terms:

$$-6x + 10y = 4.6,$$
$$6x - cy = 9.3.$$

If we add these equations, we eliminate x. If $c = \boxed{10}$, we eliminate y, as well, and are left with $0 = 13.9$. So, when $c = 10$, we have no solutions to this system. If c is not equal to 10, then when we add the equations above, we will eliminate x but not y. We will therefore have a linear equation we can solve for y, and we can then use this value for y to solve for x. So, if c is any value besides 10, then the system of equations has one solution.

5.26 Let the charge of an up quark be u and a down quark be d. From our proton, we have $d + 2u = 1$, and from the neutron, we have $2d + u = 0$. Solving the second equation for u gives $u = -2d$. Substituting this in the first equation gives $-4d + d = 1$, so $d = -1/3$. Therefore $u = -2d = \boxed{2/3}$.

5.27 Let my age now be m and my father's age now be f. Because my father's age 5 years ago plus twice my age now is 65, we have $f - 5 + 2m = 65$, or $f + 2m = 70$. Because my age 5 years ago plus three times my father's age is 130, we have $m - 5 + 3f = 130$, or $3f + m = 135$. Therefore, we have the system

$$f + 2m = 70,$$
$$3f + m = 135.$$

Subtracting twice the second equation from the first gives $-5f = -200$, so $f = \boxed{40}$.

5.28 Back in the text we already converted these words into equations. We let x be the number of northerly steps and y be the number of easterly steps, and converted the words in the riddle to the system of equations

$$3(x + y) = 4 + 4x,$$
$$5(x - 2) = 2 + 7y.$$

We simplify each equation by expanding the left sides, then moving all the variable terms to the left and the constants to the right. This gives the system

$$-x + 3y = 4,$$
$$5x - 7y = 12.$$

From the first equation, we have $x = 3y - 4$, and substituting this into the second equation gives

$$5(3y - 4) - 7y = 12 \quad \Rightarrow \quad 15y - 20 - 7y = 12 \quad \Rightarrow \quad 8y = 32 \quad \Rightarrow \quad y = 4.$$

So, $x = 3y - 4 = 12 - 4 = 8$. Therefore, we should take $\boxed{\text{8 steps north and 4 steps east}}$.

5.29 The second equation is already a simple linear equation, but the first has y in the denominator of two fractions. We multiply both sides of the first equation by y to get

$$y\left(\frac{x}{y} - 3\right) = y \cdot \frac{2}{y} \quad \Rightarrow \quad x - 3y = 2.$$

This is just a simple two-variable linear equation. Adding $3y$ to both sides gives $x = 2 + 3y$, and substituting this into our second original equation gives

$$2(2 + 3y) - 9y = -8 \quad \Rightarrow \quad 4 + 6y - 9y = -8 \quad \Rightarrow \quad -3y = -12 \quad \Rightarrow \quad y = 4.$$

Therefore, $x = 2 + 3y = 14$, so our solution is $(x, y) = \boxed{(14, 4)}$.

5.30 Adding the first and last equations eliminates p and leaves $q + 6r = -12$. Adding twice the first equation to the second also eliminates p and gives $13q - 10r = 64$. Solving $q + 6r = -12$ for q gives $q = -12 - 6r$, and substituting this into $13q - 10r = 64$ gives

$$13(-12 - 6r) - 10r = 64.$$

Because $-12 - 6r = 2(-6 - 3r)$, we can write the equation above as $13 \cdot 2(-6 - 3r) - 10r = 64$. We can divide this equation by 2 to simplify it. This gives us $13(-6 - 3r) - 5r = 32$. Expanding the left side gives $-78 - 39r - 5r = 32$, from which we find $r = -5/2$. So, $q = -12 - 6(-5/2) = -12 + 15 = 3$. From $p + 3q - 2r = 19$, we have $p + 3(3) - 2(-5/2) = 19$, from which we find $p = 5$. So, our solution is $(p, q, r) = \boxed{(5, 3, -5/2)}$.

5.31 Let the weight of an orange be o, and apple be a, and a banana be b. Then, we can use the information in the problem to create a system of equations:

$$3o + 2a + b = 15,$$
$$5o + 7a + 2b = 44,$$
$$o + 3a + 5b = 26.$$

Solving the last equation for o gives $o = 26 - 3a - 5b$. Substituting this in the first equation gives

$$3(26 - 3a - 5b) + 2a + b = 15 \quad \Rightarrow \quad 78 - 9a - 15b + 2a + b = 15 \quad \Rightarrow \quad -7a - 14b = -63 \quad \Rightarrow \quad a + 2b = 9.$$

Notice how we divide by -7 in the last step to simplify; removing the common factor of 7 makes the numbers much smaller. We substitute for o in the second equation to get

$$5(26 - 3a - 5b) + 7a + 2b = 44 \quad \Rightarrow \quad 130 - 15a - 25b + 7a + 2b = 44 \quad \Rightarrow \quad -8a - 23b = -86.$$

From $a + 2b = 9$, we have $a = 9 - 2b$, so we have

$$-8a - 23b = -86 \quad \Rightarrow \quad -8(9 - 2b) - 23b = -86 \quad \Rightarrow \quad -72 + 16b - 23b = -86 \quad \Rightarrow \quad -7b = -14 \quad \Rightarrow \quad b = 2.$$

So, $a = 9 - 2(2) = 5$, and $o = 26 - 3a - 5b = 26 - 15 - 10 = 1$. Therefore, apples weigh $\boxed{5}$, bananas weigh $\boxed{2}$, and oranges weigh $\boxed{1}$.

Challenge Problems

5.32 Let t be the tens digit and u be the units digit of my number. So, the value of my number is $10t + u$. When the digits of my number are reversed, then u is the tens digit and t is the units digit. This number has value $10u + t$. So, we have $10u + t - (10t + u) = 36$, or $9u - 9t = 36$. Dividing this equation by 9 gives $u - t = 4$.

If the tens digit of my original number is doubled and the units digit is halved, then the new tens digit is $2t$ and the new units digit is $u/2$. This number has value $2t(10) + u/2$, or $20t + u/2$. This number is 17 more than my original number, so $(20t + u/2) - (10t + u) = 17$, or $10t - u/2 = 17$.

From $u - t = 4$, we have $t = u - 4$. Substituting this into $10t - u/2 = 17$ gives

$$10(u - 4) - \frac{u}{2} = 17 \quad \Rightarrow \quad 10u - 40 - \frac{u}{2} = 17 \quad \Rightarrow \quad \frac{19u}{2} = 57 \quad \Rightarrow \quad u = 6.$$

So, we have $t = 6 - 4 = 2$, and my original number was $\boxed{26}$.

5.33 *Solution 1: Elimination.* We multiply $y = 5 + 2t$ by 2 to get $2y = 10 + 4t$. Adding this to $x = 3 - 4t$ eliminates t and gives $x + 2y = 13$. Therefore, $\boxed{x = 13 - 2y}$.

Solution 2: Substitution. We solve $y = 5 + 2t$ for t to get $t = (y - 5)/2$. Substituting this into $x = 3 - 4t$ gives

$$x = 3 - 4\left(\frac{y - 5}{2}\right) = 3 - 2(y - 5) = \boxed{13 - 2y}.$$

5.34 *Solution 1: Look at that second equation.* Since square roots must be nonnegative, the second equation, $2\sqrt{x} + \sqrt{y} = 0$, tells us that both $\sqrt{x}$ and $\sqrt{y}$ must be 0. However, if $\sqrt{x} = \sqrt{y} = 0$, then our first equation is $0 = 3$, which is not true. Because the only solution to the second equation does not satisfy the first, we have $\boxed{\text{no solutions}}$ to this system.

Solution 2: Find $\sqrt{x}$ and $\sqrt{y}$ using tactics from the text. Subtracting twice the second equation from the first gives $-3\sqrt{x} = 3$, from which we find $\sqrt{x} = -1$. But the square root of x can't be negative! So, this equation has no solutions. Therefore, the system has $\boxed{\text{no solutions}}$.

5.35 Let the numbers be x and y. We are given $xy = 2(x + y)$. We wish to evaluate $\frac{1}{x} + \frac{1}{y}$.

Solution 1: Get reciprocals into our equation. We'd like to turn the equation $xy = 2(x + y)$ into something with x and y in denominators. We start by dividing both sides by x, which gives

$$\frac{xy}{x} = \frac{2(x + y)}{x} \quad \Rightarrow \quad y = 2 + \frac{2y}{x}.$$

So, that puts x in a denominator. Let's divide by y:

$$\frac{y}{y} = \frac{2}{y} + \frac{2y/x}{y} \quad \Rightarrow \quad 1 = \frac{2}{y} + \frac{2y}{xy} \quad \Rightarrow \quad 1 = \frac{2}{y} + \frac{2}{x}.$$

Success! Dividing this equation by 2 tells us $\dfrac{1}{y} + \dfrac{1}{x} = \boxed{\dfrac{1}{2}}$.

Solution 2: Rearrange what we want. We try manipulating the expression we want, hoping that will provide clues. We write our sum of reciprocals with a common denominator:

$$\frac{1}{x} + \frac{1}{y} = \frac{y}{xy} + \frac{x}{xy} = \frac{x + y}{xy}.$$

Aha! We can use our equation to find $(x + y)/xy$. Dividing both sides of $xy = 2(x + y)$ by $2xy$ gives

$$\frac{xy}{2xy} = \frac{2(x + y)}{2xy} \quad \Rightarrow \quad \boxed{\frac{1}{2}} = \frac{x + y}{xy}.$$

5.36 Our equations look very complicated, but if we expand our left sides, we see something much simpler. For example, the first equation is:

$$\frac{3x - 4y}{xy} = -8 \quad \Rightarrow \quad \frac{3x}{xy} - \frac{4y}{xy} = -8 \quad \Rightarrow \quad \frac{3}{y} - \frac{4}{x} = -8.$$

This looks much nicer. Let's see what happens in the second equation:

$$\frac{2x + 7y}{xy} = 43 \quad \Rightarrow \quad \frac{2}{y} + \frac{7}{x} = 43.$$

Now, our system of equations looks very familiar! We have

$$\frac{3}{y} - \frac{4}{x} = -8,$$
$$\frac{2}{y} + \frac{7}{x} = 43.$$

Multiplying the first equation by 2 and the second by -3 to set up eliminating y gives

$$\frac{6}{y} - \frac{8}{x} = -16,$$
$$-\frac{6}{y} - \frac{21}{x} = -129.$$

Adding these equations gives $-\dfrac{29}{x} = -145$. Dividing both sides by -29 gives $1/x = 5$, so $x = 1/5$. Substituting this into either equation above gives $y = 1/4$, so our solution is $(x, y) = \boxed{(1/5, 1/4)}$.

5.37 We need the ratio of x to y. Our variable expressions in the equations are much more complicated than simply x/y. Solving the first equation for x in terms of y then substituting into the second equation gives us something pretty scary. Instead, we look for some way to combine the two equations that will eliminate all the variables except for an x/y expression.

Finding another way to express the variable sides of the equations in the previous problem worked pretty well. Let's try that here. Before, we expanded the expressions, "undoing" the common denominators on the left. Here, we try going the other way, writing both left sides with a common denominator. Our first equation is

$$x + \frac{1}{y} = 4 \quad \Rightarrow \quad \frac{xy}{y} + \frac{1}{y} = 4 \quad \Rightarrow \quad \frac{xy + 1}{y} = 4.$$

Our second equation gives us

$$y + \frac{1}{x} = \frac{1}{4} \quad \Rightarrow \quad \frac{xy}{x} + \frac{1}{x} = \frac{1}{4} \quad \Rightarrow \quad \frac{xy + 1}{x} = \frac{1}{4}.$$

Success! We can cancel out these ugly $xy + 1$ terms. We can either take the quotient of our two equations, or multiply the first by y and the second by x to get the system

$$xy + 1 = 4y,$$
$$xy + 1 = \frac{x}{4}.$$

The left sides are the same, so the right sides are equal. (Or, again, we could just divide the second equation by the first to get x/y.) We have $4y = x/4$, so $16y = x$, and $x/y = \boxed{16}$.

5.38 Let the amount of milk Angela drank be m and the amount of coffee she drank be c. Because she drank a total of 8 ounces, we have $m + c = 8$. Because she drank 1/4 of the milk, the whole family drank a total of $4m$ milk. Because she drank 1/6 of the coffee, the whole family drank a total of $6c$ of the coffee. Therefore, the whole family drank a total of $4m + 6c$ ounces.

Each drink is a total of 8 ounces, so if there are n people in the family (including Angela), we must have $4m + 6c = 8n$. Dividing by 2 gives $2m + 3c = 4n$. From $m + c = 8$, we have $m = 8 - c$, so we have

$$2(8 - c) + 3c = 4n \quad \Rightarrow \quad 16 + c = 4n.$$

Because we are told that Angela definitely used some coffee and some milk, c must be between 0 and 8, but cannot equal 0 or 8. Therefore, $16 + c$ is between 16 and 24, but cannot equal 16 or 24. Because $16 + c = 4n$, and n must be an integer, $16 + c$ must be a multiple of 4. The only multiple of 4 between 16 and 24, exclusive, is 20, so $16 + c = 4n = 20$. Therefore, there are $n = 20/4 = \boxed{5}$ members of the family.

5.39 We have to be careful to organize our information, and to choose an approach that's going to minimize the amount of painful algebra we have to do. We already have one equation without d, so we focus on eliminating d. Subtracting the first equation from the last gives $a + 10b - 11c = 72$. Adding twice the last equation to the third equation will also eliminate d to give $5a + 16b - 19c = 108$. Combining these two with the original second equation gives the system

$$7a + 2b - c = -28,$$
$$a + 10b - 11c = 72,$$
$$5a + 16b - 19c = 108.$$

We solve the second equation for a to get $a = 72 - 10b + 11c$. We substitute this into the other two equations to get a system of two equations with two variables. The first equation gives

$$7(72 - 10b + 11c) + 2b - c = -28 \quad \Rightarrow \quad 504 - 70b + 77c + 2b - c = -28 \quad \Rightarrow \quad -68b + 76c = -532.$$

All the constants in this equation are multiples of 4, so we divide the equation by 4 to find $-17b + 19c = -133$. Substituting our expression for a into the last equation gives

$$5(72 - 10b + 11c) + 16b - 19c = 108 \quad \Rightarrow \quad 360 - 50b + 55c + 16b - 19c = 108 \quad \Rightarrow$$
$$-34b + 36c = -252 \quad \Rightarrow \quad -17b + 18c = -126.$$

Subtracting this equation from $-17b + 19c = -133$ gives $c = -7$. Substituting this into $-17b + 18c = -126$ gives $b = 0$, so $a = 72 - 10b + 11c = 72 - 0 - 77 = -5$. Substituting $a = -5$, $b = 0$, and $c = -7$ into $2a - 3b + 5c + d = -41$ (the original first equation) gives

$$-10 + 0 - 35 + d = -41 \quad \Rightarrow \quad d = 4.$$

Therefore, our solution is $(a, b, c, d) = \boxed{(-5, 0, -7, 4)}$.

5.40 We multiply the first equation by d and the second by b to set up eliminating the y terms:

$$adx + bdy = de,$$
$$bcx + bdy = bf.$$

Subtracting the second equation from the first gives

$$(ad - bc)x = de - bf.$$

We recognize the coefficient of x and the right side of the equation as having the form of the determinants described in the problem, so we have

$$\begin{vmatrix} a & b \\ c & d \end{vmatrix} x = \begin{vmatrix} e & b \\ f & d \end{vmatrix} \quad \Rightarrow \quad x = \frac{\begin{vmatrix} e & b \\ f & d \end{vmatrix}}{\begin{vmatrix} a & b \\ c & d \end{vmatrix}}.$$

Similarly, if we multiply our first original equation by c and the second by a, we are set up to eliminate the x terms:

$$acx + bcy = ce,$$
$$acx + ady = af.$$

Subtracting the first equation from the second gives

$$(ad - bc)y = af - ce \quad \Rightarrow \quad \begin{vmatrix} a & b \\ c & d \end{vmatrix} y = \begin{vmatrix} a & e \\ c & f \end{vmatrix} \quad \Rightarrow \quad y = \frac{\begin{vmatrix} a & e \\ c & f \end{vmatrix}}{\begin{vmatrix} a & b \\ c & d \end{vmatrix}}.$$

In both our solution for x and our solution for y, we run into trouble if $ad - bc = 0$, since then our equation after elimination has no variable. We've seen this before in solving systems of equations. If combining the equations leads to an equation with no variables, then either the system has infinitely' many solutions (if the no-variable equation is true), or the system has no solutions (if the no-variable equation is false). So, we look at the two equations we found with elimination:

$$(ad - bc)x = de - bf,$$
$$(ad - bc)y = af - ce.$$

If $ad - bc = 0$, then the left sides of both are 0. Therefore, if we have $de - bf = af - ce = 0$, then we have infinitely many solutions to this system. Rearranging $de - bf = 0$ gives $b/d = e/f$, and rearranging $af - ce = 0$ gives $a/c = e/f$. Similarly, rearranging $ad - bc = 0$ gives $a/c = b/d$. So, if $a/c = b/d = e/f$, then we have infinitely many solutions. (In this case, one equation is just a constant multiple of the other.)

However, if $ad - bc = 0$ and either $de - bf$ or $af - ce$ is not zero, then there are no solutions to the system.

5.41 We look for useful ways to combine the equations. We try subtracting the first from the second, the second from the third, and the first from the third. These give us the following equations:

$$3x_1 + 5x_2 + 7x_3 + 9x_4 + 11x_5 + 13x_6 + 15x_7 = 11,$$
$$5x_1 + 7x_2 + 9x_3 + 11x_4 + 13x_5 + 15x_6 + 17x_7 = 111,$$
$$8x_1 + 12x_2 + 16x_3 + 20x_4 + 24x_5 + 28x_6 + 32x_7 = 122.$$

At the very least, these have an interesting pattern. However, we also notice that subtracting the first from the second lets us find the sum of all the variables, by giving us

$$2(x_1 + x_2 + x_3 + x_4 + x_5 + x_6 + x_7) = 100 \quad \Rightarrow \quad x_1 + x_2 + x_3 + x_4 + x_5 + x_6 + x_7 = 50.$$

Unfortunately, it's not clear how we will get to the sum we want.

Since we're a little stuck, we try working backwards. We look for ways that we could form the expression we want. To do so, we compare the expression we want to the expressions we have, and see what we would need to get from the expressions we have to the expressions we want. For example, because we know the sum of the x_i, if we could somehow evaluate

$$15x_1 + 24x_2 + 35x_3 + 48x_4 + 63x_5 + 80x_6 + 99x_7,$$

then we could just add the sum of the x_i to get what we want.

Because taking the differences of consecutive equations gave us some interesting information before, we try that here, comparing the expression we want to the last of the three expressions for which we are given a value:

$$(16x_1 + 25x_2 + 36x_3 + 49x_4 + 64x_5 + 81x_6 + 100x_7) - (9x_1 + 16x_2 + 25x_3 + 36x_4 + 49x_5 + 64x_6 + 81x_7).$$

This difference equals

$$7x_1 + 9x_2 + 11x_3 + 13x_4 + 15x_5 + 17x_6 + 19x_7.$$

If we can evaluate this sum, then we can finish the problem. We've already evaluated sums with consecutive odd numbers of coefficients, and we know the sum of the x_i, so we know how to evaluate this expression:

$$(5x_1 + 7x_2 + 9x_3 + 11x_4 + 13x_5 + 15x_6 + 17x_7) + 2(x_1 + x_2 + x_3 + x_4 + x_5 + x_6 + x_7)$$

$$= 7x_1 + 9x_2 + 11x_3 + 13x_4 + 15x_5 + 17x_6 + 19x_7,$$

so

$$7x_1 + 9x_2 + 11x_3 + 13x_4 + 15x_5 + 17x_6 + 19x_7 = 111 + 100 = 211.$$

Therefore,

$$16x_1 + 25x_2 + 36x_3 + 49x_4 + 64x_5 + 81x_6 + 100x_7$$

$$= (9x_1 + 16x_2 + 25x_3 + 36x_4 + 49x_5 + 64x_6 + 81x_7) + (7x_1 + 9x_2 + 11x_3 + 13x_4 + 15x_5 + 17x_6 + 19x_7)$$

$$= 123 + 211 = \boxed{334}.$$

6

Ratios and Percents

Exercises for Section 6.1

6.1.1 Because seven out of every ten students in the play are boys, the remaining three out of every ten are girls. Therefore, the ratio of girls to boys is $\boxed{3/7}$.

We might also solve this problem by letting the total number of students be x. Then, there are $(7/10)x$ boys, so there are $x - (7/10)x = (1 - 7/10)x = (3/10)x$ girls. Therefore, the ratio of girls to boys is

$$\frac{(3/10)x}{(7/10)x} = \frac{3}{7}.$$

6.1.2 Since the ratio of red to blue to green marbles is $1 : 5 : 3$, the ratio of green marbles to the total number of marbles is $3/(1 + 5 + 3) = 3/9 = 1/3$. Let there be x marbles in the bag. Since there are 27 green marbles, we have $27/x = 1/3$. Cross-multiplying gives $x = \boxed{81}$.

6.1.3 There are 570,000 people in 68 square miles. So, we have:

$$\frac{\text{Number of people}}{\text{Number of square miles}} = \frac{570000}{68} \approx 8382.$$

We want the number of people per 1 square mile, so we must solve the equation

$$\frac{\text{Number of people}}{1} = 8382$$

for the number of people. Therefore, the number of people per square mile, rounded to the nearest hundred, is $\boxed{8400}$.

6.1.4 Because $\frac{2}{3}$ of the people voted for Proposition 16, we know that $1 - \frac{2}{3} = \frac{1}{3}$ of the people did not. So, we know that

$$\frac{\text{Number of people who did not vote for Proposition 16}}{\text{Number of people in the town}} = \frac{1}{3}.$$

Let the number of people in my town be x. We know that 634 people didn't vote for Proposition 16, so we have

$$\frac{634}{x} = \frac{1}{3}.$$

Cross-multiplying this equation, we have $(634)(3) = (1)(x)$, so $x = \boxed{1902}$.

We also could have reasoned through this as follows. Since $\frac{1}{3}$ of the people did not vote for the proposition, we have to multiply by 3 the number of people who did not vote for it in order to find the population of the whole town.

6.1.5 Suppose Alice's original height is x feet. Because the ratio of her original height to her second height is $24 : 5$, we know that the ratio of her second height to her original height is $5 : 24$. In other words, her second height is $5/24$ of her original height, or $5x/24$ feet. Since the ratio of her second height to her third height is $1 : 12$, her third height is 12 times her second height, or $(5x/24)(12) = 5x/2$ feet. Since the ratio of her original height to her fourth height is $16 : 1$, her original height is 16 times her fourth height. Her original height is x feet, so her fourth height is $x/16$ feet. So, Alice's four heights in feet are

$$x, \frac{5x}{24}, \frac{5x}{2}, \frac{x}{16}.$$

The largest of these is $5x/2$ feet, and we know this equals 10 feet. Solving $5x/2 = 10$ gives us $x = 4$ feet. The shortest of the four heights is the last, $x/16$ feet. Since $x = 4$ feet, her shortest height is $4/16 = \boxed{0.25 \text{ feet}}$.

Exercises for Section 6.2

6.2.1 Suppose there are s people in Smalltown. We need to find expressions for the total number of men and the total number of women in Smalltown and Bigtown together in order to find our ratio. Since the ratio of men to women in Smalltown is $3 : 2$, we know that $3/5$ of the people in Smalltown are men and $2/5$ are women (make sure you see why). Since there are s people in Smalltown, there are $3s/5$ men and $2s/5$ women.

Now we have to deal with Bigtown. Since the population of Bigtown is three times that of Smalltown, there are $3s$ people in Bigtown. Because the ratio of men to women in Bigtown is $2 : 3$, we know that $2/5$ of the people in Bigtown are men and $3/5$ are women. So, the number of men in Bigtown is $(2/5)(3s) = 6s/5$ and the number of women in Bigtown is $(3/5)(3s) = 9s/5$.

Finally, we can count the men and the women overall. First the men:

$$\frac{3s}{5} + \frac{6s}{5} = \frac{9s}{5},$$

then the women:

$$\frac{2s}{5} + \frac{9s}{5} = \frac{11s}{5}.$$

Therefore, the ratio of men to women when the two cities are combined is

$$\frac{\frac{9s}{5}}{\frac{11s}{5}} = \frac{9s}{11s} = \boxed{\frac{9}{11}}.$$

6.2.2

(a) Let there be n nickels and p pennies. Since 2/7 of the coins initially in the piggy bank are nickels, we have $\dfrac{n}{n+p} = \dfrac{2}{7}$.

(b) After 84 pennies are removed, we have $p - 84$ pennies out of $n + p - 84$ coins. We are told that the ratio of pennies to total coins is 1/3, so we have

$$\frac{p - 84}{n + p - 84} = \frac{1}{3}.$$

(c) Our equations in the first two parts give us a system of equations. If we cross-multiply both equations, we have

$$7n = 2(n + p),$$
$$3(p - 84) = n + p - 84.$$

Simplifying both equations gives us

$$5n = 2p,$$
$$2p - 168 = n.$$

We can solve this system by substituting our expression for n from the second equation into the first equation, or by adding the two equations to eliminate p. Either way, we find that $n = 42$ and $p = 105$. So, there are $\boxed{42}$ nickels in the piggy bank.

6.2.3 Suppose we played x games. Since we won 2/9 of the games we played, we won $(2/9)x = 2x/9$ games. Therefore, we lost $x - 2x/9 = 7x/9$ games. Since we lost 15 more games than we won, we have

$$\frac{7x}{9} - \frac{2x}{9} = 15.$$

Simplifying the left side gives $5x/9 = 15$, and solving this equation gives $x = \boxed{27}$ games played.

6.2.4 *Solution 1: Substitution.* From the first equation, we have $a = b(3/4)$. Substituting this into the second equation gives

$$\frac{b(3/4)}{c} = 5,$$

so $\dfrac{3b}{4c} = 5$. Multiplying both sides of this by $\dfrac{4}{3}$ gives $\dfrac{b}{c} = \boxed{\dfrac{20}{3}}$.

Solution 2: Elimination. We want b/c, so take reciprocals of both sides of $a/b = 3/4$ to get $b/a = 4/3$. Multiplying this equation by $a/c = 5$ cancels the a's and leaves us with:

$$\frac{b}{c} = \frac{b}{a} \cdot \frac{a}{c} = \frac{4}{3} \cdot 5 = \boxed{\frac{20}{3}}.$$

6.2.5 Joe's coffee clearly has 2 ounces of cream, so we only have to figure out how much cream JoAnn's coffee has. She added 2 ounces of cream to her coffee, so that her cup then had $2 + 12 = 14$ ounces of liquid. Two ounces of this liquid is cream, so $2/14 = 1/7$ of her drink is cream. After she drinks two ounces, she has 12 ounces of drink left. This drink is still $1/7$ cream, since her drinking doesn't change the ratio of cream in her drink. Therefore, her drink has $12(1/7) = 12/7$ ounces of cream. So, the ratio of the amount of cream in Joe's coffee to that in JoAnn's is $2/(12/7) = \boxed{7/6}$.

6.2.6 We use Problem 6.7 in the text as a guide. In that problem, we found an expression for the minute at which the hour hand is pointing. We do the same here. Suppose the time is m minutes after 1 o'clock. As we saw in the text, during the m minutes after 1 o'clock, the hour hand moves $m/12$ minutes past 1 o'clock. Since the 1 o'clock position is the same as the 5 minute point on the clock, this means that the hour hand points at $5 + \frac{m}{12}$ minutes.

The two hands point in exactly opposite directions, so the minute at which the hour hand points must be 30 minutes smaller than the minute at which the minute hand points. Therefore, we have

$$5 + \frac{m}{12} = m - 30.$$

Solving this equation gives $m = 38\frac{2}{11}$ minutes, so the first time after 1 o'clock at which the hands of the clock point in opposite directions is $\boxed{1{:}38\dfrac{2}{11}}$.

Exercises for Section 6.3

6.3.1 Our conversion factor is $\dfrac{1 \text{ pound}}{0.4536 \text{ kg}} \approx 1$, so we have $200 \text{ k\!g} \cdot \dfrac{1 \text{ pound}}{0.4536 \text{ k\!g}} \approx \boxed{441 \text{ pounds}}$.

6.3.2 The conversion factor from inches to centimeters is $\dfrac{2.54 \text{ cm}}{1 \text{ inch}}$, so to convert your height in inches to centimeters, you multiply it by 2.54.

Conversely, the conversion factor from centimeters to inches is $\dfrac{1 \text{ inch}}{2.54 \text{ cm}}$, so to convert your height in centimeters to inches, divide by 2.54.

6.3.3 *Solution 1: Logic.* Because we're so used to converting minutes to hours, this is probably a problem we can do without conversion factors at all. Since Janie stuffs 30 envelopes in a minute, and there are 60 minutes in an hour, she stuffs $30 \cdot 60 = 1800$ envelopes an hour. So, in n hours she can stuff $\boxed{1800n}$ envelopes.

Solution 2: Conversion Factors. Janie's stuffing rate is $30 \dfrac{\text{envelopes}}{\text{min}}$. The conversion factor from minutes to hours is

$$\frac{60 \text{ min}}{1 \text{ hr}} = 1.$$

Notice that we put minutes in the numerator because we want to cancel minutes in the denominator:

$$30 \frac{\text{envelopes}}{\text{m\!i\!n}} \cdot \frac{60 \text{ m\!i\!n}}{1 \text{ hr}} = 1800 \frac{\text{envelopes}}{\text{hr}}.$$

Since Janie can stuff 1800 envelopes in an hour, she can stuff $\boxed{1800n}$ envelopes in n hours.

6.3.4 First, we determine how many euros I gave the cashier. The conversion factor from dollars to euros is $\dfrac{1 \text{ euro}}{1.32 \text{ dollars}} = 1$, so we have

$$40 \text{ dollars} \cdot \frac{1 \text{ euro}}{1.32 \text{ dollars}} \approx 30.30 \text{ euros}.$$

Since my meal costs 17 euros, I should receive $30.30 - 17 = \boxed{13.30 \text{ euros}}$ in change.

6.3.5 We wish to convert one fish into its equivalent value in bags of rice. To do so, we must first convert the fish into bread, then convert the bread into rice. The two conversion factors we then need are:

$$\frac{2 \text{ loaves of bread}}{3 \text{ fish}} = 1 \qquad \text{and} \qquad \frac{4 \text{ bags of rice}}{1 \text{ loaf of bread}} = 1.$$

Now we can perform our conversion:

$$1 \text{ fish} = 1 \text{ fish} \cdot \frac{2 \text{ loaves of bread}}{3 \text{ fish}} \cdot \frac{4 \text{ bags of rice}}{1 \text{ loaf of bread}} = \frac{2}{3} \cdot 4 \text{ bags of rice} = \boxed{\frac{8}{3} \text{ bags of rice}}.$$

6.3.6 We first must determine how much Clint pays for a cubic foot of horse feed. He pays \$10 per cubic meter, so we have to convert cubic meters to cubic feet. The conversion factor from meters to feet is

$$\frac{1 \text{ m}}{3.28 \text{ ft}} \approx 1.$$

However, we must convert m^3 to ft^3, so we must cube this conversion factor:

$$\left(\frac{1 \text{ m}}{3.28 \text{ ft}} \right)^3 \approx \frac{1 \text{ m}^3}{35.3 \text{ ft}^3} = 1.$$

So, every cubic meter is approximately 35.3 cubic feet. Since Clint pays \$10 for 35.3 cubic feet of feed, he pays $\$10/35.3 \approx \$0.283$ for each cubic foot. If he wants to double his money, he should charge $2(0.283) \approx \boxed{\$0.57}$ per cubic foot.

Exercises for Section 6.4

6.4.1

(a) The number that is 24% of 140 is $(0.24)(140) = \boxed{33.6}$.

(b) 10% of 1200 is $(0.10)(1200) = 120$, and 5% of 120 is $(0.05)(120) = \boxed{6}$.

6.4.2

(a) Suppose 66 is $x\%$ of 88. This means we have the equation $\dfrac{66}{88} = \dfrac{x}{100}$. Since $66/88 = 3/4$, our equation is $3/4 = x/100$. Solving this equation gives $x = 75$, so 66 is $\boxed{75\%}$ of 88.

(b) Suppose 210 is $x\%$ of 84. So, we must have $\dfrac{210}{84} = \dfrac{x}{100}$. Solving this equation gives $x = 250$, so 210 is $\boxed{250\%}$ of 84.

6.4.3

(a) Let the desired number be x, so we have $0.3x = 63$. Dividing by 0.3 gives $x = 63/0.3 = \boxed{210}$.

(b) Let the desired number be x, so we have $0.75x = 125$. Therefore, $x = 125/0.75 = 125/(3/4) = 500/3 = \boxed{166\frac{2}{3}}$.

6.4.4 Because 120% of 42 is $1.2(42) = 50.4$, when we increase 42 by 120%, we get $42 + 50.4 = \boxed{92.4}$. As described in the text, we could also find our answer by multiplying 42 by $1 + 1.2 = 2.2$.

6.4.5 Let our number be x. We have

$$x(1 - 0.4) = 36 \quad \Rightarrow \quad 0.6x = 36 \quad \Rightarrow \quad x = \boxed{60}.$$

Exercises for Section 6.5

6.5.1 Mike's \$2 was 10% of his bill. Since $10\% = 0.1$, Mike's bill was therefore $\$2/(0.1) = \$20$. Similarly, Joe's \$2 was 20% of his bill, so Joe's bill was $\$2/(0.2) = \$10$. Therefore, the difference between their bills was $\$20 - \$10 = \boxed{\$10}$.

6.5.2 *Solution 1: Work forwards.* Suppose there are x fish in the store at the start of Monday. Each day, 20% are sold, so the number of fish in the store is reduced to 80% of what it was at the start of the day. So, at the end of Monday, there are $0.8x$ fish. Similarly, at the end of Tuesday, there are $0.8(0.8x) = (0.8)^2x$ fish, since 20% of the fish that were available at the start of Tuesday were sold.

We are told that there are 2000 fish left at the end of Tuesday, so we must have $(0.8)^2x = 2000$. Dividing both sides by 0.8^2, we have

$$x = \frac{2000}{0.8^2} = \frac{2000}{(4/5)^2} = \frac{2000}{16/25} = 2000 \cdot \left(\frac{25}{16}\right) = \boxed{3125}.$$

Solution 2: Work backwards. Since 20% of the fish available at the start of Tuesday were sold on Tuesday, there must be $100\% - 20\% = 80\%$ of these fish remaining at the end of Tuesday. Since there are 2000 fish remaining, there must have been $2000/(0.8) = 2500$ fish at the start of Tuesday. Similarly, 2500 is 80% of the number of fish available at the start of Monday, so there were $2500/0.8 = \boxed{3125}$ fish at the start of Monday.

6.5.3 Suppose there is x of the 20% acid solution. So, there is $2x$ of the 50% acid solution, and there is $x + 2x = 3x$ of the total combined solution. The first solution is 20% acid, so it has $(0.20)x = 0.2x$ acid. The second is 50% acid, so it has $(0.50)(2x) = x$ acid. Combined, the two solutions have $0.2x + x = 1.2x$ acid. The remaining $3x - 1.2x = 1.8x$ of the solution is water. Therefore, the ratio of acid to water in the combined mixture of solutions is $(1.2x)/(1.8x) = 1.2/1.8 = \boxed{2/3}$.

6.5.4 Since t is 25% of u, we have $t = 0.25u$. To find what percent of $4t$ is $2u$, we must find the value of x such that

$$\frac{2u}{4t} = \frac{x}{100}.$$

We can use $t = 0.25u$ to find $2u/4t$. First, we divide both sides by t to get $1 = 0.25u/t$. Writing 0.25 as a fraction gives

$$1 = \frac{u}{4t}.$$

Multiplying both sides of this equation by 2 gives $2 = \frac{2u}{4t}$. Therefore, we now have $2 = \frac{x}{100}$, so $x = 200$. So, $2u$ is $\boxed{200\%}$ of $4t$.

We also could have substituted $t = 0.25u$ into $2u/(4t)$ to find

$$\frac{2u}{4t} = \frac{2u}{4(0.25u)} = \frac{2u}{u} = 2,$$

so $2u$ is 200% of $4t$.

6.5.5 Suppose there are x students in the school originally, so that $0.2x$ of them are walkers and the remaining $x - 0.2x = 0.8x$ of them are not. If the number of walkers doubles, there are $0.4x$ walkers out of a total of $0.4x + 0.8x = 1.2x$ students. So, the fraction of students who are walkers is $(0.4x)/(1.2x) = \boxed{1/3}$.

6.5.6 Suppose we use x cups of heavy cream and y cups of whole milk. Since we need 2 cups total, we have $x + y = 2$. The x cups of heavy cream have $0.36x$ cups of butterfat. The y cups of whole milk have $0.04y$ cups of butterfat. Together, we want the mixture to have $(2)(0.24) = 0.48$ cups of butterfat. Therefore, we must have $0.36x + 0.04y = 0.48$. We can get rid of the decimals by multiplying by 100 to get $36x + 4y = 48$. We then divide by 4 to get $9x + y = 12$. Subtracting $x + y = 2$ from this equation, we have $8x = 10$, so $x = \boxed{5/4}$ cups.

Challenge: See if you can find a "logic" solution to this problem! Notice that the x cups of heavy cream have $0.12x$ more butterfat than x cups of our 24% butterfat mixture, and that the y cups of whole milk have $0.2y$ less butterfat than y cups of the mixture. What does this tell us about the quantities $0.12x$ and $0.2y$?

6.5.7 We'll take two approaches:

Solution 1: Use Algebra. Since Kristin pays more than $p\%$, she must make more than \$28000. Let her income be x. Since x is greater than \$28000, we can write her total tax as

$$\left(\frac{p}{100}\right)(28000) + \left(\frac{p+2}{100}\right)(x - 28000) = 280p + \frac{(p+2)(x-28000)}{100}.$$

Seeing that nasty expression, we're already wondering if there's an easier way to do this problem. However, we could plow ahead and use the distributive property to expand the second term:

$$280p + \frac{(p+2)(x-28000)}{100} = 280p + \frac{(p)(x-28000) + 2(x-28000)}{100}$$
$$= 280p + \frac{px - 28000p + 2x - 56000}{100} = \frac{px + 2x}{100} - 560.$$

We know that this quantity is $(p + 0.25)\%$ of her total income, so we have

$$\frac{\frac{px+2x}{100} - 560}{x} = \frac{p + 0.25}{100}.$$

Cross-multiplying gives

$$100 \left(\frac{px + 2x}{100} - 560 \right) = x(p + 0.25).$$

Expanding both sides gives $px + 2x - 56000 = px + 0.25x$, from which we find $x = 32000$. So, her income is $\boxed{\$32000}$.

Solution 2: Use Logic. Rather than wade through the algebra, we think about the problem first. Kristin's total tax paid is the same as if her whole income were taxed at $(p + 0.25)\%$, so we think about how taxing her whole income at $(p + 0.25)\%$ differs from taxing part at $p\%$ and part at $(p + 2)\%$.

Suppose her whole income is taxed at $(p + 0.25)\%$. For each dollar that is taxed at $(p + 0.25)\%$ instead of $p\%$, she pays an extra 0.25%. On the other hand, for each dollar that is taxed at $(p + 0.25)\%$ instead of $(p + 2)\%$, she pays 1.75% *less*.

The total amount extra she pays on the portion originally taxed at $p\%$ must equal the size of the discount she gets on the portion originally taxed at the higher rate. The per-dollar discount on the amount originally taxed at the high rate is $1.75/0.25 = 7$ times the per-dollar tax increase on the amount originally taxed at the lower rate. Therefore, she must have had 7 times as much money taxed at the lower rate in order to keep her total tax unchanged. Since we know she has \$28000 taxed at the lower rate, she must have \$28000/7 = \$4000 taxed at the higher rate, for a total pay of \$28000 + \$4000 = \$32000.

Review Problems

6.21 Because the ratio of koi to goldfish in the pond is $3 : 11$, the ratio of koi to all fish is $3 : (11+3) = 3 : 14$. Since 3/14 of my fish are koi and I have 70 fish, I must have $(3/14)(70) = \boxed{15}$ koi. (We could also set up an equation by letting there be x koi, so that $x/70 = 3/14$.)

6.22 Since my team won 3/5 of its games, it must have lost 2/5 of its games. So, the ratio of games we won to games we lost is $(3/5)/(2/5) = 3/2$. Since we lost 24 games, we must have won $(24)(3/2) = \boxed{36}$ games. (We also could have set up an equation by letting there be x games won, so that $x/24 = 3/2$.)

6.23

(a) 15% of 26 is $(0.15)(26) = \boxed{3.9}$.

(b) 150% of 60 is $(1.5)(60) = 90$, and 20% of 90 is $(0.2)(90) = \boxed{18}$.

6.24

(a) We must find the value of x such that $\dfrac{x}{100} = \dfrac{30}{75}$. Cross-multiplying gives $75x = 3000$ and dividing by 75 gives $x = 40$, so 30 is $\boxed{40\%}$ of 75.

(b) We must find the value of x such that $\dfrac{x}{100} = \dfrac{12\frac{1}{2}}{20}$. Cross-multiplying gives $20x = (12\frac{1}{2})(100) = 1250$. Dividing by 20 gives $x = 62.5$, so $12\frac{1}{2}$ is $\boxed{62.5\%}$ of 20.

Good number sense can help us solve this part more quickly. We might notice that $12.5 = 5(2.5)$ and $20 = 8(2.5)$, so $12.5/20 = 5/8$. Since $5/8 = 0.625$, we see that 12.5 is $62\frac{1}{2}\%$ of 20.

6.25

(a) Let our desired number be x, so that we must have $\dfrac{48}{x} = \dfrac{37.5}{100}$. We can write 37.5/100 more conveniently. First, we multiply numerator and denominator by 10 to get 37.5/100 = 375/1000. Both the numerator and denominator of this fraction are divisible by 125, so 375/1000 = 3/8. (This is another example of how knowing fraction-to-decimal conversions well can help you!)

So, our equation is now $\dfrac{48}{x} = \dfrac{3}{8}$. Cross-multiplying gives (48)(8) = 3x, and solving this gives $x = 128$. So, 48 is 37.5% of $\boxed{128}$.

We also could have set up our equation by noting that because 37.5% of x is 48, we have $0.375x = 48$. Solving for x gives $x = 48/(0.375) = 48/(3/8) = 128$.

(b) Let the desired number be x, so we have $\dfrac{1/2}{x} = \dfrac{140}{100} = \dfrac{7}{5}$. Cross-multiplying gives us $7x = 5/2$, and dividing by 7 gives $x = \boxed{5/14}$.

6.26 Since there are 1000 milliliters in a liter, we have

$$\frac{\text{calories}}{\text{milliliters of soda}} = \frac{450}{1000} = \frac{45}{100} = \frac{9}{20}.$$

So, if we let there be x calories in 250 milliliters of soda, we have $x/250 = 9/20$. Solving this equation gives $x = \boxed{112.5}$ calories. (We could also note that 250 is 1/4 of 1000, so 250 mL of soda have $(1/4)(450) = 112.5$ calories.)

6.27 First we find how many minutes are in a day. There are 24 hours in a day, and 60 minutes in each hour, so there are 24(60) minutes in each day. The number of minutes in 20% of a day is $(0.20)(24)(60) = (24)(0.2)(60) = (24)(12) = 2(12)(12) = 2(144) = \boxed{288}$. Notice how we made the arithmetic easy on ourselves!

6.28 The total amount of orange juice in the two pitchers is

$$\frac{1}{3}(600) + \frac{2}{5}(600) = 200 + 240 = 440 \text{ mL}.$$

The two pitchers together have $600 + 600 = 1200$ mL of liquid, so the proportion of the mixture that is orange juice is $440/1200 = \boxed{11/30}$.

6.29 We wish to convert 4 gallons per 3 seconds to a number of gallons per hour.

Solution 1: Reason it out in steps. Since there are 60 seconds in a minute, there are $60/3 = 20$ periods of 3 seconds in a minute. Therefore, my jet needs $4(20) = 80$ gallons of jet fuel each minute. There are 60 minutes in an hour, so I need $(80)(60) = \boxed{4800}$ gallons for the whole hour.

Solution 2: Conversion factors. I need to convert seconds in the denominator to hours in the denominator, so the conversion factors I need are

$$\frac{60 \text{ s}}{1 \text{ min}} = 1 \qquad \text{and} \qquad \frac{60 \text{ min}}{1 \text{ hr}} = 1.$$

Now for the conversion:

$$\frac{4 \text{ gallons}}{3 \text{ s}} = \frac{4 \text{ gallons}}{3 \text{ s}} \cdot \frac{60 \text{ s}}{1 \text{ min}} \cdot \frac{60 \text{ min}}{1 \text{ hr}} = \frac{4}{3} \cdot 60 \cdot 60 \, \frac{\text{gallons}}{\text{hr}} = \boxed{4800 \, \frac{\text{gallons}}{\text{hr}}}.$$

6.30 There are 28.35 grams in each ounce, so we divide our 541 grams by 28.35 to find that there are approximately 19.1 ounces in 541 grams. If it isn't obvious to us that we should divide, we can use the conversion factor between ounces (oz) and grams (g):

$$\frac{1 \text{ oz}}{28.35 \text{ g}} \approx 1.$$

Multiplying 541 grams by this conversion factor, we have:

$$541 \text{ g} \approx 541 \text{ g} \cdot \frac{1 \text{ oz}}{28.35 \text{ g}} \approx 19.1 \text{ oz}.$$

6.31 We wish to convert 40 yd^2 to the equivalent number of m^2. We have the conversion factor

$$\frac{0.9144 \text{ m}}{1 \text{ yd}} \approx 1.$$

We must multiply by this conversion factor twice in order to convert yd^2 to m^2:

$$40 \text{ yd}^2 \cdot \left(\frac{0.9144 \text{ m}}{1 \text{ yd}}\right)^2 \approx 40 \text{ yd}^2 \cdot \left(\frac{0.8361 \text{ m}^2}{1 \text{ yd}^2}\right) \approx \boxed{33.45 \text{ m}^2}.$$

6.32 Suppose there are x eggs in the first carton, so there are $3x$ eggs in the second. Since 20% of the eggs in the first carton are red, $0.2x$ eggs in the first carton are red. Since 25% of the eggs in the second carton are red, and the second carton has $3x$ eggs, there are $0.25(3x) = 0.75x$ red eggs in the second carton. Therefore, out of the total of $x + 3x = 4x$ eggs in the two cartons, we have $0.2x + 0.75x = 0.95x$ red eggs, for a total percentage of

$$\frac{0.95x}{4x} = \frac{0.95}{4} = 0.2375 = \boxed{23.75\%}.$$

We can quickly check our answer by pretending there are 100 eggs in the first carton, so 20 of them are red. There are then $3(100) = 300$ eggs in the second carton, of which $(0.25)(300) = 75$ are red, for a total of 95 red eggs out of 400 total. So, $95/400 = 23.75\%$ of the eggs are red.

6.33 First we figure out how many people are seated. Since 3/4 of the chairs are used, 1/4 of them are empty. Since 6 chairs are empty, and this number represents 1/4 of the total number of chairs, we know that there are 24 chairs. Therefore, there are $(3/4)(24) = 18$ people seated. We are told that these 18 people make up 2/3 of the number of the people in the room. So, if there are y people in the room, then we have $18/y = 2/3$. Solving this equation gives us $y = \boxed{27}$ people in the room.

We could have gotten through those last few steps a little more quickly by noting that if 18 is 2/3 of the number of people in the room, then halving both tells us that 9 is equal to 1/3 of the number of people in the room. But, since 2/3 of the people are sitting, $1 - 2/3 = 1/3$ of the people are standing. Therefore, there are 9 people standing to go with the 18 who are sitting.

6.34 Let the price of the radio be x on Monday. On Tuesday, the price is reduced by 20%, so it is $100\% - 20\% = 80\%$ of the Monday price, or $0.8x$. On Wednesday, we wish for the price to be $0.5x$. This price is $(0.5x)/(0.8x) = 0.625 = 62.5\%$ of the Tuesday price. Since the Wednesday price is 62.5% of the Tuesday price, the Tuesday price must be reduced by $100\% - 62.5\% = \boxed{37.5\%}$ on Wednesday.

Notice that the actual Monday price is irrelevant, so we could work the problem by simply choosing a Monday price. Suppose the Monday price is \$100. Then, the Tuesday price is \$80 and we want the Wednesday price to be \$50. So, we must reduce the price on Wednesday by \$30, which is a $30/80 = 0.375 = 37.5\%$ reduction.

6.35 In Francesca's lemonade, there are $25 + 386 = 411$ calories. However, Francesca's lemonade has 600 grams, and we want to know the number of calories in 200 grams of lemonade. Therefore, we are only interested in the amount of calories in $200/600 = 1/3$ of Francesca's lemonade, which is $(411)(1/3) = \boxed{137}$.

6.36 Suppose there are x students. Since $x/3$ take the first test, $x/4$ take the second, $x/5$ take the third, 26 take the fourth, and each student takes one test, we must have

$$\frac{x}{3} + \frac{x}{4} + \frac{x}{5} + 26 = x.$$

Simplifying the left side gives $\frac{47}{60}x + 26 = x$. Solving this equation gives $x = \boxed{120}$.

6.37 Because p is 50% of q, we have $p = 0.5q$. Because r is 40% of q, we have $r = 0.4q$. To find what percent of r that p is, we must find the ratio p/r. We do so by dividing the equation $p = 0.5q$ by the equation $r = 0.4q$, which gives us $p/r = 0.5/0.4 = 5/4 = 1.25$. Therefore, $p/r = 125/100$, so p is $\boxed{125\%}$ of r.

6.38 Our initial solution has $0.03(50) = 1.5$ pounds of salt, so it has $50 - 1.5 = 48.5$ pounds of water. Suppose x pounds of water evaporate, so that we have 1.5 pounds of salt left in $50 - x$ pounds of solution. Since this new solution must be a 5% salt solution, we must have

$$\frac{1.5}{50 - x} = \frac{5}{100} = \frac{1}{20}.$$

Cross-multiplying gives $1.5(20) = 50 - x$, so we have $30 = 50 - x$. Solving this equation gives $x = \boxed{20}$.

6.39 Since the ratio of Corps members to Brigadiers is $3 : 1$ in the first 2400 sectors, we know that 3/4 of the first 2400 patrollers are members of the Corps. Therefore, there are $(3/4)(2400) = 1800$ Corps members among these first 2400 patrollers. Because there are an equal number of Corps members and Brigadiers in the whole galaxy, there are $3600/2 = 1800$ of each. However, we know there are 1800 Corps members among the first 2400 sectors. Therefore, there are none in the rest of the galaxy! So, the ratio of Corps members to Brigadiers in the rest of the galaxy is $\boxed{0}$.

6.40 Since we want a 20% profit above our cost of \$6.25, we need to sell the toy for 120% of our cost, or $1.2(\$6.25) = \$7.50$. When the price is reduced 25% from \$$K$, the new price is $100\% - 25\% = 75\%$ of the original price, or $0.75(\$K)$. Therefore, we must have $0.75(\$K) = \$7.50$, from which we find $K = \boxed{10}$.

6.41 Let there be r red marbles and b blue ones. Since 1/7 of the remaining marbles are red when one red marble is removed, we have

$$\frac{r-1}{b+r-1} = \frac{1}{7}.$$

Since 1/5 of the remaining marbles are red if we instead remove 2 blue marbles, we have

$$\frac{r}{b+r-2} = \frac{1}{5}.$$

Cross-multiplying both equations gives the system

$$7(r-1) = b + r - 1,$$
$$5r = b + r - 2.$$

We can eliminate b by subtracting the second equation from the first, which gives

$$7(r-1) - 5r = -1 + 2.$$

Solving this equation gives $r = 4$, and substituting this into either of our original equations gives $b = 18$. Therefore, there were originally $4 + 18 = \boxed{22}$ marbles.

6.42 There are 4 quarters in a dollar, so there are $4(25) = 100$ quarters in \$25. So, there are 100 quarters in a pound. Since 2 pennies weighs the same as 1 quarter, there are $2(100) = 200$ pennies in a pound. 200 pennies is worth $\boxed{\$2}$. (We could also slog through this with conversion factors, but this step-by-step solution is easy enough because we're used to these simple conversions.)

6.43 *Solution 1: Pick the number of students in the school.* All the information in the problem is relative information, and the answer we seek is also relative information (a percentage), so we can probably safely pick the number of students in the school. Suppose there are 1000 students in the school. (We pick 1000 to keep all of the numbers integers.) After the first 80% of the students have voted, we have 800 votes. Of these, $800(0.53) = 424$ are for Susie and $800 - 424 = 376$ are for Calvin. In order for Calvin to catch up to Susie, he must pick up at least $500 - 376 = 124$ of the final 200 votes, because he must have at least 500 votes total in the end to catch up to her. Therefore, he must receive at least $124/200 = 62/100 = \boxed{62\%}$ of the remaining votes.

Notice that if Calvin gets 62%, then Susie gets $100\% - 62\% = 38\%$ of the remaining votes. In this case, Susie's percentage of the first 800 votes is 6% more than Calvin's, and Calvin's percentage of the next 200 votes is 24% more than Susie's, and they end up tied. Hmmm... $800/200 = 24/6$. Is this a coincidence?

Solution 2: Use algebra. Let there be x students in the school. So, after $0.8x$ of them have voted, Susie has received $0.8x(0.53) = 0.424x$ votes and Calvin has received $0.8x(0.47) = 0.376x$ votes. Calvin needs a total of $0.5x$ votes overall to catch Susie, so he needs $0.5x - 0.376x = 0.124x$ of the remaining $x - 0.8x = 0.2x$ votes, or $(0.124x)/(0.2x) = 0.124/0.2 = 124/200 = 62/100 = \boxed{62\%}$. Perhaps the similarities in all the numbers in this solution and the other solution show why the other solution works!

Challenge Problems

6.44 Out of every $8 + 5 + 2 + 1 = 16$ plants, $8 + 2 = 10$ are green (because there are 8 tall green and 2 short green plants for each group of 16). So, $10/16 = 5/8$ of the plants in Gregor's garden are green. Letting x be the total number of plants Gregor has, we have $1080/x = 10/16$. Therefore, $10x = 16(1080)$, so $x = 16(1080)/10 = 16(108) = \boxed{1728}$.

6.45 We set up conversion factors:

$$\frac{0.4 \text{ spig}}{1 \text{ spoog}} = 1, \qquad \frac{0.25 \text{ speeg}}{1 \text{ spoog}} = 1, \qquad \frac{0.7 \text{ speeg}}{1 \text{ spug}} = 1.$$

We need to find the ratio of a spug to a spig to answer the question. We combine our conversion factors (or reciprocals of them) so that the speegs and spoogs cancel out and leave spigs in the denominator and spugs in the numerator:

$$1 = \frac{1 \text{ spoog}}{0.4 \text{ spig}} \cdot \frac{0.25 \text{ speeg}}{1 \text{ spoog}} \cdot \frac{1 \text{ spug}}{0.7 \text{ speeg}} = \frac{0.25 \text{ spug}}{0.4 \cdot 0.7 \text{ spig}} = \frac{0.25 \text{ spug}}{0.28 \text{ spig}}.$$

We multiply both sides by $0.28/0.25$ to find

$$\frac{1 \text{ spug}}{1 \text{ spig}} = \frac{0.28}{0.25} = 1.12.$$

So, 5 spugs is $5(1.12) = 5.6 = \boxed{560\%}$ of a spig.

6.46 We are told that

$$\frac{\text{Krzyzewski voters}}{\text{Smith voters}} = \frac{\text{Smith voters}}{\text{Williams voters}}.$$

There are 800 Krzyzewski voters and 200 Williams voters. Letting there be x Smith voters, we have

$$\frac{800}{x} = \frac{x}{200}.$$

Cross-multiplying gives us $x^2 = (200)(800) = 160000$. We take the square root of both sides (and note that x must obviously be positive) to get $x = \sqrt{160000} = \sqrt{16}\sqrt{10000} = 4(100) = 400$. Therefore, the total number of voters in the town is $800 + 400 + 200 = \boxed{1400}$.

6.47 *Solution 1: Use conversion factors.* The main difficulty in this problem is converting the square inches to square feet. Because each 25 square inch tile costs \$130, the cost per square inch is

$$\frac{\$130}{25 \text{ in}^2} = \frac{\$5.2}{1 \text{ in}^2}.$$

Because there are 12 inches in a foot, we have the conversion factor

$$\frac{12 \text{ in}}{1 \text{ ft}} = 1.$$

We put inches in the numerator because we want to cancel them in the denominator of of our earlier expression. However, we have in^2 in that expression, so we'll need two copies of our feet-to-inches conversion factor:

$$\frac{\$5.2}{1 \text{ in}^2} = \frac{\$5.2}{1 \text{ in}^2} \cdot \left(\frac{12 \text{ in}}{1 \text{ ft}}\right)^2 = \frac{\$5.2}{1 \text{ in}^2} \cdot \frac{144 \text{ in}^2}{1 \text{ ft}^2} = \frac{\$748.8}{\text{ft}^2}.$$

It costs \$748.80 to tile a square foot, so it costs $(2000)(\$748.80) = \boxed{\$1,497,600}$ to tile the whole bottom surface of the shuttle.

Solution 2: Reason it out. As before, we find that it costs $5.2 to tile a square inch. Since each foot has 12 inches, each square foot is a 12 inch by 12 inch square, which has $12^2 = 144$ square inches. Therefore, it costs $(144)(\$5.2) = \$748.80$ to tile a square foot. So, it costs $(2000)(\$748.80) = \$1,497,600$ to tile 2000 square feet. Notice that this solution is essentially exactly the same as the first solution. The conversion factors are mainly just a tool to use when we aren't sure when to multiply and when to divide.

6.48 We wish to find k, so we first use the ratio with k to make an equation. Because $1 : k = \frac{x}{yz} : \frac{y}{zx}$, we have

$$\frac{k}{1} = \frac{\frac{y}{zx}}{\frac{x}{yz}} = \frac{y}{zx} \cdot \frac{yz}{x} = \frac{y^2}{x^2}.$$

So, we need to find $\frac{y}{x}$ in order to find k. Looking at the other ratio we are given, $yz : zx : xy = 1 : 2 : 3$, we see that the ratio of the first two terms, yz and zx, is y/x, because the z's cancel out. So, we have

$$\frac{y}{x} = \frac{yz}{zx} = \frac{1}{2}.$$

So, we have $k = y^2/x^2 = (y/x)^2 = \boxed{1/4}$.

6.49 If 99% of the trees are eucalyptus trees, then 1% of them are not eucalyptus trees. Suppose x of the trees are not eucalyptus trees, so $99x$ of them are. After the cutting, 98% will be eucalyptus trees, so 2% will not. We will still have x trees that are not eucalyptus trees, but this x will be 2% of the total. So, we'll have

$$\frac{x}{\text{Total number of trees}} = \frac{2}{100}.$$

From this equation, we find that there will be $50x$ trees after the cutting. Originally, there were $100x$ trees in my neighborhood, so the plan involves removing $\boxed{50\%}$ of the existing trees. (Notice that all the information in the problem is relative information, and we seek a percentage in our answer. So, we could have found the answer by simply pretending that my neighborhood has 100 trees.)

6.50 Suppose the cost of his goods is c and the price at which he sells his goods is p. Because his profit currently is $x\%$ of his cost, the price is $x\%$ greater than his cost. So, we have

$$\frac{p}{c} = \frac{100 + x}{100}.$$

If his cost is reduced by 8%, it will be 92% of the original cost, or $0.92c$. In this case, the price, p, is $(x + 10)\%$ greater than the cost. So, we have

$$\frac{p}{0.92c} = \frac{100 + (10 + x)}{100}.$$

At first, these equations look very intimidating because they have three variables each. However, we see p/c on both left sides. This suggests two different ways to solve the problem.

Solution 1: Substitution. We can solve the second equation for p/c by multiplying both sides by 0.92:

$$\frac{p}{c} = \frac{0.92(110 + x)}{100}.$$

Combining this with our first equation, we have

$$\frac{100 + x}{100} = \frac{0.92(110 + x)}{100}.$$

Multiplying by 100 gives $100 + x = 0.92(110 + x) = 101.2 + 0.92x$. Solving this equation gives $x = \boxed{15}$.

Solution 2: Elimination. We can divide the first equation by the second to cancel out the p/c terms:

$$\frac{\dfrac{p}{c}}{\dfrac{1}{0.92} \cdot \dfrac{p}{c}} = \frac{\dfrac{100 + x}{100}}{\dfrac{100 + (10 + x)}{100}}.$$

The p/c terms cancel on the left, leaving $1/(1/0.92) = 0.92$. On the right we have $(100 + x)/(110 + x)$. Multiplying both sides by $110 + x$ gives us the same equation we had before, $100 + x = 0.92(110 + x)$, from which we find $x = \boxed{15}$.

Notice that all the information we are given in the problem is relative information. Can you find another solution to this problem by assuming that the dealer's cost is \$100?

6.51 We have to convert feet (ft) to meters (m) and seconds (s) to minutes (min), so we start by writing the two conversion factors we know we'll need:

$$\frac{3.28 \text{ ft}}{1 \text{ m}} \approx 1 = \frac{60 \text{ s}}{1 \text{ min}}.$$

Next, we write our acceleration due to gravity as a fraction with the s^2 in the denominator, since the s^2 is in the denominator in our units:

$$\frac{32.2 \text{ ft}}{1 \text{ s}^2}.$$

This tells us how to apply our conversion factors. For the feet-to-meters factor, we want the feet in the denominator to cancel with the feet in the numerator of 32.2 ft/s^2. However, we want the seconds in the numerator of the seconds–to–minutes conversion factor to cancel with the seconds in the denominator of 32.2 ft/s^2. Moreover, we'll need *two* of the seconds-to-minutes conversion factors because the seconds are squared in our denominator:

$$\frac{32.2 \text{ ft}}{1 \text{ s}^2} \approx \frac{32.2 \text{ ft}}{1 \text{ s}^2} \cdot \frac{1 \text{ m}}{3.28 \text{ ft}} \left(\frac{60 \text{ s}}{1 \text{ min}}\right)^2 = \frac{32.2 \text{ m}}{3.28 \text{ s}^2} \cdot \frac{3600 \text{ s}^2}{1 \text{ min}^2} \approx \frac{32.2 \cdot 3600}{3.28} \text{ m/min}^2 \approx \boxed{35300 \text{ m/min}^2}.$$

6.52 We tackle the whole problem in seconds, rather than in minutes and seconds. In one hour, there are 60 minutes. Each minute has 60 seconds, so there are $(60)(60) = 3600$ seconds in an hour. During each hour, Cassandra's watch moves 57 minutes and 36 seconds, which equals $57(60) + 36 = 3456$ seconds. So, the ratio of number of seconds that elapse on Cassandra's watch to the number of seconds that pass in real time is

$$\frac{\text{Cassandra's watch}}{\text{Real time}} = \frac{3456}{3600} = \frac{24}{25}.$$

(Quick computation note: I reduced that fraction quickly by noticing that $3600 - 3456 = 144$, and $144 = 36 \cdot 4$. So, $3456/3600 = 1 - 144/3600 = 1 - 4/100 = 24/25$.)

After Cassandra's watch has moved 10 hours, it has moved $10(3600) = 36000$ of its seconds (not "real time" seconds). So, we have

$$\frac{36000}{\text{Real time}} = \frac{24}{25}.$$

Solving this equation, we find that

$$\text{Real time} = \frac{25 \cdot 36000}{24} = 25 \cdot 1500.$$

Notice that we don't multiply this out. This is the time that has passed in seconds. To figure out what time it is, we first find the number of minutes that are in $25 \cdot 1500$ seconds. To do so, we divide by 60:

$$25 \cdot 1500 \; \cancel{s} \cdot \frac{1 \text{ min}}{60 \; \cancel{s}} = \frac{25 \cdot 1500}{60} \text{ min} = 25 \cdot 25 \text{ min} = 625 \text{ min}.$$

Since 625 minutes is 10 hours (600 min) and 25 minutes, it is $\boxed{10\!:\!25 \text{ PM}}$ when Cassandra's watch reads 10:00 PM.

6.53 Al believes he has won $3/(3+2+1) = 1/2$ of the candy, so he'll take half when he shows up. That means he'll leave half of the candy that is there when he arrives. Similarly, Bert believes he has won $2/(3+2+1) = 2/6 = 1/3$ of the candy, so he'll take $1/3$ of the candy he finds when he arrives. So, he'll leave $2/3$ of the candy that is present when he arrives. Finally, Carl thinks he has won $1/(3+2+1) = 1/6$ of the candy, so when he arrives, he'll take $1/6$ of what he finds and leave the other $5/6$.

Putting these together, Al will leave $1/2$ the candy he finds, Bert will leave $2/3$ of the candy he finds, and Carl will leave $5/6$. Suppose there are x pieces of candy. If they arrive in alphabetical order, there will be $x(1/2) = x/2$ pieces left after Al, then $(x/2)(2/3) = x/3$ pieces left after Carl, then $(x/3)(5/6) = 5x/18$ pieces left after Carl. But what if they come in another order?

To find the amount of candy left, we multiply by the appropriate fraction at each step. Above, we multiplied by $1/2$ for the half that Al left, then by $2/3$ for the $2/3$ that Bert left, then by $5/6$. If the students come in a different order, we just multiply by these factors in a different order. However, we are still multiplying x by $1/2$, $2/3$, and $5/6$. Since it doesn't matter in what order we multiply numbers in a product, the final amount of candy will always be $5x/18$ *no matter in which order the students arrive.* Therefore, there will be $\boxed{5/18}$ of the candy left.

6.54 The three of them together have 8 loaves of bread, so each receives $8/3 = 2\frac{2}{3}$ loaves. Therefore, Dick gives Albert $5 - 2\frac{2}{3} = 2\frac{1}{3}$ loaves and Nick gives Albert $3 - 2\frac{2}{3} = \frac{1}{3}$ loaves. To figure out how much money each of Dick and Nick should receive, we must determine what portion of Albert's bread comes from each. Of Albert's $2\frac{2}{3}$ loaves, $2\frac{1}{3}$ comes from Dick. So, Albert received

$$\frac{2\frac{1}{3}}{2\frac{2}{3}} = \frac{7/3}{8/3} = \frac{7}{8}$$

of his bread from Dick. So, Dick should get $7/8$ of the money, or $(\$8)(7/8) = \$7$, and Nick should get the remaining $\$8 - \$7 = \$1$.

6.55 Suppose the minute hand is now pointing at m. Six minutes from now, the minute hand will be pointing at $m + 6$. Three minutes ago, the minute hand was pointing at $m - 3$. At this time, the hour hand has moved $(m-3)/60$ of the way from hour 10 (minute 50) to hour 11 (minute 55). Therefore, the hour hand three minutes ago was pointing at

$$50 + 5\left(\frac{m-3}{60}\right) = 50 + \frac{m-3}{12}.$$

Since this is exactly opposite where the minute hand is 6 minutes from now, the minute hand 6 minutes from now must be pointing at a minute that is 30 minutes earlier than the minute $50 + (m-3)/12$ (which is the minute that the hour hand pointed at 3 minutes ago). So, we have

$$50 + \frac{m-3}{12} = m + 6 + 30.$$

Putting all the m terms on one side and the constants on the other, we have $11m/12 = 55/4$. Multiplying both sides by $12/11$ gives $m = 15$. So the exact time now is $\boxed{10{:}15}$.

CHAPTER 7

Proportion

Exercises for Section 7.1

7.1.1 Since x and r are directly proportional, the ratio x/r is constant. Since $x = 5$ when $r = 25$, we have $x/r = 5/25 = 1/5$. So, when $r = 40$, we have $x/40 = 1/5$. Solving this equation gives $x = \boxed{8}$.

7.1.2 The number of cookies and the number of cups of flour are directly proportional. So, if we let c be the number of cookies and f be the number of cups of flour, the ratio c/f is constant. Since $c = 30$ when $f = 2\frac{1}{4}$, we have $c/f = 30/(2\frac{1}{4}) = 30/(9/4) = 40/3$. When we have fifteen dozen cookies, we have $c = 15(12) = 180$. So, we have $180/f = 40/3$, from which we find $f = 180(3/40) = 27/2$. So, we need $\boxed{13\frac{1}{2}}$ cups of flour.

We could also have tackled this problem by noting that 180 cookies equals 6 batches of 30 cookies. So, we need 6 batches of $2\frac{1}{4}$ cups of flour, for a total of $6(2\frac{1}{4}) = 6(9/4) = 27/2$ cups of flour.

7.1.3 The drawing height is directly proportional to the actual height of an object. We are told that a 3-inch tall drawing matches an 8-foot tall object. Three inches equals $3/12 = 1/4$ foot, so we have

$$\frac{\text{Drawing height}}{\text{Actual height}} = \frac{1/4}{8} = \frac{1}{32}.$$

So, for our 50-foot tall mansion, we have

$$\frac{\text{Drawing height}}{50 \text{ feet}} = \frac{1}{32}.$$

Multiplying both sides by 50 feet, we find that the height of the drawing is $50/32 = 25/16 = \boxed{1\frac{9}{16}}$ feet. (This also equals $\frac{25}{16} \cdot 12 = 18\frac{3}{4}$ inches.)

7.1.4 The height of an object and the length of its shadow are directly proportional. Let h be the height of an object and s be the length of its shadow. From the information about the child, we have $h/s = 4/(0.8) = 5$. The flagpole with Robin on top has a height of 46 feet. Suppose its shadow has length x. Solving the equation $46/x = 5$ for x gives us $x = 46/5 = 9.2$ feet. The flagpole alone is 40 feet tall. Suppose its shadow has length y. Then, we must have $40/y = 5$. Solving this equation gives $y = 8$ feet. So, the difference in shadow lengths is $9.2 - 8 = \boxed{1.2 \text{ feet}}$.

Notice that Robin is 6 feet tall, so if his shadow is z feet long, then we must have $6/z = 5$. Solving this equation gives us $z = 1.2$ feet. But this is the same as our answer.... Is this a coincidence?

No! The difference between the lengths of the shadow of the flagpole with Robin on top and of the shadow of the flagpole alone is just the length of Robin's shadow! So all we had to do was find the length of Robin's shadow to answer the problem.

Exercises for Section 7.2

7.2.1 Since x and r are inversely proportional, their product is constant. Because $x = 5$ when $r = 25$, we have $xr = 125$. So, when $r = 40$, we have $x(40) = 125$. Dividing by 40 gives $x = 125/40 = \boxed{25/8}$.

7.2.2 The number of people mowing and the time required to mow are inversely proportional. Letting n be the number of people and t be the amount of time, we have $nt = (5)(12) = 60$ because 5 people can mow a lawn in 12 hours. If m people can mow the lawn in 3 hours, then we must have $m(3) = 60$, so $m = 20$. Therefore, we need to add $20 - 5 = \boxed{15}$ people to the job.

7.2.3 Because x and y are inversely proportional and $x = 5$ when $y = 6$, we have $xy = 30$. Because y and z are directly proportional and $y = 6$ when $z = 30$, we must have $y/z = 6/30 = 1/5$. So, when $z = 5$, we have $y/z = 1/5$, so $y = 1$. Since $xy = 30$, we have $x = 30/y = 30/1 = \boxed{30}$.

Exercises for Section 7.3

7.3.1 Doubling the number of chickens doubles the bags of scratch eaten, so the number of chickens and bags of scratch are directly proportional. If we want twice as many bags of scratch eaten but we have the same number of chickens, we must double the number of days, so the bags of scratch and the amount of time are directly proportional. Putting these together, we see that the bags of scratch is jointly proportional to the number of chickens and the amount of time. So, we see that the expression

$$\frac{\text{(Time in days)(Number of chickens)}}{\text{(Bags of scratch)}}$$

is constant. Using the given information, we find:

$$\frac{\text{(Time in days)(Number of chickens)}}{\text{(Bags of scratch)}} = \frac{(20)(5)}{10} = 10.$$

So, if we have 18 chickens and we want to find out how long it takes them to eat 100 bags of scratch, we must have

$$\frac{\text{(Time in days)(18)}}{100} = 10.$$

Multiplying both sides by 100/18, we find that the required time is $10(100/18) = 500/9 = \boxed{55\frac{5}{9} \text{ days}}$.

7.3.2 Because a is jointly proportional to b and c, there is some constant k for which $a = kbc$. Since $a = 4$ when $b = 8$ and $c = 9$, we have $4 = k(8)(9)$, so $k = 4/72 = 1/18$. So, when $b = 2$ and $c = 18$, we have $a = kbc = (1/18)(2)(18) = \boxed{2}$.

Note that in going from $b = 8$ and $c = 9$ to $b = 2$ and $c = 18$, we divide b by 4 and multiply c by 2. So, we divide a by 4 and then multiply it by 2. The result is that a is divided by 2, which takes it from 4 to 2.

7.3.3 Because the force of gravitational attraction is directly proportional to the mass of each body, when the mass of one body is doubled, the force of gravitational attraction between them is doubled. Therefore, when the masses of both bodies are doubled, the force of gravitational attraction between them is doubled twice, so it is multiplied by 4.

Because the force of gravitational attraction is inversely proportional to the square of the distance between the two bodies, the product of the force and the square of the distance between the two bodies is constant. So, when the distance between the two bodies is tripled, the square of the distance is multiplied by 9. Since the product of the force and the square of the distance between the bodies is constant, when the square of the distance is multiplied by 9, the force must be divided by 9.

Therefore, if the distance between two bodies is tripled and the mass of each of the bodies is doubled, the force is divided by 9 to account for the change in distance and multiplied by 4 to account for the change in mass. So, the force is $\boxed{\text{multiplied by 4/9}}$.

Exercises for Section 7.4

7.4.1 In the hour that Jack drives 40 mph, he covers 40 miles. In the hour that he drives 50 miles per hour, he covers 50 miles. Therefore, he covers 90 miles in 2 hours, so his average speed is $90/2 = \boxed{45 \text{ mph}}$. Make sure you see the difference between this and the question, "Jack drives 40 mph for a distance of 100 miles, then 50 mph for a distance of 100 miles. What is his average speed for the whole trip?"

7.4.2 Because it takes Brenda 5 hours to dig the whole ditch herself, she can dig 1/5 of the ditch in one hour. So, in three hours, she can dig 3/5 of the ditch. Since she and Jack can dig the whole ditch together in 3 hours, Jack must dig the other 2/5 of the ditch during those three hours. Since Jack can dig 2/5 of the ditch in 3 hours, he can dig $(2/5)/3 = 2/15$ during each hour. Therefore, if Jack needs x hours to dig the whole ditch, we must have $x(2/15) = 1$, so $x = 15/2 = \boxed{7.5 \text{ hours}}$.

We also could have let j be the number of hours Jack needs to dig the whole ditch, so that he digs $1/j$ of the ditch per hour. Since Brenda digs 1/5 of the ditch per hour, and the two together dig the whole ditch in 3 hours, we must have

$$3\left(\frac{1}{j} + \frac{1}{5}\right) = 1,$$

since together they dig $\frac{1}{j} + \frac{1}{5}$ of the ditch per hour. Solving this equation gives $j = \boxed{7.5 \text{ hours}}$, as before.

7.4.3

(a) Suppose the distance from home to work is x miles. When he drives 40 mph, it takes him $x/40$ hours to cover this distance. When he drives 60 mph, it takes him $x/60$ hours to cover this distance. We know that these two times are 6 minutes apart, so we convert these two times to minutes by multiplying both by 60 (because there are 60 minutes in an hour). So, it takes Earl $60(x/40) = 3x/2$ minutes when he drives 40 mph and $60(x/60) = x$ minutes when he drives 60 mph.

(b) Because Earl arrives at work 6 minutes earlier when driving 60 mph than he does when driving 40 mph, we have the equation

$$\frac{3x}{2} - x = 6.$$

Solving for x, we have $x = 12$.

(c) Because it takes Earl 12 minutes to get to work when he drives 60 mph, we know his house is 12 miles away from work (60 miles per hour equals 1 mile per minute). We also know that he must arrive at 8:15 to be precisely on time, because 8:12 is 3 minutes early. So, to be precisely on time, Earl must drive 12 miles in 15 minutes. Since 15 minutes is 1/4 hr, this means that Earl's speed must be $(12 \text{ miles})/(\frac{1}{4} \text{ hr}) = \boxed{48 \text{ mph}}$ to arrive on time.

7.4.4 Suppose the plane can fly p mph when there is no wind, and that the wind blows w mph. With the tailwind, the plane then flies $p + w$ mph. Since it covers 2000 miles in 5 hours with the tailwind, we have

$$5(p + w) = 2000,$$

or $p + w = 400$. When the plane flies into the wind, its rate is $p - w$ mph. Since it covers 2000 miles in 8 hours into the wind, we have

$$8(p - w) = 2000,$$

or $p - w = 250$. We can most quickly find the speed of the wind by subtracting this equation from $p + w = 400$, which gives $2w = 400 - 250 = 150$. Therefore, the speed of the wind is $w = \boxed{75 \text{ mph}}$.

7.4.5 Because Bart can cover the board in 50 minutes, he covers 1/50 of the board in each minute. Because Nelson can erase the board in 80 minutes, he can erase 1/80 of the board in each minute. So, during each minute, 1/50 of the board gets covered while 1/80 of it gets erased. The net result is that

$$\frac{1}{50} - \frac{1}{80} = \frac{8}{400} - \frac{5}{400} = \frac{3}{400}$$

of the board is covered each minute after both Bart's writing and Nelson's erasing. Since they cover 3/400 of the board each minute, the amount of time, t, needed to cover the entire board is the solution to the equation

$$\frac{3}{400} \cdot t = 1.$$

(Rate of covering times time equals amount covered.) Therefore, it will take them $t = 400/3 = \boxed{133\frac{1}{3} \text{ minutes}}$ to cover the board.

7.4.6 Let the length of the sidewalk be x feet. Since Kelsey can walk this length in 3 minutes, Kelsey walks at the rate of $x/3$ feet/min. Since she will travel the length of the sidewalk in 2 minutes if she rides the moving sidewalk without walking, the sidewalk moves at the rate of $x/2$ feet/min. Therefore, if she walks on the sidewalk, she moves at an overall rate of

$$\frac{x}{3} + \frac{x}{2} = \frac{2x}{6} + \frac{3x}{6} = \frac{5x}{6} \frac{\text{feet}}{\text{min}}.$$

Because the sidewalk is x feet long, the total time she'll spend walking on the moving sidewalk is

$$\text{Time} = \frac{\text{Distance}}{\text{Rate}} = \frac{x}{\frac{5x}{6}} = \frac{6}{5} = \boxed{1.2 \text{ minutes}}.$$

(Note that the actual value of x doesn't matter, so we could have simply chosen a convenient length for the sidewalk and solved the problem with that length to find the answer.)

7.4.7 *Solution 1: Find the time it takes to catch up.* Suppose Sunny's rate is s, so Moonbeam's is ms. Since Sunny has a head start of h meters, Moonbeam must run h meters more than Sunny. Because Moonbeam's rate is $ms - s = s(m - 1)$ greater than Sunny's, Moonbeam catches up to Sunny at a rate of $s(m - 1)$. Let t be the time it takes Moonbeam to catch up. Because we must have

(Rate at which Moonbeam catches up)(Time it take Moonbeam to catch up) =

Distance Moonbeam must make up,

we have
$$[s(m - 1)](t) = h.$$

Therefore, $t = \dfrac{h}{s(m - 1)}$. In this amount of time, Sunny runs a distance of

$$(\text{Sunny's rate})(\text{Sunny's time}) = s \cdot \frac{h}{s(m - 1)} = \frac{h}{m - 1}.$$

Moonbeam must cover this distance, plus the initial h meters head start that Sunny had, for a total of

$$h + \frac{h}{m - 1} = \frac{h(m - 1)}{m - 1} + \frac{h}{m - 1} = \frac{hm - h}{m - 1} + \frac{h}{m - 1} = \boxed{\frac{hm}{m - 1} \text{ meters}}.$$

Solution 2: Equate the times during which they are running. Let d be the distance Moonbeam must run to catch Sunny. Since Moonbeam runs at a rate of ms, Moonbeam runs a total time of $d/(ms)$ before catching Sunny. During this time, Sunny runs a total distance of $d - h$ at a rate of s. Since Sunny runs $d - h$ at a rate of s, Sunny runs for a time of $(d - h)/s$. Since Sunny and Moonbeam run for the same amount of time, we must have
$$\frac{d}{ms} = \frac{d - h}{s}.$$

Multiplying both sides of this equation by ms gives $d = m(d - h) = md - mh$. Moving all d terms to the left gives $d - md = -mh$, so $d(1 - m) = -mh$. Dividing both sides by $1 - m$ gives $d = -mh/(1 - m) = mh/(m - 1)$, as before.

Review Problems

7.17 Because s and t are directly proportional and $s = 14$ when $t = 12$, we have $s/t = 14/12 = 7/6$. Therefore, when $t = 7$, we have $s/7 = 7/6$, so $s = \boxed{49/6}$.

7.18 Because x and y are inversely proportional, the product xy is constant. If x is tripled and the product xy must remain constant, then y must be $\boxed{\text{divided by 3}}$ in order to hold the product constant:

$$(3x)\left(\frac{y}{3}\right) = (\cancel{3}x)\left(\frac{y}{\cancel{3}}\right) = xy.$$

7.19 Because a^2 and b are directly proportional and $a = 2$ when $b = 9$, we must have $a^2/b = 2^2/9 = 4/9$. So, when $a = 6$, we have $6^2/b = 4/9$. Taking the reciprocal of both sides gives $b/6^2 = 9/4$. Multiplying both sides by 36 gives $b = (9/4)(36) = \boxed{81}$.

7.20 We have:

$$\text{Time it takes to ride to work} = \frac{\text{Distance to work}}{\text{Rate of riding}} = \frac{8 \text{ miles}}{12 \text{ miles/hr}} = \frac{2}{3} \text{ hr.}$$

Two-thirds of an hour equals $(2/3)(60) = 40$ minutes. So, she must leave her house 40 minutes before 8:00 a.m., which is at $\boxed{\text{7:20 a.m.}}$

7.21 Because a is directly proportional to b and $a = 2$ when $b = 5$, we have $a/b = 2/5$. Because a is inversely proportional to c and $a = 2$ when $c = 9$, we have $ac = (2)(9) = 18$. Therefore, when $b = 3$, we have $a/3 = 2/5$, which gives us $a = (2/5)(3) = 6/5$. Since $ac = 18$, we have

$$\left(\frac{6}{5}\right)(c) = 18 \quad \Rightarrow \quad c = \frac{18}{6/5} = 18 \cdot \frac{5}{6} = \boxed{15}.$$

We also could have divided $ac = 18$ by $a/b = 2/5$ to find $bc = 45$ (so b and c are inversely proportional). Therefore, $c = 45/b = 45/3 = 15$.

7.22 After 4 minutes, Homer has peeled $3(4) = 12$ potatoes, so there are 32 left. Together, Homer and Christen peel $3 + 5 = 8$ potatoes each minute. So, it takes them $32/8 = 4$ minutes to peel the rest of the potatoes. In those 4 minutes, Christen peels $4(5) = \boxed{20}$ of the potatoes.

7.23 Distance in inches on the map is directly proportional to distance in miles in the real world, so from the given information we must have

$$\frac{\text{Distance in inches on the map}}{\text{Distance in miles in the real world}} = \frac{4}{26} = \frac{2}{13}.$$

Therefore, if a distance of x miles in the real world is represented by a distance of 11 inches on the map, we must have $11/x = 2/13$. Cross-multiplying gives $2x = 11(13) = 143$, so $x = 143/2 = \boxed{71\frac{1}{2} \text{ miles}}$.

We also could have noticed that we have to multiply 4 inches by 11/4 to get 11 inches. Because distance on the map is directly proportional to distance in the real world, when we multiply "map distance" by 11/4, we also must multiply "real world" distance by 11/4. This gives us $26(11/4) = 143/2$ miles, as before.

7.24 We assume that the number of fish tagged in a random group of fish taken from the lake is directly proportional to the total number of fish in the group. Since there are 3 tagged fish in the given sample of 75 fish, we have

$$\frac{\text{Number of total fish}}{\text{Number of tagged fish}} = \frac{75}{3} = 25.$$

Since we know there are 50 tagged fish in the whole lake, we must have

$$\frac{\text{Number of fish in lake}}{50} = 25.$$

Therefore, there are 25(50) = $\boxed{1250}$ fish in the whole lake.

7.25 While Amy waits for 60 seconds, Joey runs 10(60) = 600 feet. The track is only 400 feet long, so in this time, Joey runs all the way around the track, then 600 − 400 = 200 feet beyond where Amy starts. Amy catches up to Joey at a rate of 12 − 10 = 2 feet per second. Therefore, Amy must run for 200/2 = $\boxed{100}$ seconds before catching Joey.

7.26 Because a and b are directly proportional, their ratio is constant. So, for some constant m, we have $a/b = m$. Similarly, because b and c are directly proportional, we have $b/c = n$ for some constant n. We wish to see how a and c are related, so we eliminate b from these two equations by multiplying them. This gives us $a/c = mn$. Since m and n are constants, the ratio a/c is constant. Therefore, a and c are $\boxed{\text{directly proportional}}$ as well.

7.27 The number of chapters completed is directly proportional to the number of people, and directly proportional to the amount of time spent. So, we must have

$$\frac{(\text{Number of people})(\text{Amount of time})}{\text{Number of chapters}} = k,$$

where k is some constant. We can find k using the information that I can solve the problems in one chapter alone in 8 hours:

$$k = \frac{(1)(8)}{1} = 8.$$

So, if n people can tackle 30 chapters in 24 hours, we must have

$$\frac{(n)(24)}{30} = 8.$$

Multiplying both sides by 30/24, we have $n = 8(30/24) = \boxed{10}$ people.

We also could have reasoned our way to the answer. If I can solve 1 chapter of problems in 8 hours, then I can solve 3 chapters in 1 day (24 hours). So, I can tackle 3/30 = 1/10 of the book in a day. Therefore, we need 10 people to solve all the problems in a single day.

7.28 Each good worker can paint 1/12 of my house in an hour, so three of them together can paint 3/12 = 1/4 of my house in an hour. So, in 3 hours, the three good workers will paint 3(1/4) = 3/4 of my house. The bad workers have to paint the other 1/4 of the the house. Each bad worker paints 1/36 of the house in an hour, so each bad worker can paint 3(1/36) = 1/12 of the house in three hours. Since the bad workers together need to paint 1/4 of the house, and each bad worker can paint 1/12 of the house in three hours, I need (1/4)/(1/12) = $\boxed{3}$ bad workers.

7.29 Suppose the distance to Jayne's grandparents' house is x miles. She drives 20 mph on the way there, so it takes her $x/20$ hours to drive there. For the total round trip of $2x$ miles, her average speed is 30 mph, so her total round trip takes $2x/30 = x/15$ hours. So, the return trip must take

$x/15 - x/20 = 4x/60 - 3x/60 = x/60$ hours. Since she must drive x miles in $x/60$ hours, her rate must be $x/(x/60) = \boxed{60 \text{ miles per hour}}$.

Notice that x doesn't matter in this problem. We could have solved the problem by assuming that Jayne's grandparents' house is 60 miles away. So, her trip there took 3 hours and the round trip takes $2(60)/30 = 4$ hours, which leaves 1 hour for the return trip. Since she must drive 60 miles in an hour, she must drive 60 miles per hour.

7.30 Since Joe can build the fence alone in 4 hours, he can build 1/4 of the fence each hour. So, after 2 hours, Joe has built 1/2 of the fence. Suppose Renee can build the fence in r hours. Then, she can build $1/r$ of the fence each hour. So, Joe and Renee can build $\frac{1}{4} + \frac{1}{r}$ of the fence together each hour. Because they can build the remaining 1/2 of the fence in 90 minutes, which equals $90/60 = 3/2$ of an hour, we have

$$\frac{3}{2}\left(\frac{1}{4} + \frac{1}{r}\right) = \frac{1}{2}.$$

Multiplying both sides by $\frac{2}{3}$ gives

$$\frac{1}{4} + \frac{1}{r} = \frac{1}{2} \cdot \frac{2}{3} = \frac{1}{3},$$

so $\frac{1}{r} = \frac{1}{3} - \frac{1}{4} = \frac{1}{12}$. Taking reciprocals of both sides gives $r = \boxed{12 \text{ hours}}$.

We also could have noted that Joe works for 3.5 hours. Since he can build the whole fence in 4 hours, he builds $3.5/4 = 7/8$ of the fence in 3.5 hours. Renee then builds the other $1 - 7/8 = 1/8$ of the fence in the 1.5 hours she works. Since she can build 1/8 of the fence in 1.5 hours, it would take her $8(1.5 \text{ hours}) = \boxed{12 \text{ hours}}$ to build the whole fence herself.

7.31 Because each member of team B runs 100 meters at b meters per second, it takes each team member $100/b$ seconds to run 100 meters. Therefore, the total time for team B is $400/b$. Each of the first three members of team A run the 100 meters in

$$\frac{100}{5b/6} = 100 \cdot \frac{6}{5b} = \frac{120}{b} \text{ seconds.}$$

Therefore, the three together finish their 300 meters in $3(120/b) = 360/b$ seconds. So, the last runner on team A must run 100 meters in the remaining $400/b - 360/b = 40/b$ seconds. Since this runner's rate is a, and the runner runs 100 meters in $40/b$ seconds, we have

$$a = \frac{\text{Distance}}{\text{Time}} = \frac{100}{40/b} = 100 \cdot \frac{b}{40} = \frac{5b}{2}.$$

Dividing both sides by b gives us $a/b = \boxed{5/2}$.

7.32 Suppose each escalator moves s steps while the two ascend the escalators. Since the older sister takes 10 steps while her escalator lifts her s steps, the older sister's trip is a total of $s + 10$ steps. Since the escalator is 40 steps high and the older sister's trip of $s + 10$ steps takes her the full length of the escalator, we must have $s + 10 = 40$. So, $s = 30$.

Suppose the younger sister takes y steps. While she takes those steps, the escalator "takes away" s steps, leaving her total distance traveled equal to $y - s$ steps. This must equal the height of the escalator, so $y - s = 40$. Since $s = 30$, we have $y = s + 40 = \boxed{70 \text{ steps}}$.

Another slick way we can look at this problem is to note that in the end, the two sisters must move up $2(40) = 80$ steps. However, for every step the older sister gets from her escalator, the younger sister loses a step to hers. So, the escalators together don't give the sisters any steps at all. All 80 of these steps must come from the two sisters. One sister takes 10, so the other takes $80 - 10 = 70$.

7.33 The number of acres is directly proportional to both the number of workers and the number of days, so the expression

$$\frac{(\text{Number of workers})(\text{Number of days})}{\text{Number of acres}}$$

is constant. From the given information, this expression must equal

$$\frac{(\text{Number of workers})(\text{Number of days})}{\text{Number of acres}} = \frac{(8)(3)}{20} = \frac{6}{5}.$$

So, if it takes n people 5 days to clear 50 acres, we must have

$$\frac{(n)(5)}{50} = \frac{6}{5}.$$

Solving this equation gives $n = \boxed{12 \text{ workers}}$.

Challenge Problems

7.34 Let the man's rate be r, the time he walked be t, and the distance he walked be d. So, we have $rt = d$. We are told that if his rate were $r + \frac{1}{2}$, then his time would have been $\frac{4}{5}t$. The distance remains the same in this case, so we have

$$\left(r + \frac{1}{2}\right)\left(\frac{4}{5}t\right) = d.$$

However, we know that $d = rt$, so we have the equation

$$\left(r + \frac{1}{2}\right)\left(\frac{4}{5}t\right) = rt.$$

Dividing both sides by t gets rid of t and leaves the equation $(r + \frac{1}{2})(\frac{4}{5}) = r$. Solving this linear equation gives us $r = 2$.

We could also have reasoned our way to a solution by noting that for a fixed distance, rate and time are inversely proportional. So, if time is multiplied by $\frac{4}{5}$, we must multiply the rate by $\frac{5}{4}$ to keep the product of rate and time constant. Therefore, the new rate must be $\frac{5}{4}r$. We are told that this rate is $\frac{1}{2}$ greater than the old rate, so we must have $\frac{5}{4}r = r + \frac{1}{2}$, so $r = 2$.

Now, we find the time. If his rate were $\frac{1}{2}$ mph slower, it would be $\frac{3}{2}$ miles per hour. Then, his time would be $2\frac{1}{2}$ greater, or $t + \frac{5}{2}$. Still, our distance is the same, so we have

$$\left(\frac{3}{2}\right)\left(t + \frac{5}{2}\right) = d = rt = 2t.$$

Solving this linear equation gives us $t = 15/2$. Again, we could have reasoned our way to this. If his rate is 1/2 mph slower, his new rate is $(3/2)/2 = 3/4$ of his old rate. Because rate and time are inversely proportional, his new time must be 4/3 his old time (to keep the product of the rate and the time constant). So, the new time is $\frac{4}{3}t$. We are told this equals $t + \frac{5}{2}$, so we solve $\frac{4}{3}t = t + \frac{5}{2}$ to find $t = 15/2$.

Because the man's rate is 2 miles per hour, and he walks for 15/2 hours, he travels $2(15/2) = \boxed{15 \text{ miles}}$.

7.35   Because there are 4.2 births and 1.7 deaths every second, the population increases by $4.2 - 1.7 = 2.5$ every second. We are asked how long it will take for the population to increase by 1 billion people. We can think of this as a rate problem:

(Rate of population growth) × (Time of population growth) = Amount of population growth.

So, we have

$$\text{Time of population growth} = \frac{\text{Amount of population growth}}{\text{Rate of population growth}} = \frac{1,000,000,000 \text{ people}}{2.5 \text{ people/s}} = 400,000,000 \text{ s}.$$

That's a lot of seconds. But how many years is it? Time for our conversion factors:

$$\frac{1 \text{ min}}{60 \text{ s}} = 1, \qquad \frac{1 \text{ hr}}{60 \text{ min}} = 1, \qquad \frac{1 \text{ day}}{24 \text{ hr}} = 1, \qquad \frac{1 \text{ yr}}{365 \text{ day}} = 1.$$

So, we have

$$400,000,000 \text{ s} = 400,000,000 \text{ s} \cdot \frac{1 \text{ min}}{60 \text{ s}} \cdot \frac{1 \text{ hr}}{60 \text{ min}} \cdot \frac{1 \text{ day}}{24 \text{ hr}} \cdot \frac{1 \text{ yr}}{365 \text{ day}} \approx 12.7 \text{ years}.$$

12.7 years after July 18, 1999, will be in the year $\boxed{2012}$. (Make sure you see why it's not in the year 2011!)

7.36   Because the train is moving 60 miles per hour, the front of the train moves 1 mile every minute. Therefore, in the three minutes since the front of the train entered the tunnel, the front of the train has moved three miles. At the end of these three minutes, we know the front of the train is 1 mile beyond the end of the tunnel, because the train is one mile long and its tail is just leaving the tunnel. So, the front of the train has moved 3 miles from the beginning of the tunnel and is now 1 mile beyond the end of the tunnel. This tells us that the tunnel is $3 - 1 = \boxed{2 \text{ miles}}$ long.

7.37   The train that leaves New York at 1 p.m. will pass every train that leaves Atlanta before 1 p.m. but arrives at New York after 1 p.m. It will also pass every train that leaves Atlanta after 1 p.m. but before 11 p.m.

First, we find the earliest train that leaves Atlanta before 1 p.m. but arrives in New York after 1 p.m. The first such train arrives at 1:10 p.m., having left Atlanta 10 hours earlier at 3:10 a.m.

Next, we find the latest train that leaves Atlanta before 11 p.m. (before our 1 p.m. New York-to-Atlanta train reaches Atlanta). Because a train leaves Atlanta at 10 minutes past each hour and at 40 minutes past each hour, the latest train that leaves Atlanta before 11 p.m. is the one that leaves at 10:40 p.m..

Every train that leaves Atlanta from 3:10 a.m. to 10:40 p.m. (including both these trains) passes our 1 p.m. New York-to-Atlanta train. There are two such trains each hour from the 3 a.m.-4 a.m. hour through the 10 p.m.-11 p.m. hour. There are 20 such hours, for a total of $2(20) = \boxed{40}$ trains.

7.38 Let x be the distance Ginny skates and y be the distance she walks. Therefore, Jenna skates a distance of y and walks a distance of x. Because Ginny covers a distance of x while skating 8 mph, she skates for a total of $x/8$ hours. Similarly, she walks for $y/3$ hours. In the same way, we find that Jenna skates for $y/9$ hours and walks for $x/4$ hours. They both travel for the same amount of time, so we must have

$$\frac{x}{8} + \frac{y}{3} = \frac{y}{9} + \frac{x}{4}.$$

Subtracting $x/8$ from both sides, and subtracting $y/9$ from both sides, gives us $2y/9 = x/8$. Multiplying both sides by 8 gives $x = 16y/9$. Therefore, Ginny skates for a total of $x/8 = (16y/9)/8 = 2y/9$ hours.

Since Ginny skates at a rate of 8 mph for $2y/9$ hours, she covers $(2y/9)(8) = 16y/9$ miles in that time. Similarly, she walks at 3 mph for $y/3$ hours, thereby covering a distance of $3(y/3) = y$ miles. Since she covers a total of 20 miles, we have

$$\frac{16y}{9} + y = 20.$$

Solving this equation gives us $y = 36/5$. Therefore, we have $x = 16y/9 = 64/5$. Finally, Ginny travels for a total time of

$$\frac{x}{8} + \frac{y}{3} = \frac{8}{5} + \frac{12}{5} = 4$$

hours. So, the two get home at $\boxed{8{:}00}$.

7.39 If we can figure out how long it takes for them to meet for the first time, we then know how much time passes between each time they meet. This is because once they meet for the first time, we have essentially the same problem over again: the two are at the same place and heading in opposite directions.

Because Superman can go all the way around the world in 2.5 hours, he can go around 1/2.5 of the world in one hour. Similarly, Flash can go around 1/1.5 of the world in one hour. Suppose they meet in x hours. In x hours, Superman travels $x(1/2.5) = x/2.5$ of the world and Flash travels $x/1.5$ of the world. When they meet for the first time, Flash has gone around part of the world in one direction and Superman in the other. These paths together make a single path all the way around the world. Therefore, combining Superman's $x/2.5$ of the world with Flash's $x/1.5$ of the world gives us 1 whole trip around the world:

$$\frac{x}{2.5} + \frac{x}{1.5} = 1.$$

Solving this equation for x gives us $x = 15/16$. Therefore, Superman and Flash meet for the first time in 15/16 hours. They'll meet every 15/16 hours thereafter, so in 24 hours, they'll meet a total of $24/(15/16) = 128/5 = 25\frac{3}{5}$ times. We discard the fraction; the 3/5 just tells us that they aren't in the same spot after 24 hours (they're 3/5 of the way through one of their trips to see each other). Therefore, the two will meet $\boxed{25}$ times during 24 hours.

7.40 The time it takes to cut a lawn is directly proportional to the area of the lawn (more lawn, more time) and is inversely proportional to the speed of the mower (faster mower, less time). So, we must have

$$\frac{(\text{Speed of mower})(\text{Time of mowing})}{\text{Area of lawn}} = k,$$

where k is some constant. Therefore, we have

$$\text{Time of mowing} = \frac{k \cdot (\text{Area of lawn})}{\text{Speed of mower}}.$$

Let the area of Andy's lawn be A and the speed of Carlos' mower be c. Andy's mower is three times as fast as Carlos', so its speed is $3c$. The area of his lawn is A, so the time it takes him to mow is

$$\text{Andy's time} = \frac{k(A)}{3c} = \frac{kA}{3c}.$$

Beth's lawn is half the size of Andy's, so its area is $A/2$. Her mower is twice as fast as Carlos', so its speed is $2c$. So, her time is:

$$\text{Beth's time} = \frac{k(A/2)}{2c} = \frac{kA}{4c}.$$

Carlos' lawn is 1/3 the size of Andy's lawn, so its area is $A/3$. The speed of his mower is just c, so his time is:

$$\text{Carlos' time} = \frac{k(A/3)}{c} = \frac{kA}{3c}.$$

So, Carlos and Andy take the same amount of time, and Beth takes a little less time (since 1/4 is less than 1/3, we know $kA/4c$ is less than $kA/3c$). So, $\boxed{\text{Beth}}$ will finish first.

7.41 Because Harold beats Charlie by 1/5 mile in a 2-mile race, Charlie runs

$$1 - \frac{1/5}{2} = 1 - \frac{1}{10} = \frac{9}{10}$$

as fast as Harold. If we imagine all three are racing, we know that when Vic finishes, Harold will be 0.1 mile behind. So, Harold will have finished $2 - 0.1 = 1.9$ miles of the course. Charlie runs 0.9 times as fast as Harold, so at this point he will have finished $0.9(1.9) = 1.71$ miles of the course as Vic finishes. Therefore, Vic will beat Charlie by $2 - 1.71 = \boxed{0.29 \text{ miles}}$ in 2-mile race.

7.42 If the radius of the tires were doubled, the tires would make half as many revolutions on the return trip. So, the car would register half the distance traveled on the return trip. This makes us think that the distance shown on the instrument panel and the radius of the tires are inversely proportional. Let's try to prove it.

Let n be the number of rotations a tire on the car makes and C be the circumference (length around the outside) of the tire. Then, the distance traveled, d, is

$$d = nC.$$

So, number of rotations and circumference are inversely proportional. To figure out how far the car goes, the car's computer also multiplies the number of rotations of a tire by the circumference of the tire. However, the computer can't get out and measure the tire, so it must be programmed to think the tire has a radius of 15 inches. In other words, it always thinks the circumference of the tire is $2(15\pi) = 30\pi$. So, the car's distance measurement, m, will be given by the formula

$$m = n(30\pi).$$

Solving this for n, we have $n = m/(30\pi)$. Substituting this into our equation above, we have

$$d = \frac{mC}{30\pi}.$$

Letting the radius of the tires be r, we have $C = 2\pi r$, so

$$d = \frac{m(2\pi r)}{30\pi} = \frac{mr}{15}.$$

So, the radius of the tires and the measurement made by the car are inversely proportional because the distance of the two trips is the same (which means d is constant). So, the product of the original measurement and radius equals the product of the return trip measurement and radius. Calling this latter radius x, we have

$$(450)(15) = (440)(x),$$

so $x = (450/440)(15) \approx 15.34$. Therefore, the radius increased $\boxed{0.34 \text{ inches}}$.

7.43 Because y is jointly proportional to $x_1, x_2, \ldots, x_{100}$, we have

$$y = kx_1 x_2 x_3 \cdots x_{100}$$

for some constant k. Since we wish to leave y unchanged, we want the product of all the x_i's to be constant. When Jiri multiplies 25 of the x_i's by 3, he multiplies the entire product by 3^{25}. Vlatko wants to counter this by multiplying each of the other x_i's by some constant c. There are 75 other terms in the product, so Vlatko multiplies the whole product by c^{75}. After Jiri multiplies the whole product by 3^{25} and Vlatko multiplies it by c^{75}, we want the overall product to remain unchanged. Therefore, we must have

$$3^{25} \cdot c^{75} = 1.$$

Solving for c^{75}, we have $c^{75} = 1/3^{25}$. Raising both sides to the $1/75$ power, we have

$$c = \left(\frac{1}{3^{25}}\right)^{1/75} = \frac{1}{3^{25/75}} = \boxed{\frac{1}{3^{1/3}}}.$$

7.44 Because x and y are inversely proportional, xy is constant, so we can write $xy = k$ for some constant k. Solving this for x we have $x = k/y$.

(a) We wish to have $x^p y^q$ be constant. Letting $x = k/y$ from above, we have

$$x^p y^q = \left(\frac{k}{y}\right)^p y^q = \frac{k^p y^q}{y^p} = k^p y^{q-p}.$$

The only way this is constant for all y is if y^{q-p} is constant. This only happens if $q - p = 0$, so we must have $\boxed{p = q}$.

(b) We wish to have x^p/y^q be constant. Again letting $x = k/y$, we have

$$\frac{x^p}{y^q} = \frac{(k/y)^p}{y^q} = \frac{k^p/y^p}{y^q} = \frac{k^p}{y^p y^q} = \frac{k^p}{y^{p+q}}.$$

As before, this can only be constant if y^{p+q} stays constant, which happens only when $p + q = 0$. So, we must have $\boxed{p = -q}$.

7.45 Because Sunny beats Windy by d meters in an h meter race, Sunny runs h meters in the time Windy runs $h - d$ meters. So, if Sunny starts d meters behind Windy, they will be tied after Sunny runs h meters because in that time Windy will have run $h - d$ meters. Therefore, the two of them are tied when they are d meters from the finish line. Since Windy runs $h - d$ meters in the time Sunny runs h meters, Windy's rate is $(h - d)/h$ of Sunny's. Therefore, while Sunny runs the last d meters, Windy only covers $d[(h - d)/h]$ meters. The difference between these is the distance by which Sunny beats Windy:

$$d - \frac{d(h - d)}{h} = \frac{dh}{h} - \frac{dh - d^2}{h} = \frac{dh - dh + d^2}{h} = \boxed{\frac{d^2}{h}}.$$

7.46 Suppose the candles are lit for t hours. During that time, $t/3$ of the first candle has burned, leaving $1 - \frac{t}{3} = \frac{3-t}{3}$ of the candle left to go. Similarly, $t/4$ of the second candle has burned, leaving $1 - \frac{t}{4} = \frac{4-t}{4}$ left. There's twice as much of the second (slow-burning) candle left as of the first, so we must have

$$\frac{4 - t}{4} = 2\left(\frac{3 - t}{3}\right) = \frac{6 - 2t}{3}.$$

Cross-multiplying gives $3(4 - t) = 4(6 - 2t)$. Expanding both sides gives $12 - 3t = 24 - 8t$. Solving for t gives $t = 12/5 = 2\frac{2}{5}$. One-fifth of an hour is $60/5 = 12$ minutes, so $2\frac{2}{5}$ hours is 2 hours and 24 minutes. Two hours and 24 minutes before 4 p.m. is $\boxed{1:36 \text{ p.m.}}$

7.47 To get a handle on the problem, we draw a picture. Suppose that runner A starts at Y and B starts at Z, and that A runs clockwise (so B runs counterclockwise). Let M be the first point at which they meet and N be the second. We know that M is 100 yards from Z and that N is 60 yards from Y. We let the distance that A travels from the start to the first meeting point be x. Because Y and Z are directly opposite each other, the whole length of the track is double the length from Y to Z along the track. Since the distance from Y to Z along the track is $x + 100$, the length of the whole track is $2(x + 100)$. So, we just have to find x to find the length of the track.

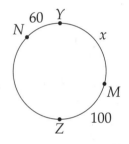

To find x, we must build an equation. We haven't used N yet, so we focus on that. As B runs from M through Y to N, runner A runs from M through Z to N. Therefore, during this time they together cover the whole track. In their initial run from their starting points to M, they together covered half the track. So, between the time they meet at M and the time they meet at N, each runs twice as far as he/she did in running to M. Specifically, runner B must run 200 yards to get from M to N. Therefore, we have $x + 60 = 200$, so $x = 140$. Finally, we know the track is $2(x + 100) = \boxed{480}$ yards long.

7.48 Suppose Anne can demolish a building in a days, Bob can do it in b days, and Carl in c days. Then, Anne can demolish $1/a$ of a building per day, Bob can demolish $1/b$, and Carl can demolish $1/c$ of the building in a day. Because Carl and Bob can demolish a building together in 6 days, together they demolish $1/6$ each day. Combining their individual demolition rates, we have

$$\frac{1}{b} + \frac{1}{c} = \frac{1}{6}.$$

Similarly, because Anne and Bob can do the job together in 3 days, we have

$$\frac{1}{a} + \frac{1}{b} = \frac{1}{3}.$$

And because Anne and Carl can do the job in 5 days, we have

$$\frac{1}{a} + \frac{1}{c} = \frac{1}{5}.$$

Because we are interested in how fast the three of them work together, we seek the expression $\frac{1}{a} + \frac{1}{b} + \frac{1}{c}$. We see each of $1/a$, $1/b$, and $1/c$ twice on the left sides of the three equations above, so we think to add all three, which gives us

$$2\left(\frac{1}{a} + \frac{1}{b} + \frac{1}{c}\right) = \frac{1}{6} + \frac{1}{3} + \frac{1}{5} = \frac{5+10+6}{30} = \frac{21}{30} = \frac{7}{10}.$$

So, we have

$$\frac{1}{a} + \frac{1}{b} + \frac{1}{c} = \frac{7}{20}.$$

Therefore, after the first day, the three of them have finished 7/20 of the job, leaving 13/20 yet to go. Then Carl quits, leaving Anne and Bob. We already know that they do 1/3 of the job each day, so they can finish the remaining 13/20 in $(13/20)/(1/3) = 39/20$ days. So, it takes them a total of $1 + \dfrac{39}{20} = \boxed{\dfrac{59}{20}}$ days.

7.49   After trying to bash this with "Rate times Time equals Distance" for a while, we get frustrated and put down our pencil. Instead, we'll have to think about it. Rather than focusing on Zuleica, let's think about Wilma. Wilma leaves at the same time as usual, but gets home 12 minutes earlier. So, she has saved 6 minutes each way. Therefore, the point where she picks up Zuleica early is 6 minutes driving time from the train station. In other words, she picks up Zuleica 6 minutes earlier than she usually does. Because Zuleica reached the train station one hour earlier than usual, she reached the train station 60 minutes earlier than she usually gets picked up. She was picked up 6 minutes earlier than usual, so she must have reached the train station $60 - 6 = 54$ minutes before she was picked up. Therefore, she was walking for $\boxed{54 \text{ minutes}}$.

CHAPTER 8

Exercises for Section 8.1

8.1.1 Since -2 is 2 steps left of the origin and the origin is 7 steps to the left of 7, the number -2 is $2 + 7 = 9$ steps to the left of 7, so the distance between -2 and 7 is $\boxed{9}$.

8.1.2

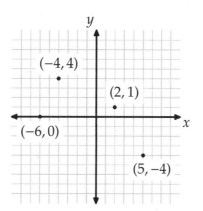

8.1.3 We use the distance formula to find that the distance is

$$\sqrt{(-5 - 7)^2 + (-2 - 3)^2} = \sqrt{144 + 25} = \boxed{13}.$$

8.1.4 The distance from a point (x, y) to the origin is $\sqrt{(x - 0)^2 + (y - 0)^2} = \sqrt{x^2 + y^2}$. Evaluating this for each of the five given points, we find that $\boxed{(6, 0)}$ is farthest from the origin.

8.1.5 We consider three cases:

- *Case 1: x is greater than y.* In this case, x is $x - y$ greater than y, so x is a distance of $x - y$ from y. Also, if x is greater than y, we have $|x - y| = x - y$. So, if x is greater than y, then x is a distance of $|x - y|$ from y.

- *Case 2: x = y.* In this case, the distance between x and y is 0, and the expression $|x - y|$ equals 0 because $x = y$. So, if $x = y$, then $|x - y|$ equals the distance between x and y.

- *Case 3: x is smaller than y.* In this case, y is $y - x$ greater than x, so y is a distance of $y - x$ from x. If y is greater than x, then $|x - y| = y - x$, since absolute value must be positive. So, if x is smaller than y, then $|x - y|$ equals the distance between x and y.

In all three cases, we find that $|x - y|$ equals the distance between x and y. These three cases cover all possible relationships between x and y, so $|x - y|$ always equals the distance between x and y.

8.1.6 Yes, the argument, and therefore the distance formula, works no matter how A and B are situated. In some configurations, $x_2 - x_1$ might be negative, but $(x_2 - x_1)^2$ is the square of the horizontal distance between A and B whether $x_2 - x_1$ is positive or not. Similarly, $(y_2 - y_1)^2$ is the square of the vertical distance between A and B regardless of whether $y_2 - y_1$ is positive or not. We can then use the Pythagorean Theorem to find AB.

For example, suppose A and B are situated as at right. Here, the leg lengths of the right triangles are $x_1 - x_2$ and $y_2 - y_1$, so the length of $\overline{AB}$ is $\sqrt{(x_1 - x_2)^2 + (y_2 - y_1)^2}$. However, because

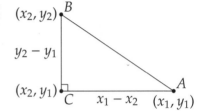

$$(x_1 - x_2)^2 = [(-1)(x_2 - x_1)]^2 = (-1)^2(x_2 - x_1)^2 = (x_2 - x_1)^2,$$

we have

$$\sqrt{(x_1 - x_2)^2 + (y_2 - y_1)^2} = \sqrt{(x_2 - x_1)^2 + (y_2 - y_1)^2}.$$

So, the distance formula we learned in the text still works for this configuration. Similarly, we can use the facts that $(x_1 - x_2)^2 = (x_2 - x_1)^2$ and $(y_1 - y_2)^2 = (y_2 - y_1)^2$ to see that the distance formula works in any configuration.

Exercises for Section 8.2

8.2.1 We have $m = \dfrac{y_2 - y_1}{x_2 - x_1} = \dfrac{-5 - 5}{2 - (-3)} = \dfrac{-10}{5} = \boxed{-2}$.

8.2.2 As discussed in the text, we know the graph is a line. We find a few points on the line by choosing values of x and solving for y. When $x = 0$, we have $-2y = -4$, so $y = 2$ and $(0, 2)$ is on the line. When $x = 2$, we have $6 - 2y = -4$, so $y = 5$ and $(2, 5)$ is on the line. When $x = 4$, we have $12 - 2y = -4$, so $y = 8$ and $(4, 8)$ is on the line. We plot these points and draw the line through them as shown at right.

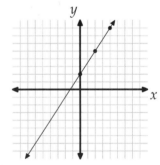

8.2.3 The graphs of the two lines are side-by-side below.

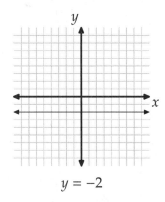

$y = -2$

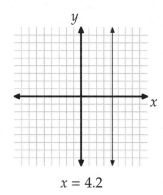

$x = 4.2$

The graph of $y = -2$ is a horizontal line and its slope is $\boxed{0}$, as described in the text. The graph of $x = 4.2$ is a vertical line, and its slope is $\boxed{\text{undefined}}$, as described in the text.

8.2.4

(a) The line goes upward as it goes from left to right, so its slope is $\boxed{\text{positive}}$. The line goes rightward faster than it goes upward (it goes right 4 units for every 3 units it goes up), so the slope is $\boxed{\text{less than 1}}$.

(b) The line is horizontal, so all points on the line have the same y-coordinate. Therefore, the slope is $\boxed{0}$.

(c) The line goes downward as it goes from left to right, so its slope is $\boxed{\text{negative}}$. It is very steep; it goes down much faster than it goes right, so its slope is $\boxed{\text{less than } -1}$.

(d) This line is vertical, so all points on the line have the same x-coordinate. Therefore, the slope is $\boxed{\text{undefined}}$.

(e) This line goes downward as it goes from left to right, so its slope is $\boxed{\text{negative}}$. It goes downward at exactly the same rate as it goes rightward, so its slope is $\boxed{-1}$.

(f) This line goes upward as it goes from left to right, so its slope is $\boxed{\text{positive}}$. It is very steep; it goes up much faster than it goes right, so its slope is $\boxed{\text{greater than 1}}$.

Exercises for Section 8.3

8.3.1 Because the three points are on the same line, the slope between the first and second equals the slope between the first and the third. This gives us the equation:

$$\frac{3 - (-5)}{(-a + 2) - 3} = \frac{2 - (-5)}{(2a + 3) - 3} \quad \Rightarrow \quad \frac{8}{-a - 1} = \frac{7}{2a} \quad \Rightarrow \quad 8(2a) = 7(-a-1) \quad \Rightarrow \quad 23a = -7 \quad \Rightarrow \quad a = \boxed{\frac{-7}{23}}.$$

8.3.2 Because the slope is 1/3, the line goes up 1 step for every 3 steps it goes to the right. So, starting from $(1, 2)$, we can find another point by going right 3 and up 1 to $(4, 3)$. We do so again to learn that point $(7, 4)$ is also on the line. Now that we have a few points, we can graph the line as shown at right.

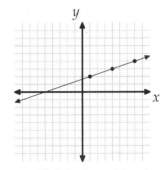

8.3.3

(a) The midpoint of $\overline{PQ}$ is $\left(\dfrac{5 + (-3)}{2}, \dfrac{3 + 6}{2} \right) = \boxed{(1, 9/2)}$.

(b) *Solution 1: Reason it out.* To get from P to Q, we go left 8 steps and up 3 steps. If Q is the midpoint of $\overline{PT}$, then the trip from Q to T is the same as the trip from P to Q, 8 steps left and 3 steps up. The point that is 8 steps left and 3 steps up from $(-3, 6)$ is $(-3 - 8, 6 + 3) = \boxed{(-11, 9)}$.

Solution 2: Use the midpoint formula. Let the coordinates of T be (x, y). Then, the midpoint of $\overline{PT}$ is $\left(\frac{x+5}{2}, \frac{y+3}{2} \right)$. Since we know the coordinates of this point are those of point Q, $(-3, 6)$, we have $(x + 5)/2 = -3$ and $(y + 3)/2 = 6$. Solving these equations gives $x = -11$ and $y = 9$, so point T is at $\boxed{(-11, 9)}$.

8.3.4 Call the new point C. As we did with the similar problem in the text, we handle the two coordinates separately. To get from A to B, we go 8 steps left and 10 steps down. Suppose we must take x steps left to get from A to C, so we must take $8 - x$ steps left to continue from C on to B. Since C is 4 times as far from B as it is from A, we must have $4x = 8 - x$. Solving this equation gives $x = 8/5$. So, C is 8/5 steps left of A. Therefore, its x-coordinate is $2 - 8/5 = 2/5$.

Let C be y steps below A, so we must take $10 - y$ steps to continue on from C to point B. Again, since C is four times as far from B as it is from A, we have $4y = 10 - y$. Solving this equation gives $y = 2$, so C is 2 steps lower than A. Therefore, its y-coordinate is $7 - 2 = 5$.

Putting our two coordinates together, we see that C is $\boxed{(2/5, 5)}$.

Suppose A is (x_1, y_1) and B is (x_2, y_2). See if you can follow the argument above to find formulas for the coordinates of the point C on $\overline{AB}$ such that $AC/BC = k$, for any positive constant k.

8.3.5 Let A be (x_1, y_1), B be (x_2, y_2), and M be $\left(\dfrac{x_1 + x_2}{2}, \dfrac{y_1 + y_2}{2} \right)$.

First, we must show that M is on the line through A and B. To do so, we show that the slope of the line through A and M equals the slope through A and B. The slope between A and B is $(y_2 - y_1)/(x_2 - x_1)$. The slope between A and M is

$$\frac{\dfrac{y_1 + y_2}{2} - y_1}{\dfrac{x_1 + x_2}{2} - x_1} = \frac{\dfrac{y_2}{2} - \dfrac{y_1}{2}}{\dfrac{x_2}{2} - \dfrac{x_1}{2}} = \frac{\dfrac{y_2}{2} - \dfrac{y_1}{2}}{\dfrac{x_2}{2} - \dfrac{x_1}{2}} \cdot \frac{2}{2} = \frac{y_2 - y_1}{x_2 - x_1}.$$

So, the slope between A and M equals the slope between A and B. Therefore, M is on the line through A and B.

Next, we must show that $AM = MB$. This is a job for the distance formula. We find:

$$AM = \sqrt{\left(\tfrac{x_1+x_2}{2} - x_1\right)^2 + \left(\tfrac{y_1+y_2}{2} - y_1\right)^2} = \sqrt{\left(\tfrac{x_2}{2} - \tfrac{x_1}{2}\right)^2 + \left(\tfrac{y_2}{2} - \tfrac{y_1}{2}\right)^2},$$
$$MB = \sqrt{\left(x_2 - \tfrac{x_1+x_2}{2}\right)^2 + \left(y_2 - \tfrac{y_1+y_2}{2}\right)^2} = \sqrt{\left(\tfrac{x_2}{2} - \tfrac{x_1}{2}\right)^2 + \left(\tfrac{y_2}{2} - \tfrac{y_1}{2}\right)^2}.$$

So, we see that $AM = MB$. Since M is on $\overleftrightarrow{AB}$ and $AM = MB$, we see that M is indeed the midpoint of $\overline{AB}$.

Exercises for Section 8.4

8.4.1 The slope between the two points is $(-2 - 4)/[1 - (-2)] = (-6)/3 = -2$. Therefore, we can use the point $(-2, 4)$ to write a point-slope form of the line: $y - 4 = -2[x - (-2)]$. Expanding the right side gives $y - 4 = -2x - 4$, and adding $2x + 4$ to both sides gives the standard form of the equation as $\boxed{2x + y = 0}$.

8.4.2 We have the slope and a point, so we can quickly write a point-slope form of the equation: $y - 5 = -3(x - 0)$. Therefore, $y - 5 = -3x$, so $\boxed{3x + y = 5}$.

8.4.3 Again, we have the slope and a point, so we have a point-slope form of our equation: $y - 7 = (1/4)(x - 2)$. In standard form, we can't have fractions as our coefficients, so we multiply both sides of this equation by 4 to get rid of the 1/4. This gives $4(y - 7) = x - 2$. Expanding the left gives $4y - 28 = x - 2$. Subtracting $4y$ from both sides and adding 2 to both sides gives us the standard form, $\boxed{x - 4y = -26}$.

8.4.4

(a) For $t = -1, 0, 1$, and 2, we have the points $(1, -3)$, $(3, 0)$, $(5, 3)$, and $(7, 6)$. These points are plotted on the graph below.

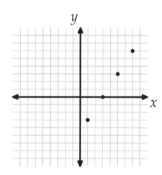

It sure looks like these points are all on a line.

(b) When we go from the point with $t = 0$ to the point with $t = 1$, we increase the x-coordinate by 2 and the y-coordinate by 3. Similarly, as we go from $t = 1$ to $t = 2$, we increase the x-coordinate by 2 and the y-coordinate by 3. And when we go from $t = 2$ to $t = 3, \ldots$ you get the picture. Each time we increase t by 1, we move in the same direction. This is very much like our description of slope in the text.

We can prove that all the points of the form $(2t + 3, 3t)$ are on a single line by showing that the slope between any two of the points is the same. We take two different points on the graph by using two different values of t, which we'll call t_1 and t_2. The two points then are $(2t_1 + 3, 3t_1)$ and $(2t_2 + 3, 3t_2)$. The slope between these two points is

$$\frac{3t_2 - 3t_1}{(2t_2 + 3) - (2t_1 + 3)} = \frac{3(t_2 - t_1)}{2t_2 - 2t_1} = \frac{3\cancel{(t_2 - t_1)}}{2\cancel{(t_2 - t_1)}} = \frac{3}{2}.$$

The slope between any two points on the graph is $3/2$, so all the points on the graph must be on the same line. We'll return to proving that each point on this line corresponds to a point $(2t + 3, 3t)$ for some t after we find the equation of the line.

(c) We found the slope, $\boxed{3/2}$ in the previous part. To find the equation of the line, we need a point on the line. We take $(3, 0)$ from the first part and we have the equation $y - 0 = (3/2)(x - 3)$. We multiply by 2 to get rid of the fraction, and we have $2y = 3x - 9$. So, in standard form, we have $\boxed{3x - 2y = 9}$.

Note that every point on this line can be written in the form $(2t + 3, 3t)$. We find the appropriate t for each point by solving $x = 2t + 3$ for t. This gives $t = (x - 3)/2$. Notice that $y = 3t = 3(x - 3)/2 = 3x/2 - 9/2$, which is the same value of y we get solving the equation $3x - 2y = 9$ for y in terms of x. In other words, the graph of the equation $3x - 2y = 9$ is the same graph as the graph consisting of all the points $(2t + 3, 3t)$ for all values of t.

8.4.5

(a) The midpoint of $\overline{BC}$ is $(\frac{-3+1}{2}, \frac{-5+1}{2}) = (-1, -2)$. The slope of the line through $(5, 9)$ and $(-1, -2)$ is $(-2 - 9)/(-1 - 5) = 11/6$. The equation of the line with slope $11/6$ through the point $(-1, -2)$ is

$$y - (-2) = \frac{11}{6}[x - (-1)] \quad \Rightarrow \quad 6(y + 2) = 11(x + 1) \quad \Rightarrow \quad \boxed{11x - 6y = 1}.$$

(b) The midpoint of $\overline{AC}$ is $(\frac{5+1}{2}, \frac{9+1}{2}) = (3, 5)$. The slope of the line through $(-3, -5)$ and $(3, 5)$ is $[5 - (-5)]/[3 - (-3)] = 10/6 = 5/3$. The equation of the line with slope $5/3$ through the point $(3, 5)$ is

$$y - 5 = \frac{5}{3}(x - 3) \quad \Rightarrow \quad 3(y - 5) = 5(x - 3) \quad \Rightarrow \quad \boxed{5x - 3y = 0}.$$

(c) To find the point that is on both lines, we must find the ordered pair (x, y) that satisfies the system

$$11x - 6y = 1,$$
$$5x - 3y = 0.$$

Multiplying the second equation by -2 gives $-10x + 6y = 0$, and adding this to the first equation gives $x = 1$. Substituting this into either of our equations gives $y = 5/3$. So, the two lines meet at $\boxed{(1, 5/3)}$. Notice that the x-coordinate of this point is one-third of the sum of the x-coordinates of A, B, and C, and the y-coordinate is one-third of the sum of the y-coordinates of A, B, and C. Is this a coincidence?

(d) Let point G be $(1, 5/3)$ and the midpoint of $\overline{AB}$ be M. Point M is $(\frac{5+(-3)}{2}, \frac{9+(-5)}{2}) = (1, 2)$. Therefore, C, M, and G all have an x-coordinate of 1, so they are all on the line $x = 1$. Furthermore, looking at their y-coordinates shows that G is between C and M, so G is on $\overline{CM}$. Therefore, the three medians of the triangle all pass through the same point. Hmmm... I wonder if this is true for all triangles?

Exercises for Section 8.5

8.5.1 The equation is already in slope-intercept form, so the coefficient of x is the slope and the constant is the y-coordinate of the y-intercept. So, the slope is $\boxed{-2}$ and the y-intercept is $\boxed{(0,1)}$. The x-intercept is the point where $y = 0$. Solving the equation $0 = -2x + 1$ gives $x = 1/2$, so the x-intercept is $\boxed{(1/2, 0)}$. To produce the graph, we can simply draw the line through the two intercepts, as shown at right.

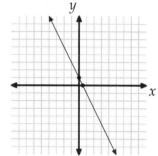

8.5.2 We can find the slope and the y-coordinate of the y-intercept quickly by putting the equation in slope-intercept form. Solving the equation $3x + 5y = 20$ for y in terms of x gives $y = -\frac{3}{5}x + 4$. So, the slope is $\boxed{-3/5}$ and the y-intercept is $\boxed{(0,4)}$. Letting $y = 0$ in $3x + 5y = 20$ gives $3x = 20$, so the x-coordinate of the x-intercept is $20/3$. So, the x-intercept is $\boxed{\left(6\frac{2}{3}, 0\right)}$.

8.5.3 The slope of $y = 2x + 3$ is 2, so our desired line has slope -2. Our line has slope -2 and y-intercept $(0, 8)$, so its equation is $\boxed{y = -2x + 8}$.

8.5.4 Make sure you see why the answer is *not* $y = 5x - 4$. We are given the x-intercept, not the y-intercept. However, we can still use point-slope form to find the equation:

$$y - 0 = 5[x - (-4)] \quad \Rightarrow \quad \boxed{y = 5x + 20}.$$

8.5.5 We put the equation in slope-intercept form by solving for y. Subtracting Ax from both sides gives $By = -Ax + C$, and dividing both sides by B gives $y = -\frac{A}{B}x + \frac{C}{B}$. Therefore, the slope is $-A/B$, as desired.

8.5.6

(a) Horizontal lines (besides the x-axis) do not ever intersect the x-axis. Therefore, these lines do not have x-intercepts. Similarly, vertical lines (besides the y-axis) do not intersect the y-axis. Therefore, these vertical lines do not have y-intercepts.

(b) First, we note that x-intercepts must have a y-coordinate of 0 and y-intercepts must have an x-coordinate of 0. Therefore, if a point is both an x-intercept and a y-intercept, it must be the origin. So, the only lines that have both an x-intercept and a y-intercept such that these two points are not different are lines through the origin.

8.5.7 *Solution 1: Use equations.* Let Mary's favorite number be m and her mother's favorite number be b. From the information about F and T, we have $6m + b = 65$ and $20m + b = 177$. Subtracting the first equation from the second gives $14m = 112$, from which we find $m = 8$. Substituting this in either equation gives $b = 17$. So, Mary's favorite number is $\boxed{8}$ and her mother's is $\boxed{17}$. Since M is the thirteenth letter in the alphabet, M becomes $8 \cdot 13 + 17 = 104 + 17 = \boxed{121}$.

Solution 2: Use graphing. Suppose we plot points on the Cartesian plane such that the x-coordinate is the letter's position in the alphabet and the y-coordinate is the encoded value of the letter. The point corresponding to F is $(6, 65)$ and the point corresponding to T is $(20, 177)$.

We can make an equation relating the position of a letter, x, to its encoded value, y, by letting Mary's favorite number be m and her mother's favorite number be b, as before. We therefore have $y = mx + b$ (make sure you see why!) Because m and b are constants, we see that the points on our graph are all on a line given by the equation $y = mx + b$. The slope of this line is Mary's favorite number and the y-coordinate of the y-intercept is her mom's favorite number.

Since the x-coordinate of point M, 13, is midway between the x-coordinates of F and T, M is the midpoint of the segment connecting points F and T. The midpoint of this segment is $(\frac{6+20}{2}, \frac{65+177}{2}) = (13, 121)$, so the encoded value of M is $\boxed{121}$.

Mary's number is the slope of the line, or $(177 - 65)/(20 - 6) = 112/14 = \boxed{8}$. So, our line is $y = 8x + b$. We know $(6, 65)$ is on this line, so we have $65 = 8 \cdot 6 + b$, from which we find $b = \boxed{17}$.

Exercises for Section 8.6

8.6.1 The slope of $2x + 3y = 5$ is $-2/3$, so a line perpendicular to this line has slope $3/2$. The x-intercept of $2x + 3y = 5$ is found by letting $y = 0$, which gives $x = 5/2$. So, we want the equation of the line with slope $3/2$ that passes through the point $(5/2, 0)$. This equation is

$$y - 0 = \frac{3}{2}\left(x - \frac{5}{2}\right) \quad \Rightarrow \quad 2y = 3x - \frac{15}{2} \quad \Rightarrow \quad 4y = 6x - 15 \quad \Rightarrow \quad \boxed{6x - 4y = 15}.$$

8.6.2 Writing the equation $x - 5 = 3y$ in slope-intercept form gives $y = (1/3)(x - 5) = (1/3)x - 5/3$, so the slope of the line is $1/3$. Any line parallel to this line must have slope $1/3$, so our line through $(2, 7)$ has slope $1/3$. Therefore, the equation of this line is

$$y - 7 = \frac{1}{3}(x - 2) \quad \Rightarrow \quad 3(y - 7) = x - 2 \quad \Rightarrow \quad \boxed{x - 3y = -19}.$$

8.6.3 If the two equations represent the same line, then when we write them with the same constant term on the right, the coefficients of x on the left must match and the coefficients of y on the left must match. We multiply the first equation by 10 and the second by 7 to give the system

$$20x - 10By = 70,$$
$$7Ax + 21y = 70.$$

These two equations are the same when $7A = 20$ and $-10B = 21$. Solving these equations for A and B gives $\boxed{A = 20/7}$ and $\boxed{B = -21/10}$.

We also could solve this problem by noting that two lines are the same if they have the same intercepts. The graph of $2x - By = 7$ has x-intercept $(7/2, 0)$, so the graph of $Ax + 3y = 10$ must also have x-intercept $(7/2, 0)$. Letting $(x, y) = (7/2, 0)$ in $Ax + 3y = 10$ gives us $7A/2 = 10$, so $A = 20/7$. Similarly, we can use the y-intercepts of the lines to find B.

8.6.4

(a) The slope of each line is $-3/(-2) = 3/2$. The x-intercept of each line is $(k/3, 0)$ and the y-intercept is $(0, -k/2)$. Using this information, we get the graphs below. The left-most line is for $k = 1$. The next line to the right is for $k = 2$, and so on.

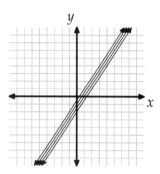

(b) The lines are parallel. The difference in the x-coordinate of the x-intercept of each line and the next is $1/3$, because k increases by 1 from each line to the next.

(c) Using the previous 2 parts as inspiration, we note that the two graphs in question are parallel lines, so the graph of $3x - 6y = 10$ is just a shifted version of the graph of $3x - 6y = 4$. But how far is it shifted? The x-coordinate of the x-intercept of $3x - 6y = 4$ is $4/3$, and the x-coordinate of the x-intercept of $3x - 6y = 10$ is $10/3$, so the second line is $10/3 - 4/3 = 6/3 = 2$ steps to the right of the first line. In the diagram at right, the graph of $3x - 6y = 4$ is dashed and the graph of $3x - 6y = 10$ is solid.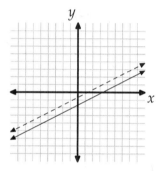

8.6.5 The slope of $3x - 5y = 7$ is $-3/(-5) = 3/5$. Because $ax + 2y = b$ is parallel to $3x - 5y = 7$, its slope must also be $3/5$. The slope of $ax + 2y = b$ is $-a/2$, so we must have $-a/2 = 3/5$. Therefore, we have $\boxed{a = -6/5}$. Now, our equation is $-6x/5 + 2y = b$. We know that $(2, 5)$ is on the graph of this equation, so we have $b = -6(2)/5 + 5 = -12/5 + 2(5) = \boxed{38/5}$.

Review Problems

8.28 The slope of the line through $(0, 3)$ and $(-8, 0)$ is $(0 - 3)/(-8 - 0) = 3/8$. If $(t, 5)$ is also on this line, then the slope of the line through $(t, 5)$ and $(0, 3)$ must also be $3/8$. Therefore, we must have

$$\frac{3 - 5}{0 - t} = \frac{3}{8} \quad \Rightarrow \quad \frac{2}{t} = \frac{3}{8} \quad \Rightarrow \quad (2)(8) = 3(t) \quad \Rightarrow \quad t = \boxed{\frac{16}{3}}.$$

8.29 We work from the inside out:

$$|1 - |2 - |3 - |4 - |5|||||| = |1 - |2 - |3 - |4 - 5|||| = |1 - |2 - |3 - |-1|||| = |1 - |2 - |3 - 1||| = |1 - |2 - 2|| = |1 - 0| = \boxed{1}.$$

8.30

(a) From the distance formula, we find that the distance between the two points is

$$\sqrt{(-5-1)^2 + [6-(-2)]^2} = \sqrt{(-6)^2 + 8^2} = \sqrt{36 + 64} = \sqrt{100} = \boxed{10}.$$

(b) The midpoint is $\left(\frac{1+(-5)}{2}, \frac{-2+6}{2}\right) = \boxed{(-2, 2)}$.

(c) The slope of the line connecting the two points is $[6-(-2)]/(-5-1) = 8/(-6) = \boxed{-4/3}$.

(d) The line passes through $(1, -2)$ and has slope $-4/3$, so a point-slope form of the equation is $y - (-2) = -\frac{4}{3}(x-1)$. Multiplying both sides by 3 gives

$$3(y+2) = -4(x-1) \quad \Rightarrow \quad 3y + 6 = -4x + 4 \quad \Rightarrow \quad \boxed{4x + 3y = -2}.$$

8.31 Because $(1, -3)$ is on the graph of $Ax - 3y = 6$, the equation must be true when $x = 1$ and $y = -3$. Therefore, we have

$$A(1) - 3(-3) = 6 \quad \Rightarrow \quad A + 9 = 6 \quad \Rightarrow \quad A = \boxed{-3}.$$

8.32 Only line ℓ goes downward as it goes from left to right, so only line ℓ has a negative slope. Therefore, line ℓ is the line with slope $-3/4$. To determine which of lines m and k has slope 2 and which has slope $1/6$, we note that line m is much steeper than line k. (In other words, line m goes upward much faster than line k.) Therefore, the slope of line m is greater than the slope of line k, so line m is the line with slope 2 and line k is the line with slope $1/6$.

8.33 We already know that line m has slope 2. We can see that it goes through $(-1, 1)$. Using this point, we find a point-slope form of the equation, and then convert this form to standard form:

$$y - 1 = 2[x - (-1)] \quad \Rightarrow \quad y - 1 = 2(x + 1) \quad \Rightarrow \quad y - 1 = 2x + 2 \quad \Rightarrow \quad \boxed{2x - y = -3}.$$

8.34 We use the distance formula to find a formula for the distance between (x_1, y_1) and $(0, 0)$:

$$\sqrt{(x_1 - 0)^2 + (y_1 - 0)^2} = \boxed{\sqrt{x_1^2 + y_1^2}}.$$

8.35 We start with a point-slope form of the line through $(4, 0)$ with slope $-1/3$, and then convert this equation to standard form:

$$y - 0 = -\frac{1}{3}(x - 4) \quad \Rightarrow \quad 3(y) = -(x - 4) \quad \Rightarrow \quad 3y = -x + 4 \quad \Rightarrow \quad \boxed{x + 3y = 4}.$$

The graph of the line is at left below.

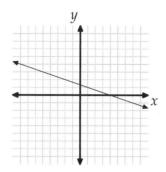

Figure 8.1: Diagram for Problem 8.35

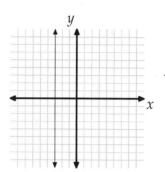

Figure 8.2: Diagram for Problem 8.36

8.36 The line $x = -3$ is a vertical line, so its slope is undefined. Its graph consists of all the points that have -3 as the x-coordinate. This graph is shown in the figure at right above.

8.37 We first find the equation of the line. The slope of the line is $[5 - (-7)]/(-3 - 3) = -2$. Using $(3, -7)$ as our point, we find that a point-slope form of the equation of the line is $y - (-7) = -2(x - 3)$. Rearranging this equation gives $2x + y = -1$. When $x = 0$, we have $y = -1$, so the y-intercept is $\boxed{(0, -1)}$. When $y = 0$, we have $x = -1/2$, so the x-intercept is $\boxed{(-1/2, 0)}$.

Notice that we could have found the y-intercept more quickly by noticing that the midpoint of segment connecting the two given points has x-coordinate 0. So, the midpoint of the segment connecting these points is the y-intercept. Can you find a similar quick way to find the x-coordinate?

8.38 In standard form, the equation is $2x - 3y = -9$. The slope of this line is $-2/(-3) = \boxed{2/3}$. When $y = 0$, we have $x = -9/2$, so the x-intercept is $(-9/2, 0) = \boxed{\left(-4\frac{1}{2}, 0\right)}$. When $x = 0$, we have $y = 3$, so the y-intercept is $\boxed{(0, 3)}$. Connecting these two points gives us our graph at right. (However, to avoid fractions, we might start with $(0, 3)$ and use our slope to go up 2 and right 3 to the point $(3, 5)$.)

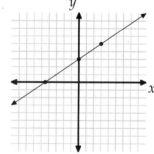

8.39 Suppose that b is greater than a. Then, we know that $a + b$ is larger than $2a$, but less than $2b$. Therefore, $(a+b)/2$ is larger than a, but less than b. Because $(a+b)/2$ is larger than a, the distance from $(a+b)/2$ to a is $(a+b)/2-a = b/2-a/2$. Because $(a + b)/2$ is smaller than b, the distance from $(a + b)/2$ to b is $b - (a + b)/2 = b/2 - a/2$. We see that these distances are the same, so $(a + b)/2$ is the same distance from a as it is from b on the number line.

If a is greater than b, we can follow essentially the same steps to show that $(a + b)/2$ is the same distance from a and from b on the number line. Finally, if $a = b$, then $(a + b)/2 = a = b$, so clearly $(a + b)/2$ is the same distance from a and b in this case.

8.40 $\boxed{\text{No}}$. It is possible that $\overleftrightarrow{AB}$ and $\overleftrightarrow{CD}$ are two different parallel lines.

8.41 *Solution 1: Find the equation of the line.* A point-slope equation of the line is $y - 0 = -2(x - 5)$, so the standard form of the equation of the line is $2x + y = 10$. When $x = 0$, we have $y = 10$, so the y-intercept is $\boxed{(0, 10)}$.

Solution 2: Use the slope without finding the equation. Because the slope of the line is -2, the line goes down 2 steps for each 1 step right. However, the point we are given on the line, $(5, 0)$, is already to the right of the y-axis, which is where the y-intercept is. Therefore, we instead think of the slope as going up 2 steps for each 1 step left. We must take 1 step left 5 times to get to the y-axis from point $(5, 0)$, so to stay on this line, we must also take 2 steps up 5 times, for a total of 10 steps. The point that is 10 steps up and 5 steps to the left of $(5, 0)$ is $\boxed{(0, 10)}$.

8.42

(a) The midpoint is $\left(\dfrac{-3 + 5}{2}, \dfrac{7 + (-12)}{2} \right) = \boxed{\left(1, -2\dfrac{1}{2} \right)}$.

(b) Because $PT/TQ = 1/3$, we know that $PT/PQ = 1/4$. (To see this, let $PT = x$, so $TQ = 3x$ and $PQ = PT + TQ = 4x$.) So, T is on $\overline{PQ}$ such that it is $1/4$ of the way from P to Q. Point Q is $5 - (-3) = 8$ to the right of P, so T is $(1/4)(8) = 2$ to the right of P. Point Q is 19 units below P, so T is $(1/4)(19) = 19/4 = 4\frac{3}{4}$ below P. The point that is 2 to the right and $4\frac{3}{4}$ below P is $\boxed{\left(-1, 2\dfrac{1}{4} \right)}$.

8.43 We find the slope of the given line by writing it in slope-intercept form. Subtracting 7 from both sides and dividing both sides by -3 gives $y = -\frac{2}{3}x + \frac{7}{3}$. So, the slope of the given line is $-2/3$. Therefore, the slope of a line perpendicular to the given line is $-1/(-2/3) = 3/2$. We find a point-slope equation of the line through $(4, 1)$ with slope $3/2$, and then convert it to standard form:

$$y - 1 = \frac{3}{2}(x - 4) \quad \Rightarrow \quad 2(y - 1) = 3(x - 4) \quad \Rightarrow \quad 2y - 2 = 3x - 12 \quad \Rightarrow \quad \boxed{3x - 2y = 10}.$$

8.44 Dividing by 3 puts the given line in slope-intercept form, $y = \frac{4}{3}x + \frac{8}{3}$. The slope of this line is $4/3$. The slope of any line parallel to this line must also be $4/3$. We find a point-slope equation of the line through $(8, -3)$ with slope $4/3$, and then convert it to standard form:

$$y - (-3) = \frac{4}{3}(x - 8) \quad \Rightarrow \quad 3(y + 3) = 4(x - 8) \quad \Rightarrow \quad 3y + 9 = 4x - 32 \quad \Rightarrow \quad \boxed{4x - 3y = 41}.$$

8.45

(a) $\boxed{\text{False}}$. The absolute value of 0 is 0.

(b) $\boxed{\text{True}}$.

(c) $\boxed{\text{False}}$. Vertical lines cannot be written in this form.

(d) $\boxed{\text{False}}$. Aside from the axes, vertical and horizontal lines do not have both an x-intercept and a y-intercept.

(e) $\boxed{\text{False}}$. The point $(0, 3)$ is the y-intercept. The x-intercept is where $y = 0$, which gives $x = -3/2$. So, the x-intercept is $(-3/2, 0)$.

(f) $\boxed{\text{True}}$.

(g) $\boxed{\text{False}}$. A vertical line and a horizontal line are perpendicular, but the product of their slopes is undefined.

(h) $\boxed{\text{True}}$.

8.46

(a) The x-intercept is the point where $y = 0$. Letting $y = 0$ gives us $x = b$, so the x-intercept is $\boxed{(b, 0)}$.

(b) We put the equation in slope-intercept form by subtracting b from both sides, then dividing by m. This gives $y = \frac{1}{m}x - \frac{b}{m}$. From this equation, we see that the slope is $\boxed{1/m}$.

8.47

(a) Because the lines are parallel, they have the same slope. The slope of $Ax + 3y = 5$ is $-A/3$ and the slope of $5x - 2y = 4$ is $-5/(-2) = 5/2$. So, we must have $-A/3 = 5/2$, which gives $A = \boxed{-15/2}$.

(b) Because the lines are perpendicular, the product of their slopes must be -1. Putting $3x = By + 2$ in slope-intercept form gives $y = \frac{3}{B}x - \frac{2}{B}$, so its slope is $3/B$. Putting $3y = -2x + 4$ in slope-intercept form gives $y = -\frac{2}{3}x + \frac{4}{3}$, so its slope is $-2/3$. Therefore, we must have

$$(3/B)(-2/3) = -1 \quad \Rightarrow \quad -2/B = -1 \quad \Rightarrow \quad \boxed{B = 2}.$$

(c) The slopes of the lines must be the same. The slope of $Ax + 3y = B$ is $-A/3$ and the slope of $2x + 6y = 7$ is $-2/6 = -1/3$. So, we must have $-A/3 = -1/3$, which gives $\boxed{A = 1}$. Since the graph of $x + 3y = B$ must be the same line as the graph of $2x + 6y = 7$, these graphs must pass through the same points. The point $(7/2, 0)$ is on the graph of $2x + 6y = 7$, so it must be on the graph of $x + 3y = B$. This gives us $(7/2) + 3(0) = B$, so $\boxed{B = 7/2}$.

We could have found the answer much more quickly. The coefficient of y in $Ax + 3y = B$ is 3, so we divide $2x + 6y = 7$ by 2 to make the coefficients of y match in the two equations. This gives $x + 3y = 7/2$. The graph of this equation gives the same graph as $Ax + 3y = B$ if and only if $A = 1$ and $B = 7/2$.

8.48 $\boxed{\text{No}}$. Every line must hit either the x-axis or the y-axis. Otherwise, the line would stay entirely within one quadrant (quarter) of the Cartesian plane, which is impossible. So, every line has either an x-intercept, or a y-intercept, or both.

Challenge Problems

8.49 Because the midpoint of the segment connecting (a, b) and (b, a) is (x, y), we have $(x, y) = \left(\frac{a+b}{2}, \frac{b+a}{2}\right)$. Therefore, we see that x and y are equal, so $\boxed{y = x}$.

8.50 Because $|x - 3| = 4$, x is 4 steps from 3 on the number line. Since -1 and 7 are the two numbers that are 4 steps from 3 on the number line, the values of x that satisfy the equation are $\boxed{-1 \text{ and } 7}$.

8.51 The slope between the given points is $\dfrac{5a^2 - 3a^2}{(3a + 4) - (2a + 4)} = \dfrac{2a^2}{a} = 2a$, so we have $2a = a + 3$. Solving this equation gives us $a = \boxed{3}$.

8.52 Because the slopes of the two lines add to 0, the slope of one line is the negative of the other line's slope. Let $(0, b)$ be the common y-intercept. From here, we offer two solutions:

Solution 1: Find the equations of the lines. Suppose these slopes are m and $-m$. The equations of the two lines are $y = mx + b$ and $y = -mx + b$. Letting $y = 0$ in each to find the x-intercepts gives $x = -b/m$ for the first equation and $x = b/m$ for the second. Therefore, the sum of the x-coordinates of the x-intercepts is $-b/m + b/m = \boxed{0}$.

Solution 2: Use the meaning of slope. Suppose the x-intercept of one line is $(c, 0)$ and the x-intercept of the other line is $(d, 0)$. Both lines go through $(0, b)$, so the slope of the first line is $(b - 0)/(0 - c) = -b/c$, and the slope of the second is $(b - 0)/(0 - d) = -b/d$. The slope of one line is the negative of the other, so we have $-b/c = -(-b/d) = b/d$, from which we find $c = -d$. (Except when $b = 0$, in which case the x-intercept of both lines is the origin.) So, the sum of the x-coordinates of the x-intercepts is $c + d = -d + d = \boxed{0}$.

8.53 If x is nonnegative, then $\sqrt{x^2} = x$. However, if x is negative, we have $\sqrt{x^2} = -x$, because the square root of a nonnegative number must be nonnegative. We saw in the text that $|x| = x$ when x is nonnegative and $|x| = -x$ when x is negative. So, because $\sqrt{x^2} = |x| = x$ when x is nonnegative and $\sqrt{x^2} = |x| = -x$ when x is negative, we have $\sqrt{x^2} = |x|$ for all real numbers.

8.54

(a) Two points on the line $y = 3x + 2$ are $(0, 2)$ and $(1, 5)$. However, Bob will mess up and plot these points as $(2, 0)$ and $(5, 1)$. Taking the leap of faith (for now) that Bob's graph will also be a line, we can produce Bob's graph at right by drawing the line through these two points. The original line is dashed in the diagram. The slopes of the lines are 3 (dashed line) and $\frac{1}{3}$ (solid line).

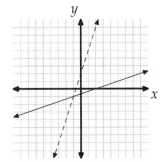

(b) Because the line Bob graphs passes through points $(2, 0)$ and $(5, 1)$, it has slope $(1 - 0)/(5 - 2) = 1/3$. Therefore, its equation is

$$y - 0 = \frac{1}{3}(x - 2) \quad \Rightarrow \quad 3y = x - 2 \quad \Rightarrow \quad x = 3y + 2.$$

Compare this equation to the equation in Bob's textbook, $y = 3x + 2$. The only difference between the two equations is that x and y are reversed! This makes sense, because Bob is just reversing x and y when he plots points. So, if Bob reverses x and y for every point on the graph of $y = 3x + 2$, he is producing the graph of the equation formed when we reverse x and y, which is $x = 3y + 2$.

(c) We use our previous part as guidance. If Bob must graph the equation $y = mx + b$, his graph will be the graph of the equation $x = my + b$. Putting this second equation in slope-intercept form gives $y = \frac{1}{m}x - \frac{b}{m}$. So, the slope of the line Bob will produce is $\boxed{1/m}$.

If Bob tries to graph $y = k$, he will produce the vertical line that is the graph of $x = k$. Similarly, if he tries to graph $x = h$, he will produce the horizontal line $y = h$.

In all cases, the graph Bob produces is the result of flipping the correct graph over the line $x = y$. (Take a look at our two lines in part (a) above; they are mirror images over the line $x = y$.)

8.55 There is a point T on $\overline{RS}$ such that $RT/TS = 3$. Suppose the coordinates of T are (x, y). Since T is on the segment connecting $(3, 5)$ to $(8, -3)$, we know x is between 3 and 8. Moreover, we know that T is 3 times as far from R as from S, so x is 3 times as far from 3 as from 8, so we have $x - 3 = 3(8 - x)$. (Remember, x is between 3 and 8, so x is $x - 3$ from 3 and $8 - x$ from 8.) Solving this equation gives $x = 27/4 = 6\frac{3}{4}$. Similarly, we have $5 - y = 3[y - (-3)]$ by considering the y-coordinates. Solving this equation gives us $y = -1$. Therefore, the point $\boxed{\left(6\frac{3}{4}, -1\right)}$ is on $\overline{RS}$ such that it is 3 times as far from R as it is from S.

Is there another point on the line through R and S that is 3 times as far from R as from S? Suppose point P is on this line. If R is between P and S, then PR/PS is less than 1, because PR is less than PS. So, no points on the line on the opposite side of R from S are three times as far from R as from S. As we move P from R to S, the ratio PR/PS goes from being very close to 0 (when P is close to R) to being very large (when P is very close to S). Somewhere in between, the ratio hits 3. We've already found that point above (point T). What happens as we move P past S?

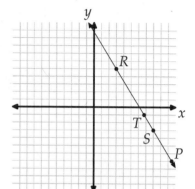

As we just move P past S, the ratio PR/PS is still very large, because PS is very close to 0. As we move P very far away from S, the ratio PR/PS gets very close to 1, since then $PR/PS = (PS + SR)/PS = 1 + SR/PS$, and SR/PS is very close to 0 when PS is huge. So, PR/PS is very large when P is near S and close to 1 when P is far past S. Therefore, for some point P in between "very close to S" and "very far away from S," we have $PR/PS = 3$. Let's find this P.

Suppose the coordinates of this P are (a, b). P is beyond S on the line through R and S. Since we must go right and down to get from R to S, we must go right and down to get from S to P. In other words, a is greater than 8 and b is less than -3. To get from R to P, we must go $a - 3$ units to the right. To get from S to P, we must go $a - 8$ units to the right. So, we have $a - 3 = 3(a - 8)$, so $a = 21/2 = 10\frac{1}{2}$. Similarly, we must go $5 - b$ units down to get from R to P and $-3 - b$ units down to get from S to P, so $5 - b = 3(-3 - b)$, which gives us $b = -7$. Therefore, the other point on the line that is three times as far from R as from S is $\boxed{\left(10\frac{1}{2}, -7\right)}$.

8.56 *Solution 1: Use slope.* Because the wire goes from 14 feet high to 11 feet high when it goes from the tall flagpole to the shorter one, the wire 'drops' 3 feet for every 12 feet along the ground it covers. So, it drops 1 foot for every 4 feet along the ground it covers. This is just like slope! To get to the ground from the top of the small flagpole, it must drop 11 feet. To do so, it must therefore cover $4(11) = \boxed{44 \text{ feet}}$ along the ground.

Solution 2: Use algebra. Let the base of the small flagpole be the origin of the coordinate axes, and the top of the small flagpole be at $(0, 11)$ because the small flagpole is 11 feet tall. Let the ground be the x-axis, so that the base of the other flagpole is at $(12, 0)$ because the bases of the flagpoles are 12 feet apart. The other flagpole is 14 feet tall, so its top is at the point $(12, 14)$. The wire is the line through the tops of these flagpoles, so the equation of the line through $(0, 11)$ and $(12, 14)$ will describe it. The slope of this line is $(14 - 11)/(12 - 0) = 3/12 = 1/4$. Therefore, the equation of this line is

$$y - 11 = \frac{1}{4}(x - 0) \quad \Rightarrow \quad 4(y - 11) = x \quad \Rightarrow \quad 4y - 44 = x.$$

The wire hits the ground where $y = 0$. Solving $4(0) - 44 = x$ gives $x = -44$, so the wire hits the ground at $(-44, 0)$. This point is a distance of 44 from the origin, which represents the base of the small flagpole. Therefore, the wire hits the ground $\boxed{44 \text{ feet}}$ from the base of the small flagpole.

8.57 Because line L passes through $(r, -3)$ and (a, b) and has slope -2, we have

$$\frac{b - (-3)}{a - r} = -2 \quad \Rightarrow \quad b + 3 = -2(a - r) \quad \Rightarrow \quad b + 3 = -2a + 2r.$$

Because line K is perpendicular to L, the slope of K is $-1/(-2) = 1/2$. Line K passes through (a, b) and $(6, r)$, so we have

$$\frac{r - b}{6 - a} = \frac{1}{2} \quad \Rightarrow \quad 2(r - b) = 6 - a \quad \Rightarrow \quad 2r - 2b = 6 - a.$$

So, we have the system of equations

$$b + 3 = -2a + 2r,$$
$$2r - 2b = 6 - a.$$

We need to find a in terms of r, so we want to eliminate b from these equations. We could solve the first equation for b and substitute the result in the second, or we could simply add 2 times the first equation to the second, which gives $2r + 6 = -4a + 4r + 6 - a$. Simplifying this equation gives us $-2r = -5a$, so $a = \boxed{2r/5}$.

8.58 First we find the graph of $(x - y + 2)(3x + y - 4) = 0$. If this equation is true for some ordered pair (x, y), then either $x - y + 2 = 0$ or $3x + y - 4 = 0$. In other words, the equation is true only if (x, y) is on either the graph of the line $x - y + 2 = 0$ or the graph of the line $3x + y - 4 = 0$. Conversely, any (x, y) that is on either of these lines is a solution to the equation $(x - y + 2)(3x + y - 4) = 0$, since such an (x, y) will make either $x - y + 2$ or $3x + y - 4$ equal to 0. So, the graph of $(x - y + 2)(3x + y - 4) = 0$ consists of the graphs of the two linear equations

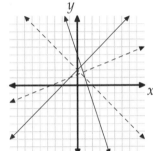

$$x - y + 2 = 0 \quad \text{and} \quad 3x + y - 4 = 0.$$

Similarly, the graph of $(x + y - 2)(2x - 5y + 7) = 0$ consists of the graphs of the two linear equations

$$x + y - 2 = 0 \quad \text{and} \quad 2x - 5y + 7 = 0.$$

Because neither of these lines is parallel to either of the lines in the graph of $(x - y + 2)(3x + y - 4) = 0$, each of these two lines meets each of the two lines in the graph of $(x - y + 2)(3x + y - 4) = 0$ once. Therefore, there are $2 \times 2 = \boxed{4}$ intersection points. (The graph shows the lines of $(x - y + 2)(3x + y - 4) = 0$ solid and those of $(x + y - 2)(2x - 5y + 7) = 0$ dashed. At how many points does each dashed line hit a solid one?)

8.59

(a) We tackle the coordinates separately. To get from P to Q, we go $x_2 - x_1$ to the right (if this quantity is negative, then we go leftward). Suppose T is the point (a, b), so that T is $a - x_1$ to the right of P. Since $PT/PQ = r$, we must have

$$\frac{a - x_1}{x_2 - x_1} = r \quad \Rightarrow \quad a - x_1 = r(x_2 - x_1) \quad \Rightarrow \quad a = r(x_2 - x_1) + x_1.$$

Similarly, we find $b = r(y_2 - y_1) + y_1$, so the coordinates of T are $\boxed{(r[x_2 - x_1] + x_1, r[y_2 - y_1] + y_1)}$.

We can also write this as $\boxed{([1 - r]x_1 + rx_2, [1 - r]y_1 + ry_2)}$.

Make sure you see why this point is between P and Q when r is between 0 and 1.

(b) We get some intuition by trying $r = 1$. When $r = 1$, our formula gives us $T = (r[x_2 - x_1] + x_1, r[y_2 - y_1] + y_1) = (x_2, y_2)$. Therefore, when $r = 1$, point T is the same as point Q.

We saw above that if T is (a, b), then $a = x_1 + r(x_2 - x_1)$. Let's look at what this means if r is greater than 1. Then, to get a, we start at x_1, the x-coordinate of P, and then add $x_2 - x_1$ *more than once*. If we just add $x_2 - x_1$ to x_1 once, we get x_2. In other words, adding $x_2 - x_1$ to the x-coordinate of P takes us to x-coordinate of Q. Then, since r is greater than 1, we add on more! So, we go beyond point Q. The same holds when we consider the y-coordinate, so when r is greater than 1, our formula still holds, and it gives us a formula for points T on the line through P and Q such that Q is between P and T.

(c) Here, we get some intuition by letting $r = 0$, which gives us $T = (r[x_2 - x_1] + x_1, r[y_2 - y_1] + y_1) = (x_1, y_1) = P$. Perhaps you see where this is heading!

Again, we focus on $a = x_1 + r(x_2 - x_1)$. When r is positive, this means that we get to T from P by moving towards Q an amount given by $r(x_2 - x_1)$. However, when r is negative, the expression $r(x_2 - x_1)$ tells us to move *away from* Q to get to T. Therefore, when r is negative, we describe points T on the line through P and Q such that P is between T and Q (i.e., T and Q are in the opposite directions from P along the line).

8.60 The expressions in the square roots look a lot like the distance formula. We can think of the first radical, $\sqrt{a^2 + 9}$ as the distance between $(0, 0)$ and $(a, 3)$. Similarly, we can think of the next radical, $\sqrt{(b - a)^2 + 4}$, as the distance between $(a, 3)$ and a point with x-coordinate equal to b. Since 4 must be the square of the difference between the y-coordinates of the two points, our other point is $(b, 5)$ or $(b, 1)$. If we use $(b, 1)$, then the expression

$$\sqrt{a^2 + 9} + \sqrt{(a - b)^2 + 4}$$

describes the length of the path from $(0, 0)$ to a point $(a, 3)$ on the line $y = 3$ then to a point $(b, 1)$ on the line $y = 1$. If we use $(b, 5)$, then the expression

$$\sqrt{a^2 + 9} + \sqrt{(b - a)^2 + 4}$$

describes the length of the path from $(0, 0)$ to a point $(a, 3)$ on the line $y = 3$ then to a point $(b, 5)$ on the line $y = 5$. The latter is much easier to deal with because such a path can be a straight line. The former path requires going from the origin up to the line $y = 3$, then down to the line $y = 1$. So, we use $(b, 5)$ instead of $(b, 1)$.

Finally, the expression $\sqrt{(8 - b)^2 + 16}$ can be thought of as the distance between $(b, 5)$ and the point $(8, 9)$. Therefore, our whole sum equals the sum:

[Distance from $(0, 0)$ to $(a, 3)$] + [Distance from $(a, 3)$ to $(b, 5)$] + [Distance from $(b, 5)$ to $(8, 9)$].

In other words, the sum gives us the total distance from the origin, to some point on the line $y = 3$, then to some point on the line $y = 5$, then on to the point $(8, 9)$. The shortest path from $(0, 0)$ to $(8, 9)$ is a straight line from $(0, 0)$ to $(8, 9)$. Moreover, this shortest path does indeed visit points on $y = 3$ and $y = 5$, as required, so the minimum value of

$$\sqrt{a^2 + 9} + \sqrt{(b - a)^2 + 4} + \sqrt{(8 - b)^2 + 16}$$

equals the length of the path from $(0, 0)$ to $(8, 9)$, which is $\sqrt{(8 - 0)^2 + (9 - 0)^2} = \sqrt{64 + 81} = \boxed{\sqrt{145}}$.

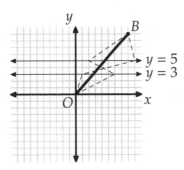

In the diagram above, we show three paths from the origin, O, to our target point, B. The shortest distance between two points is a straight line, shown in bold in the diagram. (Two other paths from O to B that hit a point on the line $y = 3$ and a point on the line $y = 5$ are also shown. We can see that these are considerably longer than the straight path.)

9

Introduction to Inequalities

Exercises for Section 9.1

9.1.1 Because $x \geq y$, we know x cannot be less than y. Because $x \leq y$, x cannot be greater than y. Therefore, the only option left is that $x = y$, which satisfies both $x \geq y$ and $x \leq y$. Therefore, $x - y = \boxed{0}$.

9.1.2 If $a \geq b$, then a is at least as large as b. Since b is strictly larger than c, a must therefore be strictly larger than c. So, $a > c$, but we cannot have $a = c$.

9.1.3 $\boxed{\text{Yes}}$. Because $x > y > 0$ and $a > b > 0$, we have $xa > yb$. Dividing both sides by ab gives us $x/b > y/a$. Make sure you see why it's important that a, b, x, and y are positive.

9.1.4 We first multiply both sides of $x < y$ by $1/x$. Since $1/x$ is negative, we reverse the inequality sign, to get $1 > y/x$. We then produce $1/y$ by multiplying both sides by $1/y$. Since $1/y$ is negative, we reverse the inequality sign when we multiply $1 > y/x$ by $1/y$, giving $1/y < 1/x$, as desired.

9.1.5

(a) We have $7^{-2} = 1/7^2 = 1/49$ and $5^{-2} = 1/5^2 = 1/25$. Since $1/49 \approx 0.0204$ and $1/25 = 0.04$, we have $1/25 > 1/49$, so, $\boxed{5^{-2}}$ is larger than 7^{-2}.

(b) We have $7^{-5} = 1/7^5 \approx 0.0000595$ and $5^{-5} = 1/5^5 = 0.00032$, so $\boxed{5^{-5}}$ is larger than 7^{-5}.

(c) Our first two parts suggest that if $x > y > 0$ and a is a negative integer, then $y^a > x^a$. We showed in the text that if b is a positive integer and $x > y > 0$, then $x^b > y^b$. When we take reciprocals of both sides of this inequality, we must reverse the direction of the inequality sign, giving, $\frac{1}{x^b} < \frac{1}{y^b}$. So, if $x > y > 0$, then $x^{-b} < y^{-b}$, where b is a positive integer. Therefore, we just let $a = -b$ and we have $x^a < \boxed{y^a}$ when a is a negative integer and $x > y > 0$.

Exercises for Section 9.2

9.2.1

(a) We square both quantities, which gives us $3\sqrt{5}$ and $6\sqrt{2}$. Squaring $3\sqrt{5}$ gives $(3\sqrt{5})^2 = 3^2(\sqrt{5})^2 = 9(5) = 45$. Squaring $6\sqrt{2}$ gives $(6\sqrt{2})^2 = 6^2(\sqrt{2})^2 = 36(2) = 72$. Clearly, $72 > 45$, so we can reverse our steps by taking the square root of both sides of this inequality two times to find:

$$72 > 45 \quad \Rightarrow \quad \sqrt{72} > \sqrt{45} \quad \Rightarrow \quad 6\sqrt{2} > 3\sqrt{5} \quad \Rightarrow \quad \sqrt{6\sqrt{2}} > \sqrt{3\sqrt{5}}.$$

So, $\boxed{\sqrt{6\sqrt{2}}}$ is larger.

(b) Just squaring both expressions won't get rid of the cube root in $\sqrt[3]{6\sqrt{2}}$, and cubing them will leave us still with the outer square root in $\sqrt{3\sqrt[3]{5}}$. However, if we write both expressions with fractional exponents, we can find a power that we can raise both expressions to in order to get rid of the fractional exponents:

$$\sqrt{3\sqrt[3]{5}} = (3 \cdot 5^{\frac{1}{3}})^{\frac{1}{2}} = 3^{\frac{1}{2}} \cdot 5^{\frac{1}{6}},$$
$$\sqrt[3]{6\sqrt{2}} = (6 \cdot 2^{\frac{1}{2}})^{\frac{1}{3}} = 6^{\frac{1}{3}} \cdot 2^{\frac{1}{6}}.$$

Aha! Raising both of these to the 6th power will knock out the fractional exponents:

$$\left(3^{\frac{1}{2}} \cdot 5^{\frac{1}{6}}\right)^6 = \left(3^{\frac{1}{2}}\right)^6 \left(5^{\frac{1}{6}}\right)^6 = 3^3 \cdot 5^1 = 135,$$
$$\left(6^{\frac{1}{3}} \cdot 2^{\frac{1}{6}}\right)^6 = \left(6^{\frac{1}{3}}\right)^6 \left(2^{\frac{1}{6}}\right)^6 = 6^2 \cdot 2 = 72.$$

Clearly $135 > 72$, so $3^3 \cdot 5 > 6^2 \cdot 2$, and we can reverse our steps to show that $\boxed{\sqrt{3\sqrt[3]{5}}}$ is larger than $\sqrt[3]{6\sqrt{2}}$:

$$3^3 \cdot 5 > 6^2 \cdot 2 \quad \Rightarrow \quad \left(3^3 \cdot 5\right)^{\frac{1}{6}} > \left(6^2 \cdot 2\right)^{\frac{1}{6}} \quad \Rightarrow \quad 3^{\frac{1}{2}}(5^{\frac{1}{3}})^{\frac{1}{2}} > 6^{\frac{1}{3}}(2^{\frac{1}{2}})^{\frac{1}{3}} \quad \Rightarrow \quad \sqrt{3\sqrt[3]{5}} > \sqrt[3]{6\sqrt{2}}.$$

9.2.2 While it isn't obvious how to compare these two fractions, it is easy to compare their reciprocals. We have

$$1 + 2 + 3 + \cdots + 99 + 100 < 2 + 3 + 4 + 5 + \cdots + 100 + 101,$$

because we form the sum on the right by adding $101 - 1$ to the sum on the left. We can then take the reciprocal of both sides of this inequality (and therefore reverse the inequality sign.) to get

$$\boxed{\frac{1}{1 + 2 + 3 + \cdots + 99 + 100}} > \frac{1}{2 + 3 + 4 + 5 + \cdots + 100 + 101}.$$

9.2.3 First, we'll compare $\sqrt{2}$ and $\sqrt[3]{3}$. We write these with fractional exponents so we can figure out what power to raise both to: $2^{\frac{1}{2}}$ and $3^{\frac{1}{3}}$. We see now that we can raise both to the 6th power to get integers:

$$\left(2^{\frac{1}{2}}\right)^6 = 2^3 = 8,$$
$$\left(3^{\frac{1}{3}}\right)^6 = 3^2 = 9.$$

Since $9 > 8$, we can reverse our steps above to see that

$$\sqrt[6]{9} > \sqrt[6]{8} \quad \Rightarrow \quad (3^2)^{\frac{1}{6}} > (2^3)^{\frac{1}{6}} \quad \Rightarrow \quad 3^{\frac{1}{3}} > 2^{\frac{1}{2}} \quad \Rightarrow \quad \sqrt[3]{3} > \sqrt{2}.$$

We can do something similar to compare $\sqrt{2}$ to $\sqrt[5]{5}$. We write them with fractional exponents: $2^{\frac{1}{2}}$ and $5^{\frac{1}{5}}$. So, we raise both to the 10th power to get rid of the fractional exponents:

$$(2^{\frac{1}{2}})^{10} = 2^5 = 32,$$
$$(5^{\frac{1}{5}})^{10} = 5^2 = 25.$$

Therefore, we see that $\sqrt{2} > \sqrt[5]{5}$. So, we can now order the three numbers from least to greatest: $\boxed{\sqrt[5]{5} < \sqrt{2} < \sqrt[3]{3}}$.

9.2.4

(a) We can compare much simpler numbers by first taking the 100th root of both of the given numbers, leaving us 2^5 and 3^3. Since $2^5 > 3^3$, we can raise both sides of this inequality to the 100th power to find $(2^5)^{100} > (3^3)^{100}$, so $\boxed{2^{500}} > 3^{300}$.

(b) What's wrong here:

> We can't compare 2^{81} and 3^{49} very easily, so we compare numbers near them: 2^{80} and 3^{48}. We take the 16th root of both, so we are comparing 2^5 and 3^3. Since $32 > 27$, we know that $2^5 > 3^3$. Raising this to the 16th power, we have $2^{80} > 3^{48}$. Therefore, we know that $2^{81} > 3^{49}$.

The explanation that $2^{80} > 3^{48}$ is perfectly valid. However, this doesn't tell us that $2^{81} > 3^{49}$. We know that $2^{81} > 2^{80}$ and $3^{49} > 3^{48}$, but we can't combine these with $2^{80} > 3^{48}$ to be able to compare 2^{81} and 3^{49}. (Note that to get from $2^{80} > 3^{48}$ to $2^{81} > 3^{49}$, we must multiply the larger side by 2 and the smaller side by 3. This is *not* a valid inequality manipulation because $2 < 3$. We must find another way.)

The problem here is that in both $2^{81} > 2^{80}$ and $3^{49} > 3^{48}$, our target numbers, 2^{81} and 3^{49} are on the same side of the inequalities (the larger side). Therefore, we won't be able to build a chain of inequalities with one on the large side and one on the small side.

However, our initial strategy of comparing our target numbers to numbers that are easier to relate to each other (2^{80} and 3^{48}) does give us an idea. We try the same thing, but start with $2^{81} > 2^{80}$ and $3^{50} > 3^{49}$. Now we compare 2^{80} and 3^{50}. Since $2^8 = 256$ and $3^5 = 243$, we have $2^8 > 3^5$, so $2^{80} > 3^{50}$. We can put these inequalities together as desired:

$$2^{81} > 2^{80} > 3^{50} > 3^{49}.$$

So, $\boxed{2^{81}}$ is larger than 3^{49}.

(c) Working with nearby powers was a winner in the last part, so we try it here. We try comparing 2^{840} to 5^{360}, because 840 and 360 have a common factor of 120. Therefore, we can take the 120th root of both, to give 2^7 and 5^3. Since $2^7 = 128$ and $5^3 = 125$, we have $2^7 > 5^3$. Raising both sides to the 120th power gives $2^{840} > 5^{360}$. But, how does this get us to 2^{845} and 5^{362}?

To get to 2^{845} from 2^{840}, we must multiply by 2^5. Similarly, we must multiply 5^{360} by 5^2 to get 5^{362}. Because $2^5 > 5^2$ and $2^{840} > 5^{360}$, the product of the left sides is greater than the product of the right sides, so $\boxed{2^{845}} > 5^{362}$.

Exercises for Section 9.3

9.3.1

(a) We subtract 2 from both sides to get $3x \geq -6$, and then divide by 3 to get $x \geq -2$. In interval notation, we have $x \in [-2, +\infty)$. The graph is shown below.

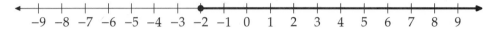

(b) Subtracting 2 from both sides, and then subtracting $2x$ from both sides gives $x < 3$. In interval notation, we have $x \in (-\infty, 3)$. The graph is shown below.

(c) We isolate $3x$ by subtracting 2 from all three parts of the inequality, which gives $-6 \leq 3x \leq 3$. Dividing all three parts by 3 isolates x and gives $-2 \leq x \leq 1$. In interval notation, we have $x \in [-2, 1]$. This is graphed below.

9.3.2 The graph consists of all the numbers greater than -1, but it does not include -1. Therefore, it represents $\boxed{x > -1}$.

9.3.3 In n months after January, 2003, the population of my town increases by $322n$, since it increases 322 each month. So, n months after January, 2003, we have $11{,}212 + 322n$ people in my town. We seek the smallest n such that $11212 + 322n > 15000$. We isolate n by subtracting 11212 to get $322n > 3788$, then by dividing by 322 to get $n > 3788/322 \approx 11.76$. So, $n = 12$ is the smallest whole number of months after which the population will be above 15,000. Therefore, the population of my town goes over 15,000 for the first time in $\boxed{\text{January, 2004}}$.

9.3.4 We isolate $-4x$ by subtracting 3 from all parts of the inequality chain, giving $-2 \leq -4x \leq 6$. We isolate x by dividing all parts of the chain by -4. Note that we must reverse the inequality signs when we perform this step, giving $\frac{1}{2} \geq x \geq -\frac{3}{2}$. The length of this interval is $(1/2) - (-3/2) = \boxed{2}$ because $1/2$ is $1/2 - (-3/2)$ to the right of $-3/2$ on the number line.

9.3.5 We'll tackle the two sides separately, and we'll first look for positive solutions. To deal with $\frac{1}{2} < \frac{n}{n+1}$, we want to get rid of the fractions. Because n is positive, so is $n + 1$, so when we multiply both sides by $n + 1$, we don't have to change the direction of the inequality sign. We also want to multiply both sides by 2 to get rid of the denominator of $\frac{1}{2}$. Multiplying both sides of $\frac{1}{2} < \frac{n}{n+1}$ by $2(n + 1)$ gives

$$2(n + 1) \cdot \frac{1}{2} < 2(n + 1) \cdot \frac{n}{n + 1} \quad \Rightarrow \quad n + 1 < 2n \quad \Rightarrow \quad 1 < n.$$

So, we have $n > 1$. What about the other side of the inequality chain?

Again, because $n + 1$ is positive, we multiply $\dfrac{n}{n + 1} < \dfrac{99}{101}$ by $101(n + 1)$ to get

$$101(n + 1) \cdot \frac{n}{n + 1} < 101(n + 1) \cdot \frac{99}{101} \quad \Rightarrow \quad 101n < 99(n + 1) \quad \Rightarrow \quad 101n < 99n + 99 \quad \Rightarrow \quad n < 49\frac{1}{2}.$$

Putting this together with $n > 1$ from before, we have $1 < n < 49\frac{1}{2}$. So, the positive integers that work are $2, 3, 4, \ldots, 49$. There are $\boxed{48}$ numbers in this list because the list has all of the first 49 positive integers except the number 1.

If n is a negative integer, we first observe that we cannot have $n = -1$, since that makes the denominator of $n/(n+1)$ equal to 0. Next, we note that for all integers n, we have $n < n + 1$. Because $n + 1$ is negative when n is a negative integer besides $n = -1$, when we divide $n < n + 1$ by $n + 1$, we reverse the inequality sign to get $\frac{n}{n+1} > 1$. Since $n/(n+1)$ must be greater than 1 when n is a negative integer besides -1, it can never be less than $99/101$. So, there are no negative integers that satisfy the inequality.

Extra challenge: Solve the problem by taking the reciprocal of the inequality chain!

9.3.6   Looking closely at the product, we see that all the terms except the first 3 in the numerators and the $k + 1$ in the denominators cancel:

$$\frac{3}{2} \cdot \frac{2}{1} \cdot \frac{1}{2} \cdot \frac{2}{3} \cdot \frac{3}{4} \cdots \frac{k}{k+1} = \frac{3}{2} \cdot \frac{\cancel{2}}{\cancel{1}} \cdot \frac{\cancel{1}}{\cancel{2}} \cdot \frac{\cancel{2}}{\cancel{3}} \cdot \frac{\cancel{3}}{\cancel{4}} \cdots \frac{\cancel{k-1}}{\cancel{k}} \cdot \frac{\cancel{k}}{k+1} = \frac{3}{k+1}.$$

This leaves us with a much simpler inequality to solve,

$$\frac{3}{k+1} \geq \frac{1}{8}.$$

Clearly, there are some positive integers k that satisfy this inequality, such as $k = 1$, so we can assume that the largest integer that satisfies it is positive. For positive values of $k + 1$, we can multiply both sides of the inequality by $k + 1$ without changing the direction of the inequality. We also multiply by 8 to get rid of the fraction on the right. These give us $3 \cdot 8 \geq k + 1$, from which we find that $23 \geq k$. So, $k = \boxed{23}$ is the largest integer that satisfies the inequality. (Note that we could have seen that since $3/(k+1) \geq 1/8$, then $3/(k+1) \geq 3/24$, which would also have given us $k + 1 \leq 24$, or $k \leq 23$.)

Exercises for Section 9.4

9.4.1   First, we graph the line $x + 2y = 6$. Because the inequality is nonstrict (meaning it is $\geq$ rather than >), we must include the line in the graph of the inequality. So, we draw a solid line rather than a dashed line. Next, we determine which side of the line to shade. The point $(0, 0)$ does not satisfy the inequality because $0 + 2(0)$ is less than 6. So, we shade the side opposite the origin.

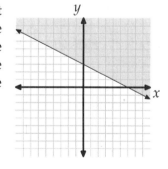

9.4.2

(a)   First we find the equation of the line. The line passes through $(2, 1)$ and $(5, -1)$, so its slope is $(-1 - 1)/(5 - 2) = -2/3$. So, the equation of the line is

$$y - 1 = -\frac{2}{3}(x - 2) \quad \Rightarrow \quad 3(y - 1) = -2(x - 2) \quad \Rightarrow \quad 3y - 3 = -2x + 4 \quad \Rightarrow \quad 2x + 3y = 7.$$

Because the line is solid, the inequality that is graphed is a nonstrict inequality; however, we must decide if it is $2x + 3y \geq 7$ or $2x + 3y \leq 7$. The origin is in the shaded region, and $2x + 3y$ is less than 7 when $(x, y) = (0, 0)$. So, the graphed inequality is $\boxed{2x + 3y \leq 7}$.

(b) Again, we find the equation of the line first. The line passes through $(5, 1)$ and $(6, 4)$, so its slope is $(4 - 1)/(6 - 5) = 3$. So, the equation of the line is

$$y - 1 = 3(x - 5) \quad \Rightarrow \quad y - 1 = 3x - 15 \quad \Rightarrow \quad 3x - y = 14.$$

The inequality is strict, because the line is dashed in the graph. The origin is not in the shaded region, so we know that the inequality is $\boxed{3x - y > 14}$.

9.4.3 We start by graphing the lines $2x + y = 0$ and $2x + y = 6$. These lines have the same slope, so they are parallel. We draw the first one solid and the second one dashed, because the inequality $0 \leq 2x + y$ is nonstrict and $2x + y < 6$ is strict. But where do we shade? Because $(x, y) = (0, 0)$ satisfies $2x + y < 6$, we shade to the left of the line $2x + y = 6$. However, the origin is on the line $2x + y = 0$, so we pick a different point to determine which side of the line $2x + y = 0$ satisfies the inequality $0 \leq 2x + y$. The point $(x, y) = (1, 0)$ satisfies this inequality, so we shade to the right of $2x + y = 0$. Therefore, the points that satisfy both inequalities are those points to the right of (or on) $2x + y = 0$ and to the left of $2x + y < 6$, as shown.

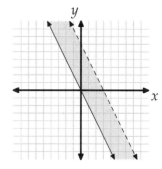

Note that we also could have determined where to shade by solving both inequalities for x. From $2x + y \geq 0$, we have $x \geq -y/2$, so we want to shade where x is greater than the line; that is, to the right of the line $2x + y = 0$. Similarly, from $2x + y < 6$, we have $x < -y/2 + 3$, so we want to shade where x is less than the corresponding points on the line $2x + y = 6$; that is, to the left of the line $2x + y = 6$.

9.4.4 We start by graphing each inequality on the same graph. We do so by first graphing the lines $2x - 4 = y$ and $y = -\frac{2}{3}x + 2$. In the graphs of the inequalities, we dash the lines, because the inequalities are strict. Because the first inequality is $y > 2x - 4$, the $>$ tells us to shade the region above the line $y = 2x - 4$. Similarly, the $<$ in the inequality $y < -\frac{2}{3}x + 2$ tells us to shade the region below the boundary line. In our graph at right, the darkest region shows those points that are common to both inequalities, and the lightly shaded region shows those points that satisfy one inequality or the other, but not both.

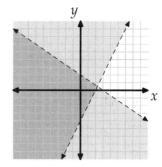

We can use this graph to create the three graphs sought in the problem. The first is the dark region, which satisfies both inequalities. This is graph (a) below. The lightly shaded regions in the original graph are reproduced as (c) below, as this depicts the region that satisfied one or the other inequality, but not both. Notice that a portion of the boundary lines are solid, because these satisfy one or the other inequality, but not both. The open circle where the lines meet indicate that this point does not satisfy either inequality.

Finally, graph (b) shows all the points that satisfy at least one if the inequalities; these are all the points that have any shading at all in our original graph. Again, portions of our boundary lines are solid, as these points satisfy at least one of the inequalities.

(a) 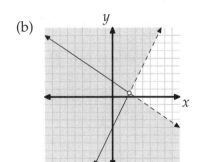   (b) 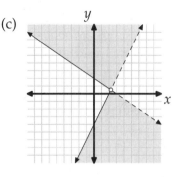   (c)

Exercises for Section 9.5

9.5.1   At noon, the pool has 0.8(10,000) = 8000 gallons of water. Each hour thereafter, the amount of water in the pool increases by 225 − 120 = 105 gallons. So, after n hours, the amount of water in the pool is $8000 + 105n$. We seek the smallest integer n for which $8000 + 105n > 10000$. Solving this inequality gives $n > 2000/105 \approx 19.05$. So, when Mitch checks the pool 19 hours after noon, it will not be overflowing, but it will be when he checks 20 hours after noon. Twenty hours after noon is $\boxed{8\ \text{AM}}$ the next day.

9.5.2   The fraction $\boxed{\dfrac{c + d}{a + b}}$ has both the largest numerator of the numbers in the list, and the smallest denominator. It is therefore the largest number in the list.

9.5.3   We first graph the region that satisfies the inequalities. This region is shown at right. As discussed in the text, we only have to test the corners of the region described by the inequalities. These corners are $(1, 2)$; $(-1, 2)$; $(-7/3, -2)$; $(7/3, -2)$. (We find these by finding the points where the lines $y = 2$ and $y = -2$ intersect the lines $y = -3x + 5$ and $y = 3x + 5$.) We evaluate $4x + y$ at each of these points and get 6, −2, −34/3, and 22/3, respectively. Therefore, the maximum value of $4x + y$ is $\boxed{22/3}$ and the minimum value is $\boxed{-34/3}$.

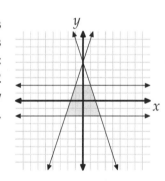

9.5.4

(a) The region is shaded below.

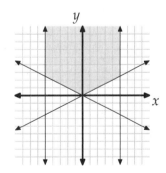

(b) At right, in addition to our shaded region, we have graphed the lines $x - y = k$ for $k = -2, 1, 5/2$, and 6. These lines are shown dashed. As k increases, the graph of the line is farther and farther to the right. The line through the boundary point $(x, y) = (5, 5/2)$ is the rightmost line that intersects the region that satisfies our inequalities. Therefore, we see that the maximum of $x - y$ is $5 - (5/2) = \boxed{5/2}$.

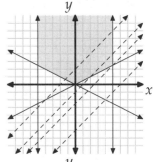

(c) Here, we have a different story. When we graph $x + y = k$ for a variety of k, we see that as k increases, the line $x + y = k$ is higher and higher. However, the shaded region extends forever upward. For any positive value of y, the point $(0, y)$ satisfies all of the given inequalities, so is in the shaded region. Therefore, we can make $x + y$ as large as we want. So, $x + y$ has $\boxed{\text{no maximum}}$.

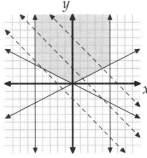

9.5.5

(a) Let x be the number of miles of highway the city builds on land and y be the number of miles the city builds on bridges. The city uses 2 truckloads of asphalt for each mile on ground and 1 truckload for each mile on bridges, and the city has 50 truckloads of asphalt. Therefore, we must have $2x + y \leq 50$. Similarly, because the city uses one gallon of paint per mile on ground and 3 gallons per mile on bridges, and the city has 80 gallons of paint, we must have $x + 3y \leq 80$. Clearly, we must also have $x \geq 0$ and $y \geq 0$, since the city can't build a negative number of miles of anything.

The number of miles the city builds is $x + y$. We wish to maximize $x + y$ under the constraints $x \geq 0$, $y \geq 0$, $2x + y \leq 50$, and $x + 3y \leq 80$. This region is graphed at right. Notice that each square in the graph is 10 by 10, instead of the usual 1 by 1.

Besides the origin, the coordinates of the boundary points are not so obvious. The line $2x + y = 50$ hits the x-axis at $x = 25$, so the boundary point on the x-axis is $(25, 0)$. Similarly, the line $x + 3y = 80$ hits the y-axis at $y = 80/3$, so $(0, 80/3)$ is another boundary point. Finally, we solve the system of equations $2x + y = 50$ and $x + 3y = 80$ to find that these two lines meet at $(14, 22)$. Evaluating $x + y$ at each of these points, we find that it is maximized at $(14, 22)$, so the greatest number of miles the city can build is $14 + 22 = \boxed{36}$ miles.

(b) We now have the additional constraint $y \geq 25$. We add this to our graph, and now we have a very tiny region that satisfies all of our constraints. Notice that the constraint $2x + y = 50$ doesn't even matter any more; our region is bounded by the y-axis, the line $y = 25$, and the line $x + 3y = 80$. These latter two lines intersect at $(5, 25)$. Solving the appropriate systems of equations tell us that the other boundary points are $(0, 80/3)$ and $(0, 25)$. Evaluating $x + y$ for each of these tells us that the city can build at most $5 + 25 = \boxed{30}$ miles of highway.

Review Problems

9.23

(a) $2 - 3x \geq 11 \quad \Rightarrow \quad -3x \geq 9 \quad \Rightarrow \quad \boxed{x \leq -3}$.

Make sure you see why we have to reverse the direction of the inequality sign in the last step. In interval notation, our solution is $\boxed{x \in (-\infty, -3]}$. The graph is below.

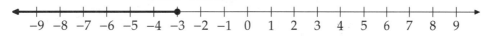

(b) $3 + 2x < 30 - 7x \quad \Rightarrow \quad 9x < 27 \quad \Rightarrow \quad \boxed{x < 3} \quad \Rightarrow \quad \boxed{x \in (-\infty, 3)}$.

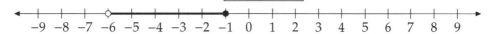

(c) Adding $2x$ to all three parts gives $8 \leq 5 - 3x < 23$. Subtracting 5 from all three parts gives $3 \leq -3x < 18$. We then divide by -3 (and change the direction of both inequality signs), to give $\boxed{-1 \geq x > -6}$. In interval notation, we have $\boxed{x \in (-6, -1]}$.

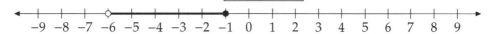

9.24

(a) $\boxed{\text{False}}$. If $a = b = c$, then we have $a \leq b$ and $b \leq c$, but it is not true that $a < c$.

(b) $\boxed{\text{True}}$. Since b must be no greater than a ($a \geq b$) and b must be no less than a ($b \geq a$), we must have $b = a$.

(c) $\boxed{\text{False}}$. If $c = 0$, then $ac = bc$. If c is negative, then $ac < bc$.

(d) $\boxed{\text{True}}$. If c is 0, then $ac = bc$. If $c < 0$, then $ac < bc$, so $ac \leq bc$.

(e) $\boxed{\text{True}}$. Subtracting a from both sides of $x + a \geq y + a$ gives $x \geq y$.

(f) $\boxed{\text{False}}$. Notice that $5 \geq 0$ and $1 \leq 2$, but it is still true that $5 + 1 \geq 0 + 2$. So, just because $x + a \geq y + b$, it is not necessarily true that $x \geq y$ and $a \geq b$.

9.25 We have $a > b$ and $c > d$, and the sum of the larger sides is larger than the sum of the smaller sides, so $a + c > b + d$. However, we can't tell which of $a + d$ and $b + c$ is larger. For example, note that $7 > 3 > 2 > 1$, and $7 + 1 > 3 + 2$, but $7 > 6 > 5 > 1$ and $7 + 1 < 6 + 5$.

9.26 We first take care of the powers of 2 by writing them all with base 2. We have $2^{36} = 2^{36}$, $4^{24} = (2^2)^{24} = 2^{48}$, and $8^4 = (2^3)^4 = 2^{12}$, so we can throw out 2^{36} and 8^4. We can compare 4^{24} to 3^{30} by comparing their sixth roots, 4^4 and 3^5. Since $4^4 = 256$ and $3^5 = 243$, we know that 4^{24} is larger than 3^{30}.

Similarly, we compare 4^{24}, 5^{18} and 6^{12} by comparing the sixth roots of all three: $4^4 = 256$, $5^3 = 125$, $6^2 = 36$. So 4^{24} is greater than both 5^{18} and 6^{12}. Finally, we take the eighth roots of 4^{24} and 7^8 to compare them: $4^3 = 64$ is much larger than 7^1, so $\boxed{4^{24}}$ is the largest of the numbers.

9.27 Suppose I buy n books. Each book costs 17 dollars, so together the books cost $17n$ dollars. Therefore, after paying for the books, I'll have $230 - 17n$ books left. I need to have at least 25 dollars left, so we must have $230 - 17n \geq 25$. Solving this inequality gives $n \leq 205/17 \approx 12.06$. Therefore, I can buy at most $\boxed{12}$ books.

9.28   Suppose I wish to compare a/b and c/d, and I know that $ad > bc$, and all the numbers are positive. We can divide both sides of $ad > bc$ by d to get $a > bc/d$, then divide this inequality by b to get $a/b > c/d$. Therefore, my magic trick will always work! If $ad > bc$ and a, b, c, and d are positive, then $a/b > c/d$.

9.29   We seek the smallest integer n such that $\sqrt{n} > 10$. We square both sides of this inequality to find $n > 100$. The smallest integer that is greater than 100 is $\boxed{101}$.

9.30

(a) We first graph the line $4x - \dfrac{y}{2} = 6$. We draw the line dashed because the inequality is strict. Then, we note that the origin, $(x, y) = (0,0)$, satisfies the inequality, so we shade the side of the graph that includes the origin. The graph is at left below.

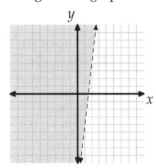
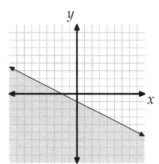

(b) We first rewrite the inequality in a form that's a little easier to graph:

$$2(y + 3) \le 4 - x \quad \Rightarrow \quad 2y + 6 \le 4 - x \quad \Rightarrow \quad x + 2y \le -2.$$

We graph the line $x + 2y = -2$ with a solid line because the inequality is nonstrict. The point $(x, y) = (0,0)$ does not satisfy the inequality, so we shade the side opposite the origin. The graph is at right above.

9.31   We first solve the inequality $7 - 3x < x - 1$. Adding $3x$ to both sides gives $7 < 4x - 1$, and adding 1 to both sides gives $8 < 4x$. Dividing both sides by 4 gives $2 < x$. Next, we take care of the other end of the inequality chain, $x - 1 \le 2x + 9$. Subtracting x from both sides, then subtracting 9 from both sides gives $-10 \le x$. The values of x that satisfy the inequality chain must satisfy both $2 < x$ and $-10 \le x$. All values of x such that $\boxed{x > 2}$ satisfy both inequalities.

9.32   We start by graphing both $3x + 2y \le 7$ and $2x + 4y \le 8$. (We simplify the latter by dividing it by 2 to get $x + 2y \le 4$.) Dashed in our diagram are several lines of the form $x + y = k$. As k gets larger, the lines are farther and farther to the right. We want the rightmost line that intersects our shaded region (or a boundary point of the shaded region). This is the line that meets the intersection point of the lines $3x + 2y = 7$ and $2x + 4y = 8$, which is $(3/2, 5/4)$. So, the maximum value of $x + y$ is $3/2 + 5/4 = \boxed{11/4}$.

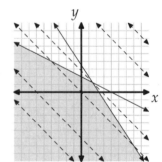

9.33   Initially, Snape's potion has 2000 mL, of which $0.30(2000) = 600$ mL is DeMuggle Juice. Every time he waves his wand, he adds 25 mL to the total potion of which $(0.75)(25) = 18.75$ mL is DeMuggle Juice. Suppose Snape waves his wand n times. Then, the potion is a total of $2000 + 25n$ mL, of which $600 + 18.75n$ is DeMuggle Juice. We want the potion to be at least 34% DeMuggle Juice, so we must have

$$\frac{600 + 18.75n}{2000 + 25n} \geq 0.34 \quad \Rightarrow \quad 600 + 18.75n \geq 0.34(2000 + 25n) \quad \Rightarrow \quad 600 + 18.75n \geq 680 + 8.5n$$

$$\Rightarrow \quad 10.25n \geq 80 \quad \Rightarrow \quad n \geq \frac{80}{10.25}.$$

Since $80/10.25 \approx 7.8$, we see that Snape must wave his wand at least $\boxed{8}$ times.

9.34 Suppose Betty buys f pounds of flour and s pounds of sugar. From the problem, we have $f \geq 6 + s/2$ and $f \leq 2s$. Putting these together, we have $2s \geq f \geq 6 + s/2$. The expression on the left end of this inequality chain must therefore be greater than or equal to the $6 + s/2$ on the right, so

$$2s \geq 6 + s/2 \quad \Rightarrow \quad 3s/2 \geq 6 \quad \Rightarrow \quad s \geq \boxed{4}.$$

9.35 We can square all three parts to isolate n, giving $81 < n < 10000$. Therefore, all the positive integers from 82 to 9999 satisfy the inequality. There are 9999 positive integers from 1 to 9999 (including 1 and 9999). We exclude the first 81 of these in our list of values of n that satisfy the inequality, giving us a total of $9999 - 81 = \boxed{9918}$ integers that satisfy the inequality chain.

9.36

(a) *Solution 1: How close is each to 1?* We have $13/17 = 1 - 4/17$ and $17/21 = 1 - 4/21$. We know how to compare $4/17$ and $4/21$. Because $17 < 21$, we can take reciprocals of both sides to give $1/17 > 1/21$ (note that we have to reverse the inequality sign). Multiplying both by -4 gives $-4/17 < -4/21$. Adding 1 to both sides gives $1 - 4/17 < 1 - 4/21$, so $13/17 < \boxed{17/21}$. (Basically, our tactic here is to note that $4/21 < 4/17$ because $1/21$ is a smaller part of 1 than $1/17$ is. So, $17/21$ is closer to 1 than $13/17$ is, which means $17/21$ is larger.)

Solution 2: Reciprocals. We first compare the reciprocals of the two numbers, $17/13$ and $21/17$. We have $17/13 = 1 + 4/13$ and $21/17 = 1 + 4/17$. Since $1/13 > 1/17$ (make sure you see why), we have $4/13 > 4/17$, so $17/13 > 21/17$. Taking the reciprocal of both sides gives $13/17 < \boxed{17/21}$. Notice that we reversed the inequality sign when we took the reciprocals of both sides.

(b) As in the second solution of part (a), we consider the reciprocals of the two numbers. We have $35/31 = 1 + 4/31$ and $41/37 = 1 + 4/37$. Because $4/31 > 4/37$, we have $35/31 > 41/37$. Taking the reciprocals of both sides of this inequality (and reversing the direction of the inequality sign) gives us $31/35 < \boxed{37/41}$.

(c) We use our first two parts as a guide:

$$\frac{100007}{100003} = 1 + \frac{4}{100003},$$

$$\frac{1000007}{1000003} = 1 + \frac{4}{1000003}.$$

So, we have

$$\frac{4}{100003} > \frac{4}{1000003} \quad \Rightarrow \quad \frac{100007}{100003} > \frac{1000007}{1000003} \quad \Rightarrow \quad \frac{100003}{100007} < \boxed{\frac{1000003}{1000007}}.$$

Challenge Problems

9.37 We start with the inequality on the left, $2r - 4 \leq r + 7$. Solving this inequality gives $r \leq 11$. We then solve the inequality on the right, $r + 7 < 3r - 15$, which gives $r > 11$. There are no values of r that satisfy both $r \leq 11$ and $r > 11$, so there are $\boxed{\text{no solutions}}$ to this inequality chain.

9.38 We isolate x^2 by subtracting 3 from both sides to get $x^2 < 9$. We can't simply take the square root of both sides and declare that $x < 3$, because we have to deal with negative numbers. If x is nonnegative, then we do indeed have $x < 3$. When x is negative, we must have $x > -3$. Combining these gives $\boxed{-3 < x < 3}$.

9.39 We notice that $\sqrt{82}$ is very close to $\sqrt{81} = 9$, so we try using this fact:

$$(\sqrt{82})^{23} > (\sqrt{81})^{23} = 9^{23} = (3^2)^{23} = 3^{46}.$$

Therefore, $\boxed{(\sqrt{82})^{23}}$ is larger.

9.40

(a) Suppose I watch x movies and buy y video games. The time information gives us $2x + 20y \leq 180$. Dividing by 2 gives $x + 10y \leq 90$. The money information gives $10x + 40y \leq 600$. Dividing this by 10 gives $x + 4y \leq 60$. We also must have $x \geq 0$ and $y \geq 0$. At right, we have our constraints graphed. (Notice the scale change on the x-axis.) The corner points are $(60, 0)$, $(40, 5)$, $(0, 9)$, and $(0, 0)$. We are asked to maximize $x + 3y$. Evaluating this expression for each of our corner points, we find that the maximum possible value is $60 + 3(0) = 60$, which occurs when I watch $\boxed{\text{60 movies}}$.

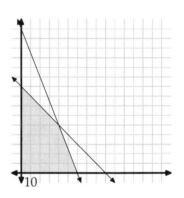

(b) Here, we want to maximize $x + 5y$. We have the same constraints as before, so we have the same corner points at which we must evaluate $x + 5y$. We find that the maximum occurs at the point $(40, 5)$, where we have $x + 5y = 40 + 5(5) = 65$. So, I must watch $\boxed{40}$ movies.

(c) Suppose I make k friends per video game, so that we must maximize $x + ky$. We still have the same constraints as before, so we must evaluate $x + ky$ at the same corner points. Clearly, $(0, 0)$ won't give us our maximum. For the other three points, we have

Point	Value of $x + ky$
$(60, 0)$	60
$(40, 5)$	$40 + 5k$
$(0, 9)$	$9k$

In order for it to make sense for me to just buy video games, $9k$ must be greater than or equal to the other two values of $x + ky$, since then it will be optimal for me to buy 9 video games and watch no movies. So, we must have $9k \geq 60$ and $9k \geq 40 + 5k$. The former gives $k \geq 20/3 = 6\frac{2}{3}$ and the latter gives $k \geq 10$. Therefore, I must make at least $\boxed{10}$ friends for each video game in order for it to make sense for me to only buy video games.

9.41 We seek the number of positive integer solutions to the inequality chain $6^4 < n^2 < 4^6$. Taking the square root of all three parts, we have $6^2 < n < 4^3$, so $36 < n < 64$. So, the solutions are the positive integers in the list 37, 38, ..., 62, 63. There are 63 integers from 1 to 63 and 36 from 1 to 36, so there are $63 - 36 = \boxed{27}$ from 37 to 63.

9.42 We can simplify $(a + b)/a$, so we think to take reciprocals of the whole chain. We reverse the inequality signs when we do so, and we have

$$\frac{2001}{4} > \frac{a+b}{a} > \frac{2001}{5} \quad \Rightarrow \quad \frac{2001}{4} > 1 + \frac{b}{a} > \frac{2001}{5} \quad \Rightarrow \quad \frac{1997}{4} > \frac{b}{a} > \frac{1996}{5}.$$

Therefore, we have $499\frac{1}{4} > \frac{b}{a} > 399\frac{1}{5}$, so $\frac{b}{a}$ could be any of the $\boxed{100}$ integers from 400 to 499 (including both 400 and 499).

9.43 First, we write our list with all the exponents evaluated:

$$2^{81}, 2^{64}, 3^{16}, 3^{16}, 4^8, 4^9.$$

We can throw out 2^{64} and 4^8. We can quickly compare 2^{81} to 4^9 by writing 4^9 with 2 as the base: $4^9 = (2^2)^9 = 2^{18}$. So, $2^{81} > 2^{18}$. All we have left is to compare 2^{81} to 3^{16}. Since $2^{81} > 2^{80} = 4^{40} > 3^{40} > 3^{16}$, we see that $\boxed{2^{3^4}}$ is the largest number in the list.

9.44 We start by graphing the region described by the restrictions. We note that two of our inequalities are nonstrict, so we use dashed lines for those two lines. We might be tempted to say that our corners are $(0,0)$, $(4,0)$, $(2,4)$, and $(0,5)$, and find the maximum value of $x + y$ for these four points. That maximum turns out to be $2 + 4 = 6$. However, the point $(2,4)$ does not satisfy the conditions of all the inequalities! It's on the dashed border, so it is not part of the region that satisfies all of the inequalities.

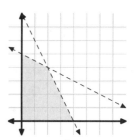

The line $x + y = 6$ passes through $(2,4)$, and doesn't pass through any point in the shaded region. If we move this line at all to the left, we will pass through a point in the shaded region. So, $x + y$ can equal values that are very, very close to 6, such as 5.9999999. In fact, $x + y$ can be as close to 6 as we want, but it can never actually equal 6. Therefore, while $x + y$ can never be 6 or greater, it has $\boxed{\text{no maximum value}}$, because there is not a value $x + y$ can actually equal but never exceed. (Remember, $x + y$ cannot equal 6.)

9.45 Because $a + b = c$ and $c + d = a$, we have $a + b + d = c + d = a$, so $b + d = 0$, which means $b = -d$. Substituting this into $b + c = d$, we have $-d + c = d$, so $c = 2d = -2b$. Putting $c = -2b$ into $a + b = c$ gives us $a + b = -2b$, which means $a = -3b$. Therefore, we have $a + b + c + d = -3b + b - 2b - b = -5b$. If b is a positive integer, the largest $-5b$ can be is $-5(1) = \boxed{-5}$.

9.46 Since we must "compute the smallest value of x," we want to create an inequality of the form

$$x \geq \text{(some number)}.$$

Unfortunately, all we have is an equation for x. We are, however, given an inequality for w: $w \geq 0$. We can substitute the given equation for w into this inequality to get

$$3z - 2001 \geq 0.$$

Solving this inequality gives $z \geq 667$. Again we can substitute, this time with $z = 2y - 2001$:

$$2y - 2001 \geq 667.$$

Solving this gives $y \geq 1334$. Finally, we substitute $y = x - 2001$ to get $x - 2001 \geq 1334$, from which we have $x \geq 3335$. Therefore, $\boxed{3335}$ is the smallest possible value of x.

9.47 We can't simply multiply both sides by $3x + 5$ to get $4x - 5 \geq 3(3x + 5)$ because if $3x + 5$ is negative, we must reverse the direction of the inequality sign. So, we break the problem into two cases.

Case 1: $3x + 5 > 0$. When $3x + 5$ is positive, we can multiply both sides by $3x + 5$ without changing the direction of the inequality sign to get

$$4x - 5 \geq 3(3x + 5) \quad \Rightarrow \quad 4x - 5 \geq 9x + 15 \quad \Rightarrow \quad -20 \geq 5x \quad \Rightarrow \quad -4 \geq x.$$

All values of x such that $x \leq -4$ make $3x + 5$ negative. This case is only valid when $3x + 5$ is positive, so there are no solutions in this case.

Case 2: $3x + 5 < 0$. When $3x + 5$ is negative, we must reverse the inequality sign when we multiply both sides by $3x + 5$:

$$4x - 5 \leq 3(3x + 5) \quad \Rightarrow \quad 4x - 5 \leq 9x + 15 \quad \Rightarrow \quad -20 \leq 5x \quad \Rightarrow \quad -4 \leq x.$$

The values of x for which $3x + 5 < 0$ are $x < -5/3$. We have just shown that all such x that also satisfy $x \geq -4$ satisfy the inequality $(4x - 5)/(3x + 5) \geq 3$. So, the values of x that satisfy $(4x - 5)/(3x + 5) \geq 3$ are $\boxed{-4 \leq x < -5/3}$.

We can also tackle the problem by bringing all terms to the left and finding a common denominator. Subtracting 3 from both sides gives

$$\frac{4x - 5}{3x + 5} - 3 \geq 0.$$

Writing the left side with a common denominator gives

$$\frac{4x - 5}{3x + 5} - 3 = \frac{4x - 5}{3x + 5} - \frac{3(3x + 5)}{3x + 5} = \frac{-5x - 20}{3x + 5}.$$

So, our inequality is now

$$\frac{-5x - 20}{3x + 5} \geq 0.$$

The left side equals zero only when $-5x - 20 = 0$, which gives $x = -4$. The quotient $(-5x - 20)/(3x + 5)$ is positive only if both $-5x - 20$ and $3x + 5$ are both positive or if they're both negative. The expression $-5x - 20$ is positive for $x < -4$ and negative for $x > -4$, and $3x + 5$ is positive for $x > -5/3$ and negative for $x < -5/3$. So, both expressions are negative when $-4 < x < -5/3$, and there are no values of x that make both expressions positive. Combining $-4 < x < -5/3$ with the solution $x = -4$, we see that the values of x that satisfy our inequality are $\boxed{-4 \leq x < -5/3}$.

9.48 If we increase M by $p\%$, we get $M\left(1 + \frac{p}{100}\right)$. If we then decrease this number by $q\%$, we get $M\left(1 + \frac{p}{100}\right)\left(1 - \frac{q}{100}\right)$. We are asked to determine when we have

$$M\left(1 + \frac{p}{100}\right)\left(1 - \frac{q}{100}\right) > M.$$

We can divide both sides by M to get

$$\left(1 + \frac{p}{100}\right)\left(1 - \frac{q}{100}\right) > 1.$$

Looking at our choices, we see that we must isolate p. We divide both sides by $\left(1 - \frac{q}{100}\right)$ and put on our algebraic manipulation hats:

$$1 + \frac{p}{100} > \frac{1}{1 - \frac{q}{100}} \quad \Rightarrow \quad 1 + \frac{p}{100} > \frac{1}{\frac{100 - q}{100}} \quad \Rightarrow \quad 1 + \frac{p}{100} > \frac{100}{100 - q}$$

$$\Rightarrow \quad \frac{p}{100} > \frac{100}{100 - q} - 1 = \frac{100}{100 - q} - \frac{100 - q}{100 - q} = \frac{q}{100 - q}.$$

Multiplying by 100 gives $p > 100q/(100 - q)$, which is choice $\boxed{\text{(E)}}$.

9.49 Since 63 and 33 are close to powers of 2, we try comparing 64^{45} to 32^{54}. We can write these both as powers of 2:

$$64^{45} = (2^6)^{45} = 2^{270},$$
$$32^{54} = (2^5)^{54} = 2^{270}.$$

We see that $64^{45} = 32^{54}$. Fortunately, one of the numbers we must compare is larger than this, and the other is smaller than it is: $33^{54} > 32^{54} = 64^{45} > 63^{45}$, so $\boxed{33^{54}}$ is larger than 63^{45}.

9.50 Suppose they sell tickets to x teachers and y students. We have $x + y \le 300$ because there are only 300 tickets. Since there must be at least 1 teacher for every 5 students, we must have $x \ge y/5$. Similarly, since we must have at least twice as many students as teachers, we have $x \le y/2$. Finally, the teachers on the bus will take $x/2$ seats and the students will take $y/3$ seats, so we must have $x/2 + y/3 \le 110$. So, our constraints are

$$x + y \le 300, \qquad x \ge \frac{y}{5}, \qquad x \le \frac{y}{2}, \qquad \frac{x}{2} + \frac{y}{3} \le 110.$$

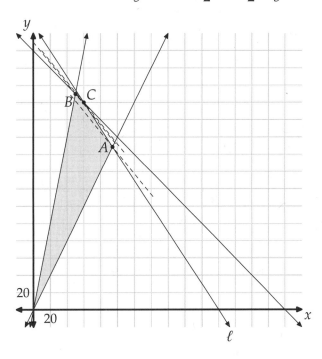

We must maximize $6x + 5y$, the revenue in dollars for the tickets. Rather than finding all four of the boundary points and evaluating $6x + 5y$ at each, we can use our understanding of slope to know which of the boundary points is the one that will maximize $6x + 5y$.

Portions of the lines of the form $6x + 5y = k$ through boundary points A, B, and C are dashed in the diagram. They're very close together! They all have slope $-6/5$. We want the one that is farthest from the origin, since this line will give us the largest value of k. It looks like the line through C is the farthest out. A quick consideration of the slopes of the lines in the diagram tells us this is correct.

The line $x/2 + y/3 = 110$, line ℓ in the diagram, has slope $-(1/2)/(1/3) = -3/2$. Therefore, line ℓ is steeper than any line of the form $6x + 5y = k$. So, as we go from right to left, the dashed line with slope $-6/5$ will go upward more slowly than line ℓ, the boundary of the shaded region. This tells us that the dashed line through A will go inside the shaded region, and therefore be below C (and the dashed line that goes through C).

Similarly, if we consider the corner point B, we see that it is along the line $x + y = 300$, which has slope -1. Since the lines of the form $6x + 5y = k$ have slope $-6/5$, these lines are steeper than $x + y = 300$, so the line of the form $6x + 5y = k$ that passes through B will go downward faster than the line $x + y = 300$. So, this line will enter the shaded region and therefore be beneath C.

Therefore, it is the line of the form $6x + 5y = k$ that passes through C that will give us our maximum value of k. Point C is the intersection point of $x + y = 300$ and $x/2 + y/3 = 110$, which is $(60, 240)$. So, they should sell tickets to $\boxed{60}$ teachers.

9.51 We don't know how to deal with square roots as exponents, but we do know how to deal with fractions as exponents. We try to find two simple fractions such that $3^{a/b} < 3^{\sqrt{3}} < 3^{c/d}$, and then hopefully we'll be able to show that $3^{a/b}$ and $3^{c/d}$ are between the same two consecutive integers.

Our first candidates for a/b and c/d might be $3/2$ and $7/4$, because $(3/2)^2 = 9/4 < 3 < 49/16 = (7/4)^2$, so $3/2 < \sqrt{3} < 7/4$, which means $3^{3/2} < 3^{\sqrt{3}} < 3^{7/4}$.

We must find the two integers that $3^{3/2}$ is between. Suppose that j is an integer such that $j < 3^{3/2} < j+1$. Squaring all three of these gives $j^2 < 3^3 < (j + 1)^2$, or $j^2 < 27 < (j + 1)^2$, so clearly $j = 5$, which tells us that $3^{3/2}$ is between 5 and 6. Now, if we can show that $3^{7/4}$ is also between 5 and 6, we're finished.

We find the two consecutive integers that $3^{7/4}$ is between in the same way. Let k be a positive integer such that $k < 3^{7/4} < k + 1$. We raise all three to the fourth power to get $k^4 < 3^7 < (k + 1)^4$, so $k^4 < 2187 < (k + 1)^4$. Uh-oh; $6^4 = 1296$ and $7^4 = 2401$, so $3^{7/4}$ is between 6 and 7. We're not finished. We'll have to find a narrower range that $3^{\sqrt{3}}$ is in.

Since $5/3 > 3/2$, we instead try $3^{5/3} < 3^{\sqrt{3}} < 3^{7/4}$, because $(5/3)^2 = 25/9 < 3$, so $5/3 < \sqrt{3}$. We already know that $3^{7/4}$ is between 6 and 7, so we focus on $3^{5/3}$.

As before, we let m be the positive integer such that $m < 3^{5/3} < m + 1$. Cubing all three gives $m^3 < 3^5 < (m + 1)^3$, so $m^3 < 243 < (m + 1)^3$. Aha! Since $6^3 = 216$ and $7^3 = 343$, we know that $m = 6$, so $3^{5/3}$ is also between 6 and 7. Therefore, we have $6 < 3^{5/3} < 3^{\sqrt{3}} < 3^{7/4} < 7$, so the greatest integer less than $3^{\sqrt{3}}$ is $\boxed{6}$.

10

Quadratic Equations – Part 1

Exercises for Section 10.1

10.1.1 We isolate x^2 by multiplying both sides by 3, to give $x^2 = 900$. Taking the square root of both sides gives $x = \pm \sqrt{900} = \pm 30$. Therefore, the values of x that satisfy the equation are $\boxed{30 \text{ and } -30}$.

10.1.2 Adding $18r$ to both sides gives $r^2 + 18r = 0$. Factoring an r out of the left side gives $r(r + 18) = 0$. The expression $r(r + 18)$ equals 0 when $r = 0$ or when $r + 18 = 0$. In the latter case, $r = -18$, so the values of r that satisfy the equation are $\boxed{0 \text{ and } -18}$.

10.1.3

(a) $(x - 7)(x + 2) = x(x + 2) - 7(x + 2) = x^2 + 2x - 7x - 14 = \boxed{x^2 - 5x - 14}$. The expression $(x - 7)(x + 2)$ equals 0 when $x - 7 = 0$ or $x + 2 = 0$. So, the values of x for which the expression equals 0 are $\boxed{7 \text{ and } -2}$.

(b) $(x - 4)(x + 4) = x(x + 4) - 4(x + 4) = x^2 + 4x - 4x - 16 = \boxed{x^2 - 16}$. The expression $(x - 4)(x + 4)$ equals 0 when $x - 4 = 0$ or $x + 4 = 0$. So, the values of x for which the expression equals 0 are $\boxed{4 \text{ and } -4}$.

(c) $\left(x - \dfrac{2}{5}\right)(x + 5) = x(x + 5) - \dfrac{2}{5}(x + 5) = x^2 + 5x - \dfrac{2}{5}x - 2 = \boxed{x^2 + \dfrac{23}{5}x - 2}$. The expression $(x - \frac{2}{5})(x + 5)$ equals 0 when $x - \frac{2}{5} = 0$ or $x + 5 = 0$. So, the values of x for which the expression equals 0 are $\boxed{\dfrac{2}{5} \text{ and } -5}$.

(d) $(x - 8)(x - 8) = x(x - 8) - 8(x - 8) = x^2 - 8x - 8x + 64 = \boxed{x^2 - 16x + 64}$. The expression $(x - 8)(x - 8)$ equals 0 only when $x - 8 = 0$. Therefore, the only value of x that makes the expression 0 is $x = \boxed{8}$.

10.1.4 Let my number be x. 100 times my number is $100x$, and the square of my number divided by 4 is $x^2/4$, so we must have $x^2/4 = 100x$. Multiplying both sides by 4 gives $x^2 = 400x$, and subtracting $400x$ from both sides gives $x^2 - 400x = 0$. Factoring out x from the left side gives $x(x - 400) = 0$, so either $x = 0$ or $x - 400 = 0$. In the latter case we have $x = 400$, so the possible values of my number are $\boxed{0 \text{ and } 400}$.

10.1.5 The expression $x - 3$ equals zero when $x = 3$ and the expression $x + 5$ equals zero when $x = -5$. Therefore, the expression $\boxed{(x - 3)(x + 5)}$ equals 0 when x equals 3 or -5.

10.1.6 By considering each of the terms in the product separately, we see that $(x-2)(x-5)(x+3) = 0$ when $x = \boxed{2}$, $\boxed{5}$, or $\boxed{-3}$. To expand $(x-2)(x-5)(x+3)$, we first expand $(x-2)(x-5)$:

$$(x-2)(x-5) = x(x-5) - 2(x-5) = x^2 - 5x - 2x + 10 = x^2 - 7x + 10.$$

Now our product is $(x^2 - 7x + 10)(x+3)$, which we can expand further by carefully distributing:

$$(x^2 - 7x + 10)(x+3) = (x^2 - 7x + 10)(x) + (x^2 - 7x + 10)(3)$$
$$= x^3 - 7x^2 + 10x + 3x^2 - 21x + 30$$
$$= \boxed{x^3 - 4x^2 - 11x + 30}.$$

Exercises for Section 10.2

10.2.1

(a) $r^2 - 9r + 18 = (r-3)(r-6)$, so we have $(r-3)(r-6) = 0$. Therefore, $r - 3 = 0$ or $r - 6 = 0$, so the two solutions for r are $\boxed{3 \text{ and } 6}$.

(b) We first get all the terms on the left side by subtracting 14 from both sides: $t^2 - 5t - 14 = 0$. Since $t^2 - 5t - 14 = (t-7)(t+2)$, we have $(t-7)(t+2) = 0$. So, our solutions are $\boxed{t = 7 \text{ and } t = -2}$.

(c) Bringing all the terms to the left gives $x^2 - 25x + 144 = 0$. We're looking for two numbers whose product is 144 and sum is -25; -9 and -16 are the numbers we want: $x^2 - 25x + 144 = (x-9)(x-16)$. Since $(x-9)(x-16) = 0$, our solutions are $\boxed{x = 9 \text{ and } x = 16}$.

(d) We isolate y^2 by adding y^2 to both sides, giving $y^2 = 200$. Taking the square root of both sides gives $y = \pm \sqrt{200} = \boxed{\pm 10\sqrt{2}}$.

(e) First we notice that all the coefficients are divisible by 3, so we can simplify the equation by dividing both sides by 3. We also write the terms on the left side starting with x^2, which gives $x^2 + 18x + 81 = 0$. Because $x^2 + 18x + 81 = (x+9)^2$, we have $(x+9)^2 = 0$. Therefore, the only value of x that satisfies the equation is $x = \boxed{-9}$.

(f) We might think to multiply through by 2 or 10 to get rid of the decimals, but that would give us a coefficient of t^2 that isn't 1, and we aren't so sure how to tackle that. So, instead, we try to factor $t^2 - 4.5t + 3.5$. We seek two numbers whose product is 3.5 and whose sum is -4.5. Clearly both numbers must be negative (positive product, negative sum), so we try the simplest possibility: -1 and -3.5. It works: $-1 - 3.5 = -4.5$ and $(-1)(-3.5) = 3.5$. So, $t^2 - 4.5t + 3.5 = (t-1)(t-3.5) = 0$. Therefore, our solutions are $\boxed{t = 1 \text{ and } t = 3.5}$.

10.2.2 As we saw in the text, if the coefficient of the quadratic term is 1, then the sum of the roots is the negative of the coefficient of the linear term and the product of the roots equals the constant term. Therefore, the sum of the roots of $x^2 + 15x - 324 = 0$ is $\boxed{-15}$ and the product is $\boxed{-324}$. (We could also have found the roots: $x^2 + 15x - 324 = (x+27)(x-12)$, so the roots are -27 and 12.)

10.2.3 To factor $x^2 - b^2$, we must find two numbers whose sum is 0 and whose product is $-b^2$. The obvious candidates are b and $-b$, and we have $x^2 - b^2 = \boxed{(x-b)(x+b)}$. This factorization is often called the difference of squares factorization, and it can be enormously useful in algebra problems. You'll see some examples later in the text.

10.2.4

(a) We let the coefficient of the quadratic term be 1. So, the coefficient of the linear term is the negative of the sum of the roots, or 3, and the constant term equals the product of the roots, −40. So, our quadratic is $\boxed{x^2 + 3x - 40}$.

(b) Factoring the quadratic gives us $x^2 + 3x - 40 = (x + 8)(x - 5)$, so the roots are $\boxed{-8 \text{ and } 5}$.

We also could have found the roots without bothering with the quadratic. Let the roots be r and s, so we are told $r + s = -3$ and $rs = -40$. This is basically the same number guessing game that factoring is. We find that r and s are −8 and 5 after a little experimentation.

10.2.5 We can factor after bringing all the terms to one side by subtracting $(4r - 10)(r^2 + 5r - 24)$ from both sides:

$$(r^2 + 5r - 24)(r^2 - 3r + 2) - (4r - 10)(r^2 + 5r - 24) = [(r^2 - 3r + 2) - (4r - 10)](r^2 + 5r - 24)$$
$$= (r^2 - 3r + 2 - 4r + 10)(r^2 + 5r - 24)$$
$$= (r^2 - 7r + 12)(r^2 + 5r - 24).$$

So, now our equation is $(r^2 - 7r + 12)(r^2 + 5r - 24) = 0$. We can factor both of these quadratics:

$$(r - 3)(r - 4)(r + 8)(r - 3) = 0.$$

The expression on the left only equals 0 when any one of the factors on the left equals 0. One of these factors, $r - 3$, is repeated, so there are only three distinct values of r that make the expression 0. These correspond to $r - 3 = 0$, $r - 4 = 0$, and $r + 8 = 0$, so our values of r that satisfy the original equation are $r = \boxed{3, 4, \text{ and } -8}$.

Exercises for Section 10.3

10.3.1

(a) We have $10x^2 - 11x + 1 = (10x - 1)(x - 1) = 0$, so we must have $10x - 1 = 0$ or $x - 1 = 0$. Therefore, our solutions are $\boxed{x = 1/10 \text{ and } x = 1}$.

(b) Moving all terms to the left, we have $11r^2 - 15r + 4 = 0$. Factoring the quadratic, we find $(11r - 4)(r - 1) = 0$, so our solutions are $\boxed{r = 4/11 \text{ and } r = 1}$.

(c) We collect all terms on one side and have $3v^2 + 34v - 24 = 0$. Factoring, we find $(3v - 2)(v + 12) = 0$, so $3v - 2 = 0$ or $v + 12 = 0$. Therefore, our solutions are $\boxed{v = 2/3 \text{ and } v = -12}$.

(d) First we collect all terms on one side to find $15x^2 + 16x + 4 = 0$. Then we factor to find $(5x + 2)(3x + 2) = 0$, so $5x + 2 = 0$ or $3x + 2 = 0$. So, our solutions are $\boxed{x = -\dfrac{2}{5} \text{ and } x = -\dfrac{2}{3}}$.

(e) We first multiply by 9 to get rid of all the fractions. This makes our equation $9t^2 + 6t + 1 = 0$. Factoring the left side gives us $(3t + 1)(3t + 1) = 0$, so our only solution occurs when $3t + 1 = 0$. So, the only solution is $\boxed{t = -1/3}$.

(f) All the coefficients are divisible by 3, so we start by dividing by 3 to get $15 - 14z - 8z^2 = 0$. Rewriting the quadratic to have a positive coefficient of z^2 (by multiplying by -1), we have $8z^2 + 14z - 15 = 0$. Factoring the left side gives us $(4z - 3)(2z + 5) = 0$, so $4z - 3 = 0$ or $2z + 5 = 0$. Therefore, our solutions are $\boxed{z = 3/4 \text{ and } z = -5/2}$.

10.3.2 This expression equals 0 when $ax + b = 0$ or $cx + d = 0$. We subtract b from both sides of $ax + b = 0$ to find $ax = -b$. Dividing both sides by a gives $x = -b/a$. Similarly, we subtract d from both sides of $cx + d = 0$ to get $cx = -d$, and dividing both sides by c yields $x = -d/c$. So, our solutions are $\boxed{-b/a \text{ and } -d/c}$.

10.3.3 The only difference between the two quadratics is the sign of the coefficient of the linear term. We know that $(6x + 7)(2x + 15) = 12x^2 + 104x + 105$. We wish to alter the binomials so that only the sign of the linear term in the quadratic changes. We can do so by simply switching $+7$ and $+15$ in the binomials to -7 and -15. This will leave the coefficient of x^2 and the constant in the quadratic unchanged, and it will just switch the sign of the coefficient of the linear term from positive to negative. If you aren't convinced, consider these two expansions:

$$(6x + 7)(2x + 15) = 6x(2x + 15) + 7(2x + 15) = 12x^2 + 90x + 14x + 105 = 12x^2 + 104x + 105,$$
$$(6x - 7)(2x - 15) = 6x(2x - 15) - 7(2x - 15) = 12x^2 - 90x - 14x + 105 = 12x^2 - 104x + 105.$$

All that changing the signs of the constants in the binomials does is change the sign of the coefficient of x in the quadratic!

So, we have $12x^2 - 104x + 105 = (6x - 7)(2x - 15) = 0$, which means $6x - 7 = 0$ or $2x - 15 = 0$. Therefore, our solutions are $\boxed{x = 7/6 \text{ and } x = 15/2}$.

10.3.4 When we factor $3x^2 + nx + 72$, our two factors are of the form $(3x + A)(x + B)$, where A and B are integers. We must have $AB = 72$, and we want $3B + A$ to be as large as possible (because $3B + A$ is the coefficient of x when $(3x + A)(x + B)$ is expanded). We make $3B + A$ as large as possible by letting $B = 72$ and $A = 1$; any other possibility reduces $3B$ much more than A increases. Therefore, the largest possible value of n is $3B + A = 3(72) + 1 = \boxed{217}$.

10.3.5

(a) $\boxed{\text{No}}$, we don't have to try them both. If we find a factorization of the form $(3x + C)(6x + D)$, then we can multiply both factors by -1 (thus multiplying the whole product by 1 and thereby leaving it unchanged) to have a factorization of the form $(-3x - C)(-6x - D)$, that is, a factorization with -3 and -6 as the coefficients of the linear terms in the binomials.

(b) We can factor $18x^2 + 21x + 5$ as $(3x + 1)(6x + 5)$ or as $(-3x - 1)(-6x - 5)$. These two expressions are equal. We can factor -1 out of both of the factors in $(-3x - 1)(-6x - 5)$ to find:

$$(-3x - 1)(-6x - 5) = [(-1)(3x + 1)][(-1)(6x + 5)] = (-1)(-1)(3x + 1)(6x + 5) = (3x + 1)(6x + 5).$$

10.3.6 The $\sqrt{3}$ in the coefficient of the linear term is a big hint. We can introduce terms with $\sqrt{3}$ in them by noticing that $9 = (\sqrt{3})(3\sqrt{3})$. With this as a guide, we factor the quadratic:

$$2x^2 + 7x\sqrt{3} + 9 = (2x + \sqrt{3})(x + 3\sqrt{3}).$$

Therefore, our quadratic equals 0 when $2x + \sqrt{3} = 0$ or $x + 3\sqrt{3} = 0$. Solving these two equations gives us the solutions $\boxed{x = -\sqrt{3}/2 \text{ and } x = -3\sqrt{3}}$.

Exercises for Section 10.4

10.4.1 If you solved this problem by finding the solutions to the equation, go back and read the section again. The sum of the roots is $-b/a$, where b is the coefficient of the linear term and a is the coefficient of the quadratic term. So, the desired sum is $-(84)/(-32) = \boxed{21/8}$.

10.4.2 As discovered in the text, the product of the roots equals the constant term divided by the coefficient of the quadratic term, or $(-500)/18 = \boxed{-250/9}$.

10.4.3

(a) Let the coefficient of the quadratic term be 1. The coefficient of the linear term then is the negative of the sum of the roots, or 1 in this case. The constant term of the quadratic equals the product of the roots, $-15/4$. So, our quadratic is $\boxed{x^2 + x - 15/4}$.

(b) Setting our quadratic equal to 0, we have $x^2 + x - 15/4 = 0$. Multiplying by 4 to get rid of the fractions gives $4x^2 + 4x - 15 = 0$. Factoring the left side gives us $(2x - 3)(2x + 5) = 0$, so our roots are $\boxed{x = 3/2 \text{ and } x = -5/2}$.

10.4.4 First we bring all the terms to one side, which gives $30z^2 - 7z - 88 = 0$. We could go ahead and try to find the roots of this equation, but that looks daunting. Instead, we expand the product we wish to find:

$$(r + 3)(s + 3) = r(s + 3) + 3(s + 3) = rs + 3r + 3s + 9 = rs + 3(r + s) + 9.$$

We know how to find rs and $r + s$. We find that $rs = -88/30$ and $r + s = -(-7)/30 = 7/30$ from the coefficients of our quadratic. Therefore, we have

$$(r + 3)(s + 3) = rs + 3(r + s) + 9 = -\frac{88}{30} + 3 \cdot \frac{7}{30} + 9 = \boxed{\frac{203}{30}}.$$

Notice that if we let $z = x - 3$ in our quadratic equation $30z^2 - 7z - 88 = 0$, the coefficient of x^2 will be 30 and the constant term will be 203, so the product of the roots of the new quadratic is $203/30$. Is this a coincidence? What are the roots of the new quadratic in terms of r and s?

10.4.5 We could approach this by choosing a coefficient of x^2 that will let us produce both fractions, such as 72. However, we know we won't make an error if we stick with the simple steps we already know. Let the coefficient of the quadratic term be 1. Then, the coefficient of the linear term is $-55/72$ and the constant term is $-25/12$. So, our quadratic equation is

$$x^2 - \frac{55}{72}x - \frac{25}{12} = 0.$$

We multiply by 72 to clear out the fractions, and we get $72x^2 - 55x - 150 = 0$. (You might have jumped straight to this quadratic from the original question.) Now we factor. Because the coefficient of x is odd, we have to multiply an odd factor of 72 by an odd factor of 150. This helps us find our factorization: $(9x + 10)(8x - 15) = 0$. From this, we find the roots $\boxed{x = -10/9 \text{ and } x = 15/8}$.

10.4.6 The sum of the roots equals the negative of the coefficient of x, so we have $m+n = -m$. Therefore, $-2m = n$. The product of the roots equals the constant term, so $mn = n$. Because $n \neq 0$, we can divide this equation by n to get $m = 1$. We can now find n, or note that the sum of the roots equals $-m$, so $m + n = -m = \boxed{-1}$.

10.4.7 If we expand the left side, we have

$$(x - p)(x - q) = x(x - q) - p(x - q) = x^2 - qx - px + pq = x^2 - (p + q)x + pq.$$

The other side of the equation is a constant, since there isn't an x term. So, if we view the equation as a quadratic in x, the sum of the roots is $-[-(p + q)] = p + q$. We know that one of the roots is r, so if the other is s, we have $r + s = p + q$, so $s = \boxed{p + q - r}$.

Exercises for Section 10.5

10.5.1 First, we try $k = 5$:
$$\frac{x - 1}{x - 2} = \frac{x - 5}{x - 6}.$$

Cross-multiplying gives $(x-1)(x-6) = (x-2)(x-5)$. Expanding both sides gives $x^2 - 7x + 6 = x^2 - 7x + 10$. Subtracting $x^2 - 7x$ from both sides gives us $6 = 10$, which clearly has no solutions. So, $k = 5$ clearly gives us no solutions for x.

Next, we try $k = 6$:
$$\frac{x - 1}{x - 2} = \frac{x - 6}{x - 6}.$$

The right side is just 1 (except when $x = 6$, which we cannot allow). So, our equation is $(x - 1)/(x - 2) = 1$. Multiplying both sides by $x - 2$ gives $x - 1 = x - 2$, and subtracting x from both sides of this equation gives $-1 = -2$, which also clearly has no solutions. So, $k = 6$ gives us no solutions for x, either.

10.5.2 We first square both sides to get rid of the square root sign. This gives $x + 1$ on the left and

$$(1 - x)^2 = (1 - x)(1 - x) = 1(1 - x) - x(1 - x) = 1 - x - x + x^2 = 1 - 2x + x^2$$

on the right. So, our equation is $x+1 = 1-2x+x^2$. Putting all terms on the same side, we have $x^2 - 3x = 0$. Factoring the quadratic gives $x(x - 3) = 0$, so $x = 0$ or $x = 3$.

Because we squared the equation as a step, we must check for extraneous solutions. When $x = 0$, we have $\sqrt{0 + 1} = 1 - 0$, which is true. When $x = 3$, we have $\sqrt{3 + 1} = 1 - 3$, which is not true. So, $x = 3$ is an extraneous solution and we must discard it. The only solution is $\boxed{x = 0}$.

10.5.3 It's a quadratic in disguise! Let $x = t^2$. Then our equation becomes $x^2 - 11x + 18 = 0$, or $(x - 2)(x - 9) = 0$. So, our solutions are $x = 2$ and $x = 9$. But we want t, not x. Since $t^2 = x$, when $x = 2$, we have $t^2 = 2$, which gives solutions $t = \pm\sqrt{2}$. When $x = 9$, we have $t^2 = 9$, which gives solutions $t = 3$ and $t = -3$. So, we have four solutions: $\boxed{\sqrt{2}, -\sqrt{2}, 3, \text{ and } -3}$.

10.5.4 Let the square originally have side length s, so that its area is s^2. After the side length is increased by 12, the side length is $s + 12$, so the area is $(s + 12)^2$. The difference in these areas is 200, so we have

$$(s + 12)^2 - s^2 = 200.$$

Simplifying the left side gives $(s+12)^2 - s^2 = s^2 + 24s + 144 - s^2 = 24s + 144$, so our equation is $24s + 144 = 200$. Therefore, we have $24s = 56$, so $s = \boxed{7/3 \text{ inches}}$.

10.5.5 Solving $2ab = -30$ for b gives us $b = -15/a$. Substituting this into our first equation gives

$$a^2 + 2\left(\frac{225}{a^2}\right) = 43.$$

Multiplying both sides of this equation by a^2 to get rid of the fraction gives $a^4 + 450 = 43a^2$. Bringing all the terms to one side gives $a^4 - 43a^2 + 450 = 0$. Letting $x = a^2$ gives us $x^2 - 43x + 450 = 0$. The quadratic factors as $(x - 25)(x - 18) = 0$, so we have $x = 25$ and $x = 18$.

When $x = 25$, we find $a^2 = 25$, so $a = \pm 5$. When $a = 5$, we have $b = -15/a = -3$. Similarly, when $a = -5$, we have $b = -15/a = 3$. So, $\boxed{(5, -3)}$ and $\boxed{(-5, 3)}$ are solutions to the system.

When $x = 18$, we find $a^2 = 18$, so $a = \pm 3\sqrt{2}$. When $a = 3\sqrt{2}$, we find $b = -15/a = -5/\sqrt{2} = -5\sqrt{2}/2$. Similarly, when $a = -3\sqrt{2}$, we find $b = 5\sqrt{2}/2$. So, $\boxed{(3\sqrt{2}, -5\sqrt{2}/2)}$ and $\boxed{(-3\sqrt{2}, 5\sqrt{2}/2)}$ are also solutions to the system of equations. Notice that in the text, we never identified or addressed this set of solutions to the system of equations. Why didn't we? (Let a and b have these values in the expression $a + b\sqrt{2}$ and you should see why!)

10.5.6 Let the population in 1996 be x. We are told that the expression

$$\frac{(\text{Population in year } n + 2) - (\text{Population in year } n)}{\text{Population in year } n + 1}$$

is constant. So, taking n to be the first year (1994), then taking n to be the second year (1995), we can evaluate our expression for two different years and set the results equal:

$$\frac{(\text{Pop. in year 1996}) - (\text{Pop. in year 1994})}{\text{Pop. in year 1995}} = \frac{(\text{Pop. in year 1997}) - (\text{Pop. in year 1995})}{\text{Pop. in year 1996}}.$$

In order, the populations for our 4 years are 39, 60, x, and 123, so our equation is

$$\frac{x - 39}{60} = \frac{123 - 60}{x} = \frac{63}{x}.$$

Cross-multiplying gives $x(x - 39) = 60(63)$, or $x^2 - 39x - 3780 = 0$. We can factor this by looking back at where the 3780 came from: $(60)(63)$. We need the factors of 3780 to be 39 apart, so we need them to be more than 3 apart. We try multiplying 60 by 3/4 and 63 by 4/3 (thus keeping the product of the factors constant), and we have $(60)(63) = (45)(84)$. Aha! We have

$$x^2 - 39x - 3780 = (x - 84)(x + 45).$$

Clearly, population must be positive, so we take the positive root. The population in 1996 is $\boxed{84}$.

Review Problems

10.31

(a) Factoring the left side of $r^2 - 7r = 0$ gives $r(r - 7) = 0$, so our solutions are $\boxed{r = 0 \text{ and } r = 7}$.

(b) Rearranging $x^2 + 3x = 7x - x^2$ gives $2x^2 - 4x = 0$, dividing this by 2 gives $x^2 - 2x = 0$, and factoring gives $x(x - 2) = 0$. So, our solutions are $\boxed{x = 0, 2}$.

(c) $2t^2 = 242 \implies t^2 = 121 \implies t = \pm\sqrt{121} = \boxed{\pm 11}$.

(d) $16 - y^2 = -4 \implies y^2 = 20 \implies y = \pm\sqrt{20} = \boxed{\pm 2\sqrt{5}}$.

10.32

(a) Expanding the product, we have $(x - 6)(x) = \boxed{x^2 - 6x}$. We have $(x - 6)(x) = 0$ when $x - 6 = 0$ or $x = 0$, so $\boxed{x = 6 \text{ and } x = 0}$ make $(x - 6)(x)$ equal to zero.

(b) Expanding the product, we find $(8-x)(x-8) = 8(x-8)-x(x-8) = 8x-64-x^2+8x = \boxed{-x^2 + 16x - 64}$. The product $(8 - x)(x - 8)$ equals 0 when $8 - x = 0$ or $x - 8 = 0$. Both of these give us $x = \boxed{8}$.

(c) $(x-7)(x+2) = x(x+2)-7(x+2) = x^2+2x-7x-14 = \boxed{x^2 - 5x - 14}$. The solutions to $(x-7)(x+2) = 0$ are $\boxed{x = 7 \text{ and } x = -2}$.

(d) $(x + 8)(x - 8) = x(x - 8) + 8(x - 8) = x^2 - 8x + 8x - 64 = \boxed{x^2 - 64}$. The solutions to $(x + 8)(x - 8) = 0$ are $\boxed{x = -8 \text{ and } x = 8}$.

10.33 Cross-multiplying, we have $x^2 = 64$. Taking the square root of both sides, we find the two solutions $x = \pm\sqrt{64} = \boxed{\pm 8}$.

10.34

(a) Factoring the left side gives $(t - 7)(t - 1) = 0$, so our solutions are $\boxed{t = 7 \text{ and } t = 1}$.

(b) Rearranging gives $x^2 - 6x - 72 = 0$ and factoring gives $(x - 12)(x + 6) = 0$, so our solutions are $\boxed{x = 12 \text{ and } x = -6}$.

(c) Dividing by 2 gives $r^2-2r = 35$, rearranging gives $r^2-2r-35 = 0$, and factoring gives $(r-7)(r+5) = 0$, so the solutions are $\boxed{r = 7 \text{ and } r = -5}$.

(d) Expanding both sides gives $x^2 + 10x = -100 - 10x$, rearranging gives $x^2 + 20x + 100 = 0$, and factoring gives $(x + 10)(x + 10) = 0$. So, our only solution is $\boxed{x = -10}$.

10.35 Because $x = 1$ is a solution, we can let x equal 1 in the equation $ax^2 + bx + 2 = 0$. This gives $a + b + 2 = 0$. Similarly, letting $x = -1$ in the equation gives us $a - b + 2 = 0$. Adding $a - b + 2 = 0$ to $a + b + 2 = 0$ gives $2a + 4 = 0$, so $a = -2$. Substituting this into either equation above gives $b = 0$, so $\boxed{a = -2 \text{ and } b = 0}$.

We also could have noted that the quadratic $(x - 1)(x + 1) = 0$ has solutions $x = -1$ and $x = 1$. Expanding the product of binomials gives $x^2 - 1 = 0$. We need the constant to be 2, so we multiply both sides by -2 to find $-2x^2 + 2 = 0$. The coefficients of this quadratic give us $a = -2$ and $b = 0$.

10.36 $x^2 + 5x + 6 = x^2 + 19x + 34 \implies 5x + 6 = 19x + 34 \implies -14x = 28 \implies x = \boxed{-2}$.

10.37

(a) $(3y-4)(y+6) = 3y(y+6) - 4(y+6) = 3y^2 + 18y - 4y - 24 = \boxed{3y^2 + 14y - 24}$. If $(3y-4)(y+6) = 0$, then $3y - 4 = 0$ or $y + 6 = 0$. So, our solutions are $\boxed{y = 4/3 \text{ and } y = -6}$.

(b) $(6y-1)(6y-1) = 6y(6y-1) - 1(6y-1) = 36y^2 - 6y - 6y + 1 = \boxed{36y^2 - 12y + 1}$. We have $(6y-1)(6y-1) = 0$ only if $6y - 1 = 0$, which only happens if $\boxed{y = 1/6}$.

(c) $(2y-7)(2y+7) = 2y(2y+7) - 7(2y+7) = 4y^2 + 14y - 14y - 49 = \boxed{4y^2 - 49}$. We have $(2y-7)(2y+7) = 0$ only if $2y - 7 = 0$ or $2y + 7 = 0$. So, our solutions are $\boxed{y = 7/2 \text{ and } y = -7/2}$.

(d) $(5y-3)(4y+9) = 5y(4y+9) - 3(4y+9) = 20y^2 + 45y - 12y - 27 = \boxed{20y^2 + 33y - 27}$. We have $(5y-3)(4y+9) = 0$ only if $5y - 3 = 0$ or $4y + 9 = 0$. So, our solutions are $\boxed{y = 3/5 \text{ and } y = -9/4}$.

10.38 We don't have to try $(2x-4)(4x-1)$ because when this is expanded, the coefficient of x will clearly be even (because both coefficients in the first binomial are even). Instead, we have $8x^2 - 33x + 4 = \boxed{(8x-1)(x-4)}$.

10.39 We could multiply both products out and add the results, but instead we notice that there is a factor of $2x + 3$ in each product, so we can factor it out:

$$(2x+3)(x-4) + (2x+3)(x-6) = (2x+3)[(x-4) + (x-6)] = (2x+3)(2x-10).$$

So, our equation is $(2x+3)(2x-10) = 0$, which tells us $2x + 3 = 0$ or $2x - 10 = 0$. These give us the solutions $x = -3/2$ and $x = 5$, and the sum of these is $\boxed{7/2}$.

10.40

(a) Rearranging gives $2x^2 - 11x + 12 = 0$ and factoring gives $(2x-3)(x-4) = 0$, so our solutions are $\boxed{x = 3/2 \text{ and } x = 4}$.

(b) Expanding the left side gives $12x^2 - 52x + 36 = 1$ and rearranging gives $12x^2 - 52x + 35 = 0$. Factoring the left side gives $(6x-5)(2x-7) = 0$, so our solutions are $\boxed{x = 5/6 \text{ and } x = 7/2}$.

(c) Multiplying by 3 to get rid of the fractions gives $27t^2 = 6t + 1$, and rearranging this gives $27t^2 - 6t - 1 = 0$. Factoring then gives us $(9t+1)(3t-1) = 0$, so our solutions are $\boxed{t = -1/9 \text{ and } t = 1/3}$.

(d) Rearranging gives $12r^2 + 139r - 60 = 0$, and factoring gives $(12r-5)(r+12) = 0$, so our solutions are $\boxed{r = 5/12 \text{ and } r = -12}$. (We find this factorization by noticing we'll have to multiply an odd factor of 12 by an odd factor of 60 to make the coefficient of r odd, and by seeing that the coefficient of r is large, so we'll have to pair up a big factor of 12 with a big factor of 60.)

10.41 We have
$$(y+13)(y+a) = y(y+a) + 13(y+a) = y^2 + ay + 13y + 13a.$$

Because there is no y term in the expansion, we must have $ay + 13y = 0$, so $ay = -13y$ and $\boxed{a = -13}$. Hmm... the constant in the binomial $y + 13$ is positive 13. Is this a coincidence?

10.42   Because $x = 1$ is a solution to the quadratic, we have $a + 5 = 3$, so $a = -2$. Therefore, the quadratic is $-2x^2 + 5x = 3$. We rearrange and factor:

$$-2x^2 + 5x = 3 \quad \Rightarrow \quad 2x^2 - 5x + 3 = 0 \quad \Rightarrow \quad (x - 1)(2x - 3) = 0,$$

which tells us the other solution is $\boxed{x = 3/2}$. We could also be crafty and note that the sum of the solutions to $-2x^2 + 5x = 3$ is $-5/(-2) = 5/2$, so the other solution is $5/2 - 1 = 3/2$. (See if you can find another solution by considering the product of the solutions of the quadratic.)

10.43   Rearranging the equation gives $35x^2 - 18x - 60 = 0$. The sum of the roots is $-(-18)/35 = \boxed{18/35}$ and the product of the roots is $(-60)/35 = \boxed{-12/7}$.

10.44   *Solution 1: Set the quadratic equal to 0 for $x = 4$.* Heather knows that the quadratic is $4x^2 + bx - 24$ for some value of b. Because 4 is a root of the quadratic, we know the quadratic equals 0 when $x = 4$. So, we must have $4 \cdot 4^2 + b(4) - 24 = 0$. Solving this equation gives us $b = -10$, so the linear term is $\boxed{-10x}$.

Solution 2: Use the sum and product of roots. Because the constant is -24 and the quadratic term is $4x^2$, Heather knows that the product of the roots is $(-24)/4 = -6$. Noel tells her that one of the roots is 4. Heather then lets the other root be r, and she knows that $4r = -6$. Therefore, $r = -6/4 = -3/2$. So, the roots are 4 and $-3/2$. There are a couple ways Heather could finish from here.

She could note that the quadratic equals zero when $x = 4$ or $x = -3/2$. One such quadratic is

$$(x - 4)\left(x + \frac{3}{2}\right) = x^2 - \frac{5}{2}x - 6.$$

However, this quadratic doesn't have the right quadratic term or the right constant. Heather fixes this by multiplying the quadratic by 4 to get $4x^2 - 10x - 24$, so the linear term is $\boxed{-10x}$.

She also could have noted that the sum of the roots is $4 + (-3/2) = 5/2$. Letting b be the coefficient of the linear term, Heather could then use this sum of roots to note that $-b/4 = 5/2$. So, $b = -10$ and the linear term is again $-10x$.

10.45   We guess that the answer is of the form $a + b\sqrt{3}$, so we have

$$a + b\sqrt{3} = \sqrt{61 - 28\sqrt{3}}.$$

Squaring both sides gives us $61 - 28\sqrt{3}$ on the right and

$$(a + b\sqrt{3})^2 = (a + b\sqrt{3})(a + b\sqrt{3}) = a^2 + 2ab\sqrt{3} + 3b^2$$

on the left, so our equation is

$$a^2 + 3b^2 + 2ab\sqrt{3} = 61 - 28\sqrt{3}.$$

We guess (and hope) that a and b are integers, so we have the system of equations

$$a^2 + 3b^2 = 61,$$
$$2ab\sqrt{3} = -28\sqrt{3}.$$

The second equation gives us $ab = -14$, which reduces the possibilities quite a bit. We find that $(a, b) = (7, -2)$ and $(a, b) = (-7, 2)$ satisfy our system of equations. So, which is the right answer, $7 - 2\sqrt{3}$

or $-7 + 2\sqrt{3}$? Because $\sqrt{49} > \sqrt{12}$, we have $7 > 2\sqrt{3}$. So, $\boxed{7 - 2\sqrt{3}}$ is the one that is positive, and therefore is our answer.

10.46 Suppose the shop increases the price by exactly x dollars, to $10 + x$ dollars. They will then sell $2x$ fewer bracelets, for a total of $50 - 2x$. Therefore, their revenue is $(10 + x)(50 - 2x)$. We need this to equal 600 dollars, so we have the equation

$$(10 + x)(50 - 2x) = 600.$$

We can factor a 2 out of the second factor on the left, then divide both sides by two.

$$(10 + x)(50 - 2x) = 600 \quad \Rightarrow \quad (10 + x)(2)(25 - x) = 600 \quad \Rightarrow \quad (10 + x)(25 - x) = 300.$$

Now, we expand, rearrange, and factor:

$$250 - 10x + 25x - x^2 = 300 \quad \Rightarrow \quad x^2 - 15x + 50 = 0 \quad \Rightarrow \quad (x - 5)(x - 10) = 0.$$

The solutions to this equation are $x = 5$ and $x = 10$, so the shop should increase the price by \$5 or \$10. Therefore, they should set the price at $\boxed{\$15 \text{ or } \$20}$.

10.47 We can get the x's out of the denominators by multiplying both sides by x^2. On the left we have

$$x^2 \left(\frac{10}{x^2} + \frac{22}{x} + 4 \right) = 10 + 22x + 4x^2.$$

So, our equation is

$$10 + 22x + 4x^2 = 0 \quad \Rightarrow \quad 2x^2 + 11x + 5 = 0 \quad \Rightarrow \quad (2x + 1)(x + 5) = 0.$$

Therefore, our solutions are $\boxed{x = -1/2 \text{ and } x = -5}$.

10.48 It's a quadratic in disguise! Let $x = w^2$, so our equation becomes $2x^2 - 5x + 2 = 0$. Factoring gives us $(2x - 1)(x - 2) = 0$, which has solutions $x = 1/2$ and $x = 2$. From $x = 1/2$, we have $w^2 = 1/2$, so $w = \pm\sqrt{1/2} = \pm 1/\sqrt{2} = \pm\sqrt{2}/2$. From $x = 2$, we have $w^2 = 2$, so $w = \pm\sqrt{2}$. Therefore, we have four solutions:

$$\boxed{w = \frac{\sqrt{2}}{2}, -\frac{\sqrt{2}}{2}, \sqrt{2}, -\sqrt{2}}.$$

10.49 We start by cross-multiplying, then we expand and rearrange:

$$(x - 6)(x - 2) = 4(x - 5) \quad \Rightarrow \quad x^2 - 8x + 12 = 4x - 20 \quad \Rightarrow \quad x^2 - 12x + 32 = 0 \quad \Rightarrow \quad (x - 8)(x - 4) = 0.$$

So, we find that $\boxed{x = 4 \text{ and } x = 8}$ are our solutions. (Neither makes a denominator equal to zero, so neither solution is extraneous.)

10.50 We can factor the quadratics in the numerators on the left side, and lo and behold, we can simplify the fractions:

$$\frac{y^2 - 9y + 8}{y - 1} + \frac{3y^2 + 16y - 12}{3y - 2} = \frac{\cancel{(y - 1)}(y - 8)}{\cancel{y - 1}} + \frac{\cancel{(3y - 2)}(y + 6)}{\cancel{3y - 2}}$$

$$= y - 8 + y + 6.$$

So, our equation is $2y - 2 = -3$, which gives $y = \boxed{-1/2}$. (A quick check shows that this solution is not extraneous.)

Challenge Problems

10.51 If Chewbacca loses one pack of cherry gum, the ratio of the number of pieces of cherry gum he has to the number of pieces of grape gum is $(20 - x)/30$. If he instead finds 5 packs of grape gum, this ratio will be $20/(30 + 5x)$. These ratios must be equal, so we must have

$$\frac{20 - x}{30} = \frac{20}{30 + 5x} \quad \Rightarrow \quad (20 - x)(30 + 5x) = (30)(20) \quad \Rightarrow \quad (20 - x)(5)(6 + x) = (30)(20).$$

Dividing both sides by 5 gives $(20 - x)(6 + x) = (30)(4)$ and expanding the left side of this gives $120 + 14x - x^2 = 120$. Therefore, $x^2 - 14x = 0$, so $x(x - 14) = 0$. We can't have $x = 0$, so we must have $x = \boxed{14}$.

10.52 LuAnn has the correct quadratic and linear terms, so the sum of her roots will match the sum of the roots of the correct quadratic. The sum of her roots is 3, so the sum of the roots of the correct equation is also 3. Similarly, Bobby has the correct quadratic and constant terms, so he has the right product of roots, which equals $(-4)(10) = -40$. Since the correct quadratic has roots which add to 3 and multiply to -40, the roots of the correct quadratic are the roots of $x^2 - 3x - 40 = 0$. Since $x^2 - 3x - 40 = (x - 8)(x + 5)$, the correct roots are $\boxed{x = 8 \text{ and } x = -5}$.

10.53

(a) For guidance, we first pretend $b = 1$. This gives us $x^2 + 7x + 10 = 0$, so $(x + 2)(x + 5) = 0$. Returning to $x^2 + 7bx + 10b^2 = 0$, we see that all we have to do is tack on a b to the constants of our earlier factorization:
$$x^2 + 7bx + 10b^2 = 0 \quad \Rightarrow \quad (x + 2b)(x + 5b) = 0.$$
So, our solutions are $\boxed{x = -2b \text{ and } x = -5b}$.

(b) We already know that solutions to the equation are $x = -2b$ and $x = -5b$. So, if $x = 25$, we have either $25 = -2b$ or $25 = -5b$, so the two possible values of b are $\boxed{b = -25/2 \text{ and } b = -5}$.

10.54 First, we isolate the radical: $\sqrt{x + 7} = 13 - x$. Then, we square both sides to get rid of the radical. This gives us
$$x + 7 = (13 - x)^2 = (13 - x)(13 - x) = 169 - 26x + x^2.$$

Rearranging gives $x^2 - 27x + 162 = 0$ and factoring gives $(x - 18)(x - 9) = 0$, so we have $x = 18$ and $x = 9$ as our solutions. However, we must check whether or not they are extraneous. Since $\sqrt{9 + 7} + 9 = 4 + 9 = 13$, the solution $x = 9$ is OK; however, the solution $x = 18$ is extraneous, since it gives $\sqrt{18 + 7} + 18 = 23$, not 13. So, $\boxed{x = 9}$ is the only solution.

10.55 We could multiply everything out (yuck). Or we could be a little clever by combining the last two terms before multiplying out:

$$(2x + 1)(9x + 1) + (9x + 1)(17x - 3) = (9x + 1)[(2x + 1) + (17x - 3)].$$

That little bit of cleverness makes us look at the first term to see if there is any other clever factoring to do. We can at least factor a 4 out of one factor in that first term, which gives

$$4(2x - 1)(2x + 1) + (2x + 1)(9x + 1) + (9x + 1)(17x - 3).$$

Now, we can combine the first two products by factoring out $2x + 1$:

$$4(2x - 1)(2x + 1) + (2x + 1)(9x + 1) + (9x + 1)(17x - 3) = (2x + 1)[4(2x - 1) + (9x + 1)] + (9x + 1)(17x - 3)$$
$$= (2x + 1)(17x - 3) + (9x + 1)(17x - 3).$$

Aha! We can keep factoring and combining terms:

$$(2x + 1)(17x - 3) + (9x + 1)(17x - 3) = [(2x + 1) + (9x + 1)](17x - 3) = (11x + 2)(17x - 3),$$

so our equation is $(11x + 2)(17x - 3) = 0$. Therefore, our roots are $\boxed{x = -2/11 \text{ and } x = 3/17}$.

10.56 We start with $(2 + x)^3 = (2 + x)(2 + x)(2 + x)$. First, we multiply the first two to find $(2 + x)(2 + x) = 4 + 4x + x^2$. So, we have

$$(2 + x)^3 = (4 + 4x + x^2)(2 + x) = (4 + 4x + x^2)(2) + (4 + 4x + x^2)(x) = 8 + 12x + 6x^2 + x^3.$$

$(2 + x)^3$ has a couple more terms than just $2^3 + x^3$ due to the distributive property.

10.57 Those square roots are a pain, so let's compare the squares of the two numbers. $(\sqrt{2})^2$ is simply 2, while

$$(2\sqrt{3} - 2)^2 = (2\sqrt{3} - 2)(2\sqrt{3} - 2) = 12 - 2(2)(2\sqrt{3}) + 4 = 16 - 8\sqrt{3}.$$

So, we now must compare 2 and $16 - 8\sqrt{3}$. If $8\sqrt{3} > 14$, then $16 - 8\sqrt{3} < 2$ (make sure you see why). To compare $8\sqrt{3}$ and 14, we square both. Since $(8\sqrt{3})^2 = 8^2 \cdot 3 = 192$ and $14^2 = 196$, we know that $8\sqrt{3} < 14$. Therefore, 14 is closer to 16 than $8\sqrt{3}$ is, so $16 - 8\sqrt{3} > 2$. Since $16 - 8\sqrt{3} = (2\sqrt{3} - 2)^2$, we can take the square root of both sides of $16 - 8\sqrt{3} > 2$ to determine that $\boxed{2\sqrt{3} - 2}$ is greater than $\sqrt{2}$.

10.58

(a) This product equals 0 when $x - 1 = 0$, $x + 2 = 0$, or $x - 5 = 0$, so the values for which the product equals 0 are $\boxed{1, -2, \text{ and } 5}$.

(b) The sum is $1 - 2 + 5 = \boxed{4}$ and the product is $(1)(-2)(5) = \boxed{-10}$.

(c) Since $(x - 1)(x + 2) = x^2 + x - 2$, we have

$$(x - 1)(x + 2)(x - 5) = (x^2 + x - 2)(x - 5) = (x^2 + x - 2)(x) + (x^2 + x - 2)(-5) = \boxed{x^3 - 4x^2 - 7x + 10}.$$

(d) The negative of the sum of our solutions is the coefficient of x^2 and the negative of the product of our solutions is the constant term. Hmm...

(e) We check to see if the relationship from part (d) is a coincidence by replacing our roots with constants p, q, and r: $(x - p)(x - q)(x - r)$. The values of x for which this product is 0 are $x = p$, q, and r. We now expand the product:

$$(x - p)(x - q)(x - r) = [(x - p)(x) + (x - p)(-q)](x - r)$$
$$= (x^2 - px - qx + pq)(x - r)$$
$$= (x^2 - px - qx + pq)(x) + (x^2 - px - qx + pq)(-r)$$
$$= x^3 - px^2 - qx^2 + pqx - rx^2 + rpx + qrx - pqr$$
$$= x^3 - (p + q + r)x^2 + (pq + qr + rp)x - pqr.$$

Sure enough, the coefficient of x^2 is the negative of the sum of the solutions of $(x-p)(x-q)(x-r) = 0$, and the constant term is the negative of the product. Also, look at the interesting form of the coefficient of x. For future study: what if the coefficient of x^3 is not 1? What if there are 4 or 5 or 6 terms in the initial product?

10.59 Let the original side length be x. So, the area of each face of the original cube is x^2, which means the total area of all 6 faces is $6x^2$. The edge length of the new cube is $x + 4$, so the area of each face is $(x + 4)^2$, which means the area of all 6 faces is $6(x + 4)^2$. We therefore have the equation

$$6(x + 4)^2 - 6x^2 = 576.$$

We first divide by 6, then expand and simplify:

$$(x + 4)^2 - x^2 = 96 \quad \Rightarrow \quad x^2 + 8x + 16 - x^2 = 96 \quad \Rightarrow \quad 8x + 16 = 96 \quad \Rightarrow \quad x = 10.$$

Since the original cube has sides of length 10, its volume is $10^3 = \boxed{1000}$.

10.60 We know that $r + s = -b$ and $rs = c$, so we try to write $r^2 s + s^2 r$ in terms of rs and $r + s$. Factoring out rs does the trick: $rs(r + s) = 10$. So, we have $-bc = 10$. Because 10 can be factored as $1 \cdot 10$ or $2 \cdot 5$, there are 8 possible ordered pairs:

$$\boxed{(b, c) = (2, -5);\ (-2, 5);\ (5, -2);\ (-5, 2);\ (10, -1);\ (-10, 1);\ (1, -10);\ (-1, 10)}$$

10.61

(a) $\boxed{\text{No}}$! The roots might not be rational numbers. For example, $y^2 - 2 = 0$ has roots of $\pm \sqrt{2}$.

(b) If we let $y = 0$, we have $3y^2 - y - 12 = -12$. If we let $y = 3$, we have $3y^2 - y - 12 = 12$. So, when y is 0, the quadratic is negative, but when $y = 3$, the quadratic is positive. So, if we increase y gradually from 0 to 3, the quadratic will change gradually from a negative number to a positive number. Somewhere along the way, it will hit 0, because it must change smoothly.

(c) *Solution 1: Sum and product of roots.* The sum of the roots of our original quadratic is $r + s = -(-1)/3 = 1/3$, and the product of these roots is $rs = (-12)/3 = -4$. The sum of the roots of the new quadratic is $r + 2 + s + 2 = r + s + 4 = 1/3 + 4 = 13/3$. The product of the roots of the new quadratic is

$$(r + 2)(s + 2) = rs + 2r + 2s + 4 = rs + 2(r + s) + 4 = -4 + \frac{2}{3} + 4 = \frac{2}{3}.$$

We have the sum and the product of the roots of the new quadratic, so we have the coefficients of the new quadratic, which is $\boxed{z^2 - \frac{13}{3}z + \frac{2}{3}}$. (The result of multiplying this quadratic by any nonzero constant is also an acceptable answer.)

Solution 2: Transform the roots. The old quadratic has roots r and s and the new one has roots $r + 2$ and $s + 2$. So, we try to transform the old quadratic into the new one. If we let $y = z - 2$, notice that $y = r$ when $z = r + 2$ and $y = s$ when $z = s + 2$. So, if the quadratic $3y^2 - y - 12$ equals 0 when $y = r$ or $y = s$, then the quadratic

$$3(z - 2)^2 - (z - 2) - 12$$

equals 0 when $z = r + 2$ or $z = s + 2$. Simplifying the quadratic gives $3z^2 - 13z + 2$. Dividing this quadratic by 3 gives the same quadratic as we found in our first solution.

10.62 Suppose the man's rate is r and the current's rate is c. We have

$$\text{Time downstream} = \frac{\text{Distance downstream}}{\text{Rate downstream}} = \frac{15}{r+c}$$

and

$$\text{Time upstream} = \frac{\text{Distance upstream}}{\text{Rate upstream}} = \frac{15}{r-c}.$$

We are told that the downstream trip is 5 hours shorter than the upstream trip, so

$$\frac{15}{r-c} - \frac{15}{r+c} = 5.$$

Multiplying both sides by $(r-c)(r+c)$, we have $15(r+c) - 15(r-c) = 5(r-c)(r+c)$, or

$$30c = 5r^2 - 5c^2.$$

Similarly, if the man doubles his rate, we have

$$\text{Time downstream} = \frac{\text{Distance downstream}}{\text{Rate downstream}} = \frac{15}{2r+c}$$

and

$$\text{Time upstream} = \frac{\text{Distance upstream}}{\text{Rate upstream}} = \frac{15}{2r-c}.$$

This downstream trip is only 1 hour shorter than the upstream trip, so

$$\frac{15}{2r-c} - \frac{15}{2r+c} = 1.$$

Multiplying both sides by $(2r-c)(2r+c)$, we have $15(2r+c) - 15(2r-c) = (2r-c)(2r+c)$, or

$$30c = 4r^2 - c^2.$$

Subtracting this equation from $30c = 5r^2 - 5c^2$, we have $0 = r^2 - 4c^2$, so $r^2 = 4c^2$. Substituting this into $30c = 4r^2 - c^2$ gives $30c = 15c^2$. Since c must be positive, we can divide by $15c$ to get $c = \boxed{2 \text{ mph}}$.

10.63 We can equate the left sides of the first two equations. Expanding and rearranging gives us

$$a(b+c-5) = b(a+c-5) \quad \Rightarrow \quad ab + ac - 5a = ab + bc - 5b \quad \Rightarrow \quad ac - 5a = bc - 5b \quad \Rightarrow \quad a(c-5) = b(c-5).$$

Subtracting $b(c-5)$ from both sides gives

$$a(c-5) - b(c-5) = (a-b)(c-5) = 0.$$

So, either $a = b$ or $c = 5$. We tackle these two cases separately.

Letting $a = b$ in our final equation gives $2a^2 = 50$, so $a = \pm 5$. Substituting $a = b = 5$ into the first equation gives $5c = 7$, so $c = 7/5$. Similarly, letting $a = b = -5$, we have $-5(-5 + c - 5) = 7$. Solving this equation gives us $c = 43/5$.

Letting $c = 5$ in either of the first two equations gives $ab = 7$. Solving for a gives $a = 7/b$. Substituting this into our last equation gives

$$\frac{49}{b^2} + b^2 = 50 \quad \Rightarrow \quad 49 + b^4 = 50b^2 \quad \Rightarrow \quad b^4 - 50b^2 + 49 = 0.$$

Letting $x = b^2$, we have $x^2 - 50x + 49 = 0$, and factoring gives us $(x - 1)(x - 49) = 0$. So, $x = 1$ or $x = 49$, which gives us the possible values for b of ± 1 or ± 7. Using $a = 7/b$, we can then compute values for a.

We therefore have a total of 6 solutions:

$$(a, b, c) = \left(5, 5, \frac{7}{5}\right) ; \left(-5, -5, \frac{43}{5}\right) ; (7, 1, 5); (-7, -1, 5); (1, 7, 5); (-1, -7, 5).$$

10.64 We let the expression equal x:

$$x = \sqrt{90 + \sqrt{90 + \sqrt{90 + \cdots}}}.$$

We square both sides to get rid of one radical:

$$x^2 = 90 + \sqrt{90 + \sqrt{90 + \cdots}}.$$

We still have that big nasty infinite radical. However, we already said the nasty radical is equal to x! So, we can substitute x in place of the radical to get $x^2 = 90 + x$. So, $x^2 - x - 90 = 0$, and factoring the left side gives us $(x - 10)(x + 9) = 0$. Since x is clearly positive, we have $x = \boxed{10}$.

10.65 $\boxed{\text{Yes}}$. Suppose Joel's quadratic is $ax^2 + bx + c$. Suppose the quadratic equals k_1 when $x = 1$, k_2 when $x = 2$, and k_3 when $x = 3$. We then have three equations:

$$a + b + c = k_1,$$
$$4a + 2b + c = k_2,$$
$$9a + 3b + c = k_3.$$

The values of k_1, k_2, and k_3 are constant; these are just the numbers Joel (and Eve) found when letting x equal 1, 2, and 3 in their quadratics. Subtracting the first equation from the second and then subtracting the first equation from the third, gives us the system

$$3a + b = k_2 - k_1,$$
$$8a + 2b = k_3 - k_1.$$

Subtracting the twice the first of these equations from the second gives $2a = k_1 - 2k_2 + k_3$, so we have $a = (k_1 - 2k_2 + k_3)/2$. We can substitute this into our equation $3a + b = k_2 - k_1$ to find b in terms of k_1, k_2, and k_3. Then we can use a and b to find c in terms of k_1, k_2, and k_3. Therefore, once we know what the quadratic $ax^2 + bx + c$ gives us when $x = 1$, $x = 2$, and $x = 3$, we can find the coefficients of the quadratic. Joel's and Eve's quadratics give the same results for $x = 1$, $x = 2$, and $x = 3$, so they must have the same coefficients. So, they must be the same quadratics.

CHAPTER 11

Special Factorizations

Exercises for Section 11.1

11.1.1 For each part, we use the expansion $(a + b)^2 = a^2 + 2ab + b^2$.

(a) $(x - 5)^2 = x^2 + 2(x)(-5) + (-5)^2 = \boxed{x^2 - 10x + 25}$.

(b) $(x + 5)^2 = x^2 + 2(x)(5) + 5^2 = \boxed{x^2 + 10x + 25}$.

(c) $(3y - 7)^2 = (3y)^2 + 2(3y)(-7) + (-7)^2 = \boxed{9y^2 - 42y + 49}$.

(d) $(3y + 7)^2 = (3y)^2 + 2(3y)(7) + 7^2 = \boxed{9y^2 + 42y + 49}$.

11.1.2

(a) The quadratic term is a perfect square, but the constant term is negative, so this quadratic is not the square of a binomial.

(b) The quadratic term is the square of r and the constant term is 5^2. The linear term is $2(r)(5)$, so we see that $r^2 + 10r + 25 = \boxed{(r + 5)^2}$.

(c) The quadratic term is the square of x and the constant is the square of 20 (or the square of -20). However, the linear term is not equal to $2(x)(20)$ or $2(x)(-20)$. This quadratic is not the square of a binomial.

(d) The quadratic is the square of $3y$, the constant term is the square of -5, and the linear term equals $2(3y)(-5)$, so we have $9y^2 - 30y + 25 = \boxed{(3y - 5)^2}$.

(e) The first term is the square of $t^2/3$, the last term is the square of 6, and the middle term is $2(t^2/3)(6) = 4t^2$, so we have

$$\frac{t^4}{9} + 4t^2 + 36 = \boxed{\left(\frac{t^2}{3} + 6\right)^2}.$$

(f) At first, we might think that this quadratic is not the square of a binomial, because neither the quadratic term nor the linear term appears to be a perfect square. However, we notice that all the coefficients are multiples of 2, so we can factor out a 2 from the entire expression:

$$2y^2 - 28y + 98 = 2(y^2 - 14y + 49).$$

The expression in parentheses is the square of a binomial! So, we have

$$2y^2 - 28y + 98 = 2(y^2 - 14y + 49) = 2(y - 7)^2.$$

Since $2 = (\sqrt{2})^2$, we can write our quadratic as the square of a binomial:

$$2y^2 - 28y + 98 = 2(y - 7)^2 = \left(\sqrt{2}\right)^2 (y - 7)^2 = \boxed{\left(y\sqrt{2} - 7\sqrt{2}\right)^2}.$$

11.1.3   The whole figure is a square with side length $a + b$, so its area is $(a + b)^2$. We can also find the area of the square by adding up the four pieces that together make up the square. The bottom left piece is a square with side length a, so its area is a^2. The upper left and bottom right pieces are each a by b rectangles, so each has area ab. The upper right piece is a square with side length b so its area is b^2. The area of each piece is shown in the diagram at right. The sum of the areas of the pieces is $a^2 + ab + ab + b^2 = a^2 + 2ab + b^2$. 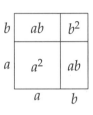 Since these pieces together make up the entire big square, and the big square has area $(a + b)^2$, we have $(a + b)^2 = a^2 + 2ab + b^2$.

11.1.4   As described in the text, we use the expansion $(a+b)^2 = a^2 + 2ab + b^2$ to perform both computations. First, we let $a = 30$ and $b = 1$ to compute 31^2:

$$31^2 = (30 + 1)^2 = 30^2 + 2(30)(1) + 1^2 = 900 + 60 + 1 = \boxed{961}.$$

Then, we let $a = 300$ and $b = -1$ to compute 299^2:

$$299^2 = (300 - 1)^2 = 300^2 + 2(300)(-1) + (-1)^2 = 90000 - 600 + 1 = \boxed{89401}.$$

11.1.5

(a) We have $4 \cdot 5 = 20$, so we tack 25 on the end of 20 to get $45^2 = \boxed{2025}$. We have $9 \cdot 10 = 90$, so we tack 25 on the end of 90 to find $95^2 = \boxed{9025}$. We have $11 \cdot 12 = 132$, so we tack 25 on the end of 132 to find $115^2 = \boxed{13225}$.

(b) Each number that ends in 5 can be written as $10n + 5$ for some value of n. The number $10n$ results from tacking a 0 onto the end of n, then adding 5 just changes the 0 to a 5. Therefore, when we erase the 5 from the end of $10n + 5$, we are basically reversing these steps to get n back. So, the number that results when we erase the 5 from the end of $10n + 5$ is just n.

When we square $10n + 5$, we have

$$(10n + 5)^2 = (10n)^2 + 2(10n)(5) + 5^2 = 100n^2 + 100n + 25 = 100(n^2 + n) + 25.$$

The product $100(n^2 + n)$ will end in 00, so when we add on 25, we'll get a number that ends in 25. So, we see where we get the "write the digits 25 at the end of the product" from. But what about that $100(n^2 + n)$? We can factor $n^2 + n$ as $n(n + 1)$, so we have

$$(10n + 5)^2 = 100(n^2 + n) + 25 = 100n(n + 1) + 25.$$

Aha! Now, we see where our steps come from. When we erase the 5 from the end of $10n + 5$, we get the number n. Our rules say to multiply this n by $n + 1$, which gives us $n(n + 1)$. When we tack 25 onto the end of this, we are first multiplying $n(n + 1)$ by 100 (in order to move this number two digits to the left, to make room for the 25), then, we are adding 25. So, tacking 25 onto the end of $n(n + 1)$ gives us $100n(n + 1) + 25$. This number looks familiar! We can reverse our algebra steps above to see that

$$100n(n + 1) + 25 = 100n^2 + 100n + 25 = (10n)^2 + 2(10n)(5) + 5^2 = (10n + 5)^2.$$

11.1.6   Instead of squaring those nasty numbers, we let $x = 199919981991$. Then, we can use algebra:

$$199919981997^2 - 2 \cdot 199919981994^2 + 199919981991^2 = (x + 6)^2 - 2(x + 3)^2 + x^2$$
$$= x^2 + 12x + 36 - 2(x^2 + 6x + 9) + x^2$$
$$= x^2 + 12x + 36 - 2x^2 - 12x - 18 + x^2$$
$$= \boxed{18}.$$

That's a lot easier than squaring 12-digit numbers!

Exercises for Section 11.2

11.2.1   Each expression in the first three parts is the difference of two perfect squares, so we can use our difference of squares factorization, $x^2 - y^2 = (x - y)(x + y)$.

(a)   We have $t^2 - 49 = t^2 - 7^2 = \boxed{(t - 7)(t + 7)}$.

(b)   We have $36 - 9x^2 = 6^2 - (3x)^2 = (6 - 3x)(6 + 3x)$. We can factor a 3 out of each of $6 - 3x$ and $6 + 3x$ to give $3 \cdot (2 - x) \cdot 3 \cdot (2 + x) = \boxed{9(2 - x)(2 + x)}$. (We could also have factored out a 9 at the beginning: $36 - 9x^2 = 9(4 - x^2) = 9(2 - x)(2 + x)$.)

(c)   We have $121a^2b^4 - c^2 = (11ab^2)^2 - c^2 = \boxed{(11ab^2 - c)(11ab^2 + c)}$.

(d)   At first, it doesn't look like we have a difference of squares factorization because neither term looks like a perfect square. However, we can factor $8x^2$ out of both terms, and then we see the difference of squares:

$$800x^4 - 72x^2y^2 = 8x^2(100x^2 - 9y^2) = 8x^2[(10x)^2 - (3y)^2] = \boxed{8x^2(10x - 3y)(10x + 3y)}.$$

11.2.2   In the text, we learned that the difference between $(n + 1)^2$ and n^2 is $n + (n + 1)$. Therefore, to find 55556^2 given 55555^2, we simply add 55555 and 55556 to 55555^2:

$$55556^2 = 55555^2 + 55555 + 55556 = 3086358025 + 111111 = \boxed{3086469136}.$$

11.2.3

(a)   Because $(3^4)^2 = 3^8$ and $(2^3)^2 = 2^6$, both 3^8 and 2^6 are perfect squares.

(b) Because 3^8 and 2^6 are perfect squares, we can use the difference of squares factorization to find

$$3^8 - 2^6 = (3^4)^2 - (2^3)^2 = (3^4 - 2^3)(3^4 + 2^3) = (81 - 8)(81 + 8) = \boxed{(73)(89)}.$$

11.2.4 Since w^4 and 16 are both perfect squares, we can use our difference of squares factorization:

$$w^4 - 16 = (w^2)^2 - 4^2 = (w^2 - 4)(w^2 + 4).$$

We're not finished! The expression $w^2 - 4$ is also a difference of squares, which we can factor as $w^2 - 4 = (w - 2)(w + 2)$. So, we have

$$w^4 - 16 = (w^2 - 4)(w^2 + 4) = \boxed{(w - 2)(w + 2)(w^2 + 4)}.$$

11.2.5 The whole large square has area a^2. The little piece we cut out has area b^2. So, the area of the region left after we cut out the little square is $a^2 - b^2$. We can show this equals $(a - b)(a + b)$ if we can show that this leftover region has area $(a - b)(a + b)$. We know that a rectangle with consecutive sides of length $a - b$ and $a + b$ has area $(a - b)(a + b)$, so we try to find a way to rearrange our leftover region into such a rectangle.

In the diagram at left below, we have all the dimensions of the leftover piece labeled.

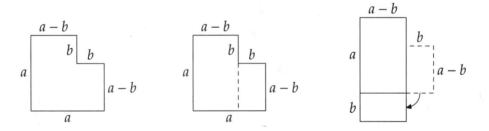

The sides of length $a - b$ give us a clue as to how to build our rectangle. We let the top side of length $a - b$ be one side of our rectangle, then cut off the right piece of the leftover region along the dashed line shown in the middle diagram. Then, we slide this piece down to the bottom as shown in the final diagram, thus completing our $a - b$ by $a + b$ rectangle. Since our leftover region after cutting a square of side length b from a square of side length a has total area $a^2 - b^2$ when computed by our first method, and total area $(a - b)(a + b)$ when computed by our second method, we have $a^2 - b^2 = (a - b)(a + b)$.

11.2.6 We can rearrange our first equation to find $x^4 - y^4 = 18\sqrt{3}$. We do so because we know we can factor the left side as the difference of squares, and we know the value of one of these factors. Since $x^4 - y^4 = (x^2)^2 - (y^2)^2 = (x^2 - y^2)(x^2 + y^2)$, and we are given that $x^2 + y^2 = 6$, we have

$$(x^2 - y^2)(x^2 + y^2) = 18\sqrt{3} \quad \Rightarrow \quad (x^2 - y^2)(6) = 18\sqrt{3} \quad \Rightarrow \quad x^2 - y^2 = 3\sqrt{3}.$$

That strategy worked once; let's try it again. We have $x^2 - y^2 = (x - y)(x + y)$ and we are given $x + y = 3$, so we have

$$(x - y)(x + y) = 3\sqrt{3} \quad \Rightarrow \quad (x - y)(3) = 3\sqrt{3} \quad \Rightarrow \quad x - y = \boxed{\sqrt{3}}.$$

Exercises for Section 11.3

11.3.1

(a) Since x^3 and 1000 are perfect cubes, we can use our difference of cubes factorization to factor $x^3 - 1000$:

$$x^3 - 1000 = x^3 - 10^3 = (x - 10)[x^2 + (x)(10) + 10^2] = \boxed{(x - 10)(x^2 + 10x + 100)}.$$

(b) We have $27r^3 + 64 = (3r)^3 + 4^3 = (3r + 4)[(3r)^2 - (3r)(4) + 4^2] = \boxed{(3r + 4)(9r^2 - 12r + 16)}.$

(c) We start by factoring $2z^3$ out of both terms, and then use the difference of cubes factorization:

$$2z^3 - 16z^6 = 2z^3(1 - 8z^3) = 2z^3[1 - (2z)^3] = 2z^3(1 - 2z)[1 + (1)(2z) + (2z)^2] = \boxed{2z^3(1 - 2z)(1 + 2z + 4z^2)}.$$

(d) We might see this as the sum of the cubes $-a^3b^3$ and -8, or we might start by factoring out -1 to see that

$$-a^3b^3 - 8 = -(a^3b^3 + 8) = -[(ab)^3 + 2^3] = -(ab + 2)[(ab)^2 - (ab)(2) + 2^2] = \boxed{-(ab + 2)(a^2b^2 - 2ab + 4)}.$$

11.3.2 Both 3^6 and 2^9 are cubes, so we can factor $3^6 + 2^9$ as the sum of cubes:

$$3^6 + 2^9 = (3^2)^3 + (2^3)^3 = (3^2 + 2^3)[(3^2)^2 - (3^2)(2^3) + (2^3)^2]$$
$$= (9 + 8)[9^2 - (9)(8) + 8^2] = 17(81 - 72 + 64) = \boxed{17 \cdot 73}.$$

11.3.3 We first factor $x^6 - 64$ as the difference of squares:

$$x^6 - 64 = (x^3)^2 - 8^2 = (x^3 - 8)(x^3 + 8).$$

Our first factor is a difference of cubes and the second is a sum of cubes, so we have

$$x^6 - 64 = (x^3 - 8)(x^3 + 8) = \boxed{(x - 2)(x^2 + 2x + 4)(x + 2)(x^2 - 2x + 4)}.$$

11.3.4

(a) We begin as we did with $x^3 - y^3$. In order to get an x^4 term and a $-y^4$ term by multiplying $x - y$ by another factor, this second factor must have both an x^3 term and a y^3 term. So, we start by trying $(x - y)(x^3 + y^3)$:

$$(x - y)(x^3 + y^3) = x(x^3 + y^3) - y(x^3 + y^3) = x^4 + xy^3 - x^3y - y^4.$$

We succeeded in forming $x^4 - y^4$, but we have some extra terms. So, we have to add more terms to $x^3 + y^3$ to cancel these terms out. One of the extra terms in our product above is xy^3. In order to get rid of it, we can include an xy^2 term into our second factor, since the product of this with $-y$ is $-xy^3$, which cancels the $+xy^3$:

$$(x - y)(x^3 + xy^2 + y^3) = x(x^3 + xy^2 + y^3) - y(x^3 + xy^2 + y^3)$$
$$= x^4 + x^2y^2 + xy^3 - x^3y - xy^3 - y^4 = x^4 + x^2y^2 - x^3y - y^4.$$

Uh-oh. We got rid of xy^3, but now we have another extra term, x^2y^2. We can get rid of this by including an x^2y in our second factor, since the product of this with $-y$ in the first factor is $-x^2y^2$, which cancels the x^2y^2:

$$(x-y)(x^3 + x^2y + xy^2 + y^3) = x(x^3 + x^2y + xy^2 + y^3) - y(x^3 + x^2y + xy^2 + y^3)$$
$$= x^4 + x^3y + x^2y^2 + xy^3 - x^3y - x^2y^2 - xy^3 - y^4 = x^4 - y^4.$$

We got rid of all our extra terms! We have $x^4 - y^4 = \boxed{(x-y)(x^3 + x^2y + xy^2 + y^3)}$. See if you can find another solution using the fact that $x^4 - y^4$ is a difference of squares.

(b) Since $x^5 - y^5$ is a difference of perfect powers, we look at the factorizations we have found of other similar differences of powers:

$$x^2 - y^2 = (x-y)(x+y),$$
$$x^3 - y^3 = (x-y)(x^2 + xy + y^2),$$
$$x^4 - y^4 = (x-y)(x^3 + x^2y + xy^2 + y^3).$$

We see a pattern! In the factorization of each expression of the form $x^k - y^k$, one factor is always $x - y$ and the other factor is $x^{k-1} + x^{k-2}y + x^{k-3}y^2 + \cdots + xy^{k-2} + y^{k-1}$. Let's see if it works when $k = 5$:

$$(x-y)(x^4 + x^3y + x^2y^2 + xy^3 + y^4) = x(x^4 + x^3y + x^2y^2 + xy^3 + y^4) - y(x^4 + x^3y + x^2y^2 + xy^3 + y^4)$$
$$= x^5 + x^4y + x^3y^2 + x^2y^3 + xy^4 - x^4y - x^3y^2 - x^2y^3 - xy^4 - y^5$$
$$= x^5 - y^5.$$

So, we have $x^5 - y^5 = \boxed{(x-y)(x^4 + x^3y + x^2y^2 + xy^3 + y^4)}$.

(c) The previous part leads the way:

$$(x-y)(x^{k-1} + x^{k-2}y + \cdots + xy^{k-2} + y^{k-1})$$
$$= x(x^{k-1} + x^{k-2}y + \cdots + xy^{k-2} + y^{k-1}) - y(x^{k-1} + x^{k-2}y + \cdots + xy^{k-2} + y^{k-1})$$
$$= x^k + x^{k-1}y + \cdots + x^2y^{k-2} + xy^{k-1} - x^{k-1}y - x^{k-2}y^2 - \cdots - xy^{k-1} - y^k$$
$$= x^k - y^k.$$

Make sure you see how all the other terms cancel. This shows that

$$x^k - y^k = \boxed{(x-y)(x^{k-1} + x^{k-2}y + \cdots + xy^{k-2} + y^{k-1})}.$$

11.3.5 We could solve for y in the first equation and substitute the result in the second, but that looks terrifying. We'll have to find a better way to combine the equations.

In general, factoring is a very useful tool, and we know how to factor the sum of cubes, so we try adding the equations, which gives
$$x^3 + y^3 = 19x + 19y.$$

This is very promising, because $x + y$ is a factor of both sides:

$$(x + y)(x^2 - xy + y^2) = 19(x + y).$$

Furthermore, because $|x| \neq |y|$, we know that $x + y \neq 0$, so we can divide by it, giving

$$x^2 - xy + y^2 = 19.$$

We definitely think we're close, because this is only one extra term, $-xy$, from what we want to find.

Using a cubes factorization worked well once, so we try it again, this time subtracting the second equation from the first to give

$$x^3 - y^3 = 11x - 11y.$$

Once again, our two sides have a common factor:

$$(x - y)(x^2 + xy + y^2) = 11(x - y).$$

We know that $x - y \neq 0$, so we can divide by $(x - y)$ to give

$$x^2 + xy + y^2 = 11.$$

Now we have a system of equations that twice includes the expression we want, $x^2 + y^2$:

$$x^2 - xy + y^2 = 19,$$
$$x^2 + xy + y^2 = 11.$$

We eliminate the terms we don't want by adding the equations, yielding $2x^2 + 2y^2 = 30$. Dividing by 2 gives us our result: $x^2 + y^2 = \boxed{15}$.

Exercises for Section 11.4

11.4.1

(a) We can make the denominator rational by multiplying it by $\sqrt{6}$, so we have

$$\frac{3}{\sqrt{6}} = \frac{3}{\sqrt{6}} \cdot \frac{\sqrt{6}}{\sqrt{6}} = \frac{3\sqrt{6}}{6} = \boxed{\frac{\sqrt{6}}{2}}.$$

(b) We could rationalize the denominator by multiplying by $\sqrt[5]{9^4}$, but before we do so, we note that $\sqrt[5]{9} = \sqrt[5]{3^2}$, so we can rationalize the denominator by multiplying by $\sqrt[5]{3^3}$:

$$\frac{2}{\sqrt[5]{9}} = \frac{2}{\sqrt[5]{3^2}} \cdot \frac{\sqrt[5]{3^3}}{\sqrt[5]{3^3}} = \frac{2\sqrt[5]{3^3}}{\sqrt[5]{3^5}} = \boxed{\frac{2\sqrt[5]{27}}{3}}.$$

11.4.2   As we learned in the text, we rationalize a denominator of the form $a\sqrt{b} + c\sqrt{d}$ by multiplying by $a\sqrt{b} - c\sqrt{d}$.

(a) $\dfrac{2}{1+\sqrt{5}} = \dfrac{2}{1+\sqrt{5}} \cdot \dfrac{1-\sqrt{5}}{1-\sqrt{5}} = \dfrac{2-2\sqrt{5}}{1+\sqrt{5}-\sqrt{5}-5} = \dfrac{2-2\sqrt{5}}{-4} = \boxed{\dfrac{-1+\sqrt{5}}{2}}.$

(b) $\dfrac{\sqrt{7}}{3-2\sqrt{7}} = \dfrac{\sqrt{7}}{3-2\sqrt{7}} \cdot \dfrac{3+2\sqrt{7}}{3+2\sqrt{7}} = \dfrac{3\sqrt{7}+2\cdot 7}{9-6\sqrt{7}+6\sqrt{7}-2^2\cdot 7} = \boxed{-\dfrac{14+3\sqrt{7}}{19}}.$

11.4.3 The denominator looks a lot like one of the factors in the factorization of a sum of cubes. If we write the 9's as 3^2 we see it even more clearly:

$$9 - 3\sqrt[3]{3} + \sqrt[3]{9} = 3^2 - 3\sqrt[3]{3} + (\sqrt[3]{3})^2.$$

This is one of the factors of $3^3 + (\sqrt[3]{3})^3$. The other factor is $3 + \sqrt[3]{3}$, so we multiply the numerator and denominator of our fraction by this:

$$\frac{4}{9 - 3\sqrt[3]{3} + \sqrt[3]{9}} = \frac{4}{9 - 3\sqrt[3]{3} + \sqrt[3]{9}} \cdot \frac{3 + \sqrt[3]{3}}{3 + \sqrt[3]{3}} = \frac{4(3 + \sqrt[3]{3})}{3^3 + (\sqrt[3]{3})^3}$$

$$= \frac{4(3 + \sqrt[3]{3})}{30} = \boxed{\frac{6 + 2\sqrt[3]{3}}{15}}.$$

11.4.4 If we let $a = \sqrt{2} - \sqrt{5}$, we have $\dfrac{2}{a+\sqrt{7}}$, and we see that we can get rid of the $\sqrt{7}$ by multiplying by $a - \sqrt{7}$:

$$\frac{2}{\sqrt{2} - \sqrt{5} + \sqrt{7}} = \frac{2}{a + \sqrt{7}} \cdot \frac{a - \sqrt{7}}{a - \sqrt{7}} = \frac{2a - 2\sqrt{7}}{a^2 - a\sqrt{7} + a\sqrt{7} - 7} = \frac{2a - 2\sqrt{7}}{a^2 - 7}.$$

Now, we substitute $a = \sqrt{2} - \sqrt{5}$ back in to find

$$\frac{2a - 2\sqrt{7}}{a^2 - 7} = \frac{2\sqrt{2} - 2\sqrt{5} - 2\sqrt{7}}{(\sqrt{2} - \sqrt{5})^2 - 7} = \frac{2\sqrt{2} - 2\sqrt{5} - 2\sqrt{7}}{2 - 2\sqrt{10} + 5 - 7}$$

$$= \frac{2\sqrt{2} - 2\sqrt{5} - 2\sqrt{7}}{-2\sqrt{10}} = \frac{-\sqrt{2} + \sqrt{5} + \sqrt{7}}{\sqrt{10}}.$$

We can rationalize the denominator of this expression by multiplying by $\sqrt{10}$:

$$\frac{-\sqrt{2} + \sqrt{5} + \sqrt{7}}{\sqrt{10}} = \frac{-\sqrt{2} + \sqrt{5} + \sqrt{7}}{\sqrt{10}} \cdot \frac{\sqrt{10}}{\sqrt{10}}$$

$$= \frac{-\sqrt{20} + \sqrt{50} + \sqrt{70}}{10} = \boxed{\frac{-2\sqrt{5} + 5\sqrt{2} + \sqrt{70}}{10}}.$$

11.4.5 Each fraction is of the form $\dfrac{1}{\sqrt{a} + \sqrt{a-1}}$. We rationalize the denominator of such a fraction by multiplying by $\sqrt{a} - \sqrt{a-1}$:

$$\frac{1}{\sqrt{a} + \sqrt{a-1}} = \frac{1}{\sqrt{a} + \sqrt{a-1}} \cdot \frac{\sqrt{a} - \sqrt{a-1}}{\sqrt{a} - \sqrt{a-1}} = \frac{\sqrt{a} - \sqrt{a-1}}{a - (a-1)} = \sqrt{a} - \sqrt{a-1}.$$

So, our expression equals

$$\left(\sqrt{100} - \sqrt{99}\right) + \left(\sqrt{99} - \sqrt{98}\right) + \left(\sqrt{98} - \sqrt{97}\right) + \cdots + \left(\sqrt{3} - \sqrt{2}\right) + \left(\sqrt{2} - \sqrt{1}\right).$$

All the terms except the first and the last cancel, leaving $\sqrt{100} - \sqrt{1} = 10 - 1 = \boxed{9}$.

Exercises for Section 11.5

11.5.1 We have $ab + 5b + 2a + 10 = ab + 5b + 2a + 2 \cdot 5$, so we have a straightforward application of Simon's Favorite Factoring Trick:
$$ab + 5b + 2a + 10 = \boxed{(a + 5)(b + 2)}.$$

11.5.2 We have $xy + 8x - 3y - 24 = xy + 8x + (-3)y + (8)(-3)$, so we can apply Simon's Favorite Factoring Trick:
$$xy + 8x - 3y - 24 = \boxed{(x - 3)(y + 8)}.$$

11.5.3 We can write the expression as
$$19 \cdot 13 - 7 \cdot 13 + 3 \cdot 19 - 21 = 19 \cdot 13 + (-7) \cdot 13 + 3 \cdot 19 + (-7) \cdot 3,$$

so we can use Simon's Favorite Factoring Trick:

$$19 \cdot 13 + (-7) \cdot 13 + 3 \cdot 19 + (-7) \cdot 3 = (19 - 7)(13 + 3) = 12(16) = \boxed{2^6 \cdot 3}.$$

11.5.4 We seek the number of pairs of numbers x and y such that
$$\frac{1}{x} + \frac{1}{y} = \frac{1}{6}.$$

We get rid of the fractions by multiplying both sides by $6xy$, which gives us $6x + 6y = xy$. Putting all the terms on one side gives us $xy - 6x - 6y = 0$. From the coefficients of x and y, we see that we can add $(-6)(-6) = 36$ to both sides in order to be able to use Simon's Favorite Factoring Trick:

$$xy - 6x - 6y + 36 = 36 \quad \Rightarrow \quad (x - 6)(y - 6) = 36.$$

Since $36 = 1 \cdot 36 = 2 \cdot 18 = 3 \cdot 12 = 4 \cdot 9 = 6 \cdot 6$, we have 5 cases to worry about:

$x - 6 = 1$	$x - 6 = 2$	$x - 6 = 3$	$x - 6 = 4$	$x - 6 = 6$
$y - 6 = 36$	$y - 6 = 18$	$y - 6 = 12$	$y - 6 = 9$	$y - 6 = 6$

Make sure you see why we don't worry about reversing $x - 6$ and $y - 6$ in each of our cases: while $x = 10$, $y = 15$ and $x = 15$, $y = 10$ do both satisfy our original equation, they both lead to the same sum of distinct unit fractions. We also don't have to worry about negative numbers that multiply to 36, since all such cases will make either x or y nonpositive.

Each of these cases gives us one pair of unit fractions whose sum is $1/6$. Taking each case in turn, we find
$$\frac{1}{7} + \frac{1}{42} = \frac{1}{8} + \frac{1}{24} = \frac{1}{9} + \frac{1}{18} = \frac{1}{10} + \frac{1}{15} = \frac{1}{12} + \frac{1}{12} = \frac{1}{6}.$$
However, the problem specifies that the unit fractions that sum to $\frac{1}{6}$ must be distinct, so we must exclude $\frac{1}{12} + \frac{1}{12}$ from our count. There are $\boxed{4}$ pairs of fractions that fit the problem.

11.5.5 The coefficients -3 and 5 give us our clue what to add to both sides to allow us to factor the left side. We add $(-3)(5)$ to both sides, which gives
$$pq - 3p + 5q - 15 = -15 \quad \Rightarrow \quad (p+5)(q-3) = -15.$$
Since $-15 = (1)(-15) = (-1)(15) = (3)(-5) = (-3)(5)$, we have eight cases to consider:

$$
\begin{array}{llll}
p + 5 = 1 & p + 5 = -1 & p + 5 = 3 & p + 5 = -3 \\
q - 3 = -15 & q - 3 = 15 & q - 3 = -5 & q - 3 = 5
\end{array}
$$

$$
\begin{array}{llll}
p + 5 = -15 & p + 5 = 15 & p + 5 = -5 & p + 5 = 5 \\
q - 3 = 1 & q - 3 = -1 & q - 3 = 3 & q - 3 = -3
\end{array}
$$

These 8 cases give us 8 solutions:
$$(p, q) = \boxed{(-4, -12); \ (-6, 18); \ (-2, -2); \ (-8, 8); \ (-20, 4); \ (10, 2); \ (-10, 6); \ (0, 0)}.$$

11.5.6

(a) Don't get intimidated by the exponents on the variables. This is still just Simon's Favorite Factoring Trick:
$$x^2 y^2 - 4y^2 - x^2 + 4 = (x^2)(y^2) + (-4)(y^2) + (-1)(x^2) + (-4)(-1) = (x^2 - 4)(y^2 - 1).$$
If you don't see how that works, let $a = x^2$ and $b = y^2$ in the original expression. We can continue our factorization using difference of squares:
$$x^2 y^2 - 4y^2 - x^2 + 4 = (x^2 - 4)(y^2 - 1) = \boxed{(x - 2)(x + 2)(y - 1)(y + 1)}.$$

(b) The coefficient of the product cd makes this expression different from the others we have factored with Simon's Favorite Factoring Trick. However, the cd, c, and d terms sure suggest trying to find a way to use Simon's Trick. We can get rid of the coefficient by letting $a = 2c$ in our expression. This gives us
$$2cd - 3d - 14c + 21 = ad - 3d - 7a + 21.$$
This is just like the expressions we have already factored, so we use Simon's Favorite Factoring Trick to find
$$ad - 3d - 7a + 21 = (a - 3)(d - 7).$$
Letting $a = 2c$ in this tells us
$$2cd - 3d - 14c + 21 = (a - 3)(d - 7) = \boxed{(2c - 3)(d - 7)}.$$

Review Problems

11.26

(a) $(7 - x)^2 = 7^2 + 2(7)(-x) + (-x)^2 = 49 - 14x + x^2 = \boxed{x^2 - 14x + 49}$.

(b) $(2t - 9)^2 = (2t)^2 + 2(2t)(-9) + (-9)^2 = \boxed{4t^2 - 36t + 81}$.

11.27

(a) $45^2 = (40 + 5)^2 = 40^2 + 2(40)(5) + 5^2 = 1600 + 400 + 25 = \boxed{2025}$.

(b) $91^2 = (90 + 1)^2 = 90^2 + 2(90)(1) + 1^2 = 8100 + 180 + 1 = \boxed{8281}$. We also could have found this by computing $91^2 = 90^2 + 90 + 91 = 8100 + 181 = 8281$.

(c) $401^2 = (400 + 1)^2 = 400^2 + 2(400)(1) + 1^2 = 160000 + 800 + 1 = \boxed{160801}$. We also could have found this by computing $401^2 = 400^2 + 400 + 401 = 160000 + 801 = 160801$.

(d) $199^2 = (200 - 1)^2 = 200^2 + 2(200)(-1) + (-1)^2 = 40000 - 400 + 1 = \boxed{39601}$.

11.28   If $9x^2 + 24x + a$ is the square of a binomial, then the binomial has the form $3x + b$ for some number b, because $(3x)^2 = 9x^2$. So, we compare $(3x + b)^2$ to $9x^2 + 24x + a$. Expanding $(3x + b)^2$ gives

$$(3x + b)^2 = (3x)^2 + 2(3x)(b) + b^2 = 9x^2 + 6bx + b^2.$$

Equating the linear term of this to the linear term of $9x^2 + 24x + a$, we have $6bx = 24x$, so $b = 4$. Equating the constant term of $9x^2 + 6bx + b^2$ to that of $9x^2 + 24x + a$ gives us $a = b^2 = \boxed{16}$.

11.29

(a) $r^2 - 121 = r^2 - 11^2 = \boxed{(r - 11)(r + 11)}$.

(b) $-32t^2 + 50 = 2(-16t^2 + 25) = 2(25 - 16t^2) = \boxed{2(5 - 4t)(5 + 4t)}$. We might also write this expression as $-2(4t - 5)(4t + 5)$.

11.30   We use the difference of squares factorization:

$$111^2 - 89^2 = (111 - 89)(111 + 89) = 22(200) = \boxed{4400}.$$

11.31   We might expand both of the squares of binomials first, but it's a little faster to factor the difference of squares first:

$$(207 + 100)^2 - (207 - 100)^2 = [(207 + 100) + (207 - 100)][(207 + 100) - (207 - 100)]$$
$$= (207 + 207)(100 + 100) = (414)(200) = \boxed{82800}.$$

11.32   Multiplying the two numbers out is scary. Instead, we let $x = 4050607$, so that $4050607^2 = x^2$ and $(4050608)(4050606) = (x + 1)(x - 1) = x^2 - 1$. Since x^2 is 1 more than $x^2 - 1$, we see that $\boxed{4050607^2}$ is bigger than $(4050608)(4050606)$.

11.33   We could square those expressions for x and y, but it's much more convenient to factor $x^2 - y^2$ as $(x + y)(x - y)$. Since $x + y = 2(2001^{1002})$ and $x - y = -2(2001^{-1002})$, we have

$$x^2 - y^2 = (x + y)(x - y) = 2(2001^{1002}) \cdot (-2)(2001^{-1002}) = [2 \cdot (-2)](2001^{1002} \cdot 2001^{-1002}) = \boxed{-4}.$$

11.34   As discussed in the text, the difference between x^2 and $(x + 1)^2$ is $x + x + 1 = 2x + 1$. Therefore, we have $2x + 1 = 63$, so $x = 31$. So, the two consecutive perfect squares with a difference of 63 are 31^2 and 32^2, or $\boxed{961 \text{ and } 1024}$.

11.35

(a) $a^3 + 27 = a^3 + 3^3 = (a + 3)(a^2 - a \cdot 3 + 3^2) = \boxed{(a + 3)(a^2 - 3a + 9)}$.

(b) $a^3b^3 + 8c^3 = (ab)^3 + (2c)^3 = (ab + 2c)[(ab)^2 - (ab)(2c) + (2c)^2] = \boxed{(ab + 2c)(a^2b^2 - 2abc + 4c^2)}$.

(c) $2r^3 - 16 = 2(r^3 - 8) = 2(r^3 - 2^3) = 2(r - 2)(r^2 + r \cdot 2 + 2^2) = \boxed{2(r - 2)(r^2 + 2r + 4)}$.

(d) $1000 - x^6y^3 = 10^3 - (x^2y)^3 = (10 - x^2y)[10^2 + (10)(x^2y) + (x^2y)^2] = \boxed{(10 - x^2y)(100 + 10x^2y + x^4y^2)}$.

11.36   We can just multiply it out:

$$(\sqrt[3]{t} - \sqrt[3]{u})(\sqrt[3]{t^2} + \sqrt[3]{tu} + \sqrt[3]{u^2}) = \sqrt[3]{t}(\sqrt[3]{t^2} + \sqrt[3]{tu} + \sqrt[3]{u^2}) - \sqrt[3]{u}(\sqrt[3]{t^2} + \sqrt[3]{tu} + \sqrt[3]{u^2})$$
$$= \sqrt[3]{t^3} + \sqrt[3]{t^2u} + \sqrt[3]{tu^2} - \sqrt[3]{t^2u} - \sqrt[3]{tu^2} - \sqrt[3]{u^3} = \boxed{t - u}.$$

With such a simple answer, we wonder if there's another way we could have seen it. Seeing all those cube roots makes us think of our difference of cubes and sum of cubes factorizations. If we view t as $(\sqrt[3]{t})^3$ and u as $(\sqrt[3]{u})^3$, we see that the given expression is the difference of cubes factorization of $t - u = (\sqrt[3]{t})^3 - (\sqrt[3]{u})^3$.

11.37

(a) $\dfrac{21}{\sqrt{21}} = \dfrac{21}{\sqrt{21}} \cdot \dfrac{\sqrt{21}}{\sqrt{21}} = \dfrac{21\sqrt{21}}{21} = \boxed{\sqrt{21}}$.

(b) $\dfrac{20}{\sqrt[3]{25}} = \dfrac{20}{\sqrt[3]{5^2}} = \dfrac{20}{\sqrt[3]{5^2}} \cdot \dfrac{\sqrt[3]{5}}{\sqrt[3]{5}} = \dfrac{20\sqrt[3]{5}}{5} = \boxed{4\sqrt[3]{5}}$.

(c) We first write $18 = 2 \cdot 3^2$. We take care of the 2's and the 3's in the denominator separately:

$$\frac{4}{\sqrt[4]{18}} = \frac{4}{\sqrt[4]{2}\sqrt[4]{3^2}} = \frac{4}{\sqrt[4]{2}\sqrt[4]{3^2}} \cdot \frac{\sqrt[4]{2^3}\sqrt[4]{3^2}}{\sqrt[4]{2^3}\sqrt[4]{3^2}}$$
$$= \frac{4\sqrt[4]{2^3 \cdot 3^2}}{\sqrt[4]{2^4}\sqrt[4]{3^4}} = \frac{4\sqrt[4]{72}}{2 \cdot 3} = \boxed{\frac{2\sqrt[4]{72}}{3}}.$$

11.38

(a) $\dfrac{8}{\sqrt{15} - \sqrt{7}} = \dfrac{8}{\sqrt{15} - \sqrt{7}} \cdot \dfrac{\sqrt{15} + \sqrt{7}}{\sqrt{15} + \sqrt{7}} = \dfrac{8(\sqrt{15} + \sqrt{7})}{15 - 7} = \boxed{\sqrt{15} + \sqrt{7}}$.

(b) $\dfrac{9}{3\sqrt{2} - 2\sqrt{5}} = \dfrac{9}{3\sqrt{2} - 2\sqrt{5}} \cdot \dfrac{3\sqrt{2} + 2\sqrt{5}}{3\sqrt{2} + 2\sqrt{5}} = \dfrac{27\sqrt{2} + 18\sqrt{5}}{(3\sqrt{2})^2 - (2\sqrt{5})^2} = \dfrac{27\sqrt{2} + 18\sqrt{5}}{18 - 20} = \boxed{-\dfrac{27\sqrt{2} + 18\sqrt{5}}{2}}$.

11.39

(a) We use Simon's Favorite Factoring Trick:

$$pq - 7p + 9q - 63 = pq + (-7)(p) + 9q + (-7)(9) = \boxed{(p + 9)(q - 7)}.$$

(b)   After factoring out -1, we have $-(rs - 5r - 2s + 10)$. We can factor $rs - 5r - 2s + 10$:

$$-(rs - 5r - 2s + 10) = -[rs + (-5)(r) + (-2)(s) + (-5)(-2)] = \boxed{-(r - 2)(s - 5)}.$$

11.40   We can view the denominator of this fraction as one of the factors of the difference of cubes factorization of $1^3 - (2\sqrt[3]{2})^3$. The other factor is $1^2 + (1)(2\sqrt[3]{2}) + (2\sqrt[3]{2})^2 = 1 + 2\sqrt[3]{2} + 4\sqrt[3]{4}$. Multiplying the numerator and denominator of our fraction by this factor gives

$$\frac{2}{1 - 2\sqrt[3]{2}} = \frac{2}{1 - 2\sqrt[3]{2}} \cdot \frac{1 + 2\sqrt[3]{2} + 4\sqrt[3]{4}}{1 + 2\sqrt[3]{2} + 4\sqrt[3]{4}} = \frac{2 + 4\sqrt[3]{2} + 8\sqrt[3]{4}}{1^3 - (2\sqrt[3]{2})^3}$$

$$= \frac{2 + 4\sqrt[3]{2} + 8\sqrt[3]{4}}{1 - 2^3 \cdot 2} = \boxed{-\frac{2 + 4\sqrt[3]{2} + 8\sqrt[3]{4}}{15}}.$$

11.41   We would like to factor the left side. To do so, we add $(-2)(7) = -14$ to both sides, so that we can use Simon's Favorite Factoring Trick on the left:

$$xy - 2x + 7y - 14 = 49 - 14 \quad \Rightarrow \quad xy + (-2)x + 7y + (-2)(7) = 35 \quad \Rightarrow \quad (x + 7)(y - 2) = 35.$$

We have $35 = 1 \cdot 35 = 5 \cdot 7 = (-1)(-35) = (-5)(-7)$, so we have the following possibilities:

$x + 7 = 1$	$x + 7 = 5$	$x + 7 = -1$	$x + 7 = -5$
$y - 2 = 35$	$y - 2 = 7$	$y - 2 = -35$	$y - 2 = -7$
$x + 7 = 35$	$x + 7 = 7$	$x + 7 = -35$	$x + 7 = -7$
$y - 2 = 1$	$y - 2 = 5$	$y - 2 = -1$	$y - 2 = -5$

These give us the following 8 solutions:

$$(x, y) = \boxed{(-6, 37);\ (-2, 9);\ (-8, -33);\ (-12, -5);\ (28, 3);\ (0, 7);\ (-42, 1);\ (-14, -3)}.$$

11.42   Writing 3 as $\sqrt{9}$ and 2 as $\sqrt{4}$, we see that each term has the form $\dfrac{1}{\sqrt{a} - \sqrt{a - 1}}$. We rationalize the denominator of such fractions as follows:

$$\frac{1}{\sqrt{a} - \sqrt{a - 1}} \cdot \frac{\sqrt{a} + \sqrt{a - 1}}{\sqrt{a} + \sqrt{a - 1}} = \frac{\sqrt{a} + \sqrt{a - 1}}{a - (a - 1)} = \sqrt{a} + \sqrt{a - 1}.$$

So, our expression equals

$$(\sqrt{9} + \sqrt{8}) - (\sqrt{8} + \sqrt{7}) + (\sqrt{7} + \sqrt{6}) - (\sqrt{6} + \sqrt{5}) + (\sqrt{5} + \sqrt{4}).$$

Everything cancels except $\sqrt{9} + \sqrt{4} = 3 + 2 = \boxed{5}$.

Challenge Problems

11.43 *Solution 1: Find a, b, and c.* Let $a = 2n - 1$ and $b = 2n + 1$. We have

$$b^2 - a^2 = (2n + 1)^2 - (2n - 1)^2 = [(2n + 1) + (2n - 1)][(2n + 1) - (2n - 1)] = (4n)(2) = 8n.$$

Therefore, we have $8n = 344$, so $n = 43$. So, $a = 2n - 1 = 85$, $b = 87$, and $c = 89$, so $c^2 - b^2 = 89^2 - 87^2 = (89 - 87)(89 + 87) = \boxed{352}$.

Solution 2: Use what we know about consecutive squares. Again, let $a = 2n - 1$, so $b = 2n + 1$ and $c = 2n + 3$. The difference between $a^2 = (2n - 1)^2$ and the next square, $(2n)^2$, is $(2n - 1) + 2n = 4n - 1$. The difference between $(2n)^2$ and the next square, $b^2 = (2n + 1)^2$, is $2n + (2n + 1) = 4n + 1$. So, the difference between a^2 and b^2 is

$$(2n - 1) + 2n + 2n + (2n + 1) = 8n.$$

Similarly, the difference between $b^2 = (2n + 1)^2$ and $c^2 = (2n + 3)^2$ is

$$(2n + 1) + (2n + 2) + (2n + 2) + (2n + 3) = 8n + 8.$$

Since we are told that $b^2 - a^2 = 8n = 344$, we have $c^2 - b^2 = 8n + 8 = \boxed{352}$.

11.44

(a) $29^2 = (30 - 1)^2 = 30^2 + 2(30)(-1) + (-1)^2 = 900 - 60 + 1 = \boxed{841}$.

(b) $299^2 = (300 - 1)^2 = 300^2 + 2(300)(-1) + (-1)^2 = 90000 - 600 + 1 = \boxed{89401}$.

(c) $2999^2 = (3000 - 1)^2 = 3000^2 + 2(3000)(-1) + (-1)^2 = 9000000 - 6000 + 1 = \boxed{8994001}$.

(d) The first few parts give us a pattern. It looks tough to deal with 29999999^2, but we do know how to deal with 30000000^2, so we write 29999999^2 as

$$(30000000 - 1)^2 = 30000000^2 + 2(30000000)(-1) + (-1)^2 = (3 \cdot 10^7)^2 - 6 \cdot 10^7 + 1$$
$$= 9 \cdot 10^{14} - 6 \cdot 10^7 + 1 = 899999940000001.$$

The sum of the digits in this number is $8 + 6 \cdot 9 + 4 + 1 = \boxed{67}$.

11.45 We have a whole bunch of differences of squares. We factor and factor and factor, and we have

$$\frac{(1998 - 1996)(1998 + 1996)(1998 - 1995)(1998 + 1995) \cdots (1998 - 0)(1998 + 0)}{(1997 - 1996)(1997 + 1996)(1997 - 1995)(1997 + 1995) \cdots (1997 - 0)(1997 + 0)}.$$

This equals
$$\frac{(2)(3994)(3)(3993)(4)(3992) \cdots (1997)(1999)(1998)(1998)}{(1)(3993)(2)(3992)(3)(3991) \cdots (1996)(1998)(1997)(1997)}.$$

The numerator consists of 1998 times the product of all the numbers from 2 to 3994, while the denominator consists of 1997 times the product of all the numbers from 1 to 3993. Therefore, everything cancels except $1998 \cdot 3994$ in the numerator and the 1997 in the denominator, so our expression equals $1998 \cdot 3994/1997 = 1998 \cdot 2 = \boxed{3996}$.

11.46 If we view $\sqrt[4]{2}$ as $\sqrt{\sqrt{2}}$, we can use the same tactic we used in the text for problems involving denominators of the form $a - \sqrt{b}$:

$$\frac{2}{2 - \sqrt[4]{2}} = \frac{2}{2 - \sqrt{\sqrt{2}}} = \frac{2}{2 - \sqrt{\sqrt{2}}} \cdot \frac{2 + \sqrt{\sqrt{2}}}{2 + \sqrt{\sqrt{2}}} = \frac{4 + 2\sqrt{\sqrt{2}}}{4 - \left(\sqrt{\sqrt{2}}\right)^2} = \frac{4 + 2\sqrt[4]{2}}{4 - \sqrt{2}}.$$

We know how to rationalize the denominator of this fraction:

$$\frac{4 + 2\sqrt[4]{2}}{4 - \sqrt{2}} = \frac{4 + 2\sqrt[4]{2}}{4 - \sqrt{2}} \cdot \frac{4 + \sqrt{2}}{4 + \sqrt{2}} = \frac{16 + 4\sqrt{2} + 8\sqrt[4]{2} + 2\sqrt[4]{2}\sqrt{2}}{16 - 2}$$

$$= \frac{16 + 4\sqrt{2} + 8\sqrt[4]{2} + 2\sqrt[4]{2}\sqrt[4]{4}}{14} = \boxed{\frac{8 + 2\sqrt{2} + 4\sqrt[4]{2} + \sqrt[4]{8}}{7}}.$$

11.47 The expression looks like the expansion of the square of a binomial. If we write $5^{12} = (5^6)^2$, $10^6 = (5 \cdot 2)^6 = 5^6 \cdot 2^6$, and $2^{12} = (2^6)^2$, we have

$$5^{12} - 2 \cdot 10^6 + 2^{12} = (5^6)^2 - 2(5^6)(2^6) + (2^6)^2 = (5^6 - 2^6)^2.$$

Our binomial is a difference of squares, so we can factor more. We have

$$(5^6 - 2^6)^2 = [(5^3)^2 - (2^3)^2]^2 = [(5^3 - 2^3)(5^3 + 2^3)]^2.$$

We might evaluate each of these factors, or factor them further using our sum of cubes and difference of cubes factorizations:

$$[(5^3 - 2^3)(5^3 + 2^3)]^2 = [(5 - 2)(5^2 + 5 \cdot 2 + 2^2)(5 + 2)(5^2 - 5 \cdot 2 + 2^2)]^2 = [(3)(39)(7)(19)]^2.$$

Since $39 = 3 \cdot 13$, we see that the largest prime divisor is $\boxed{19}$.

11.48 We write the expression we seek with a common denominator, ab:

$$\frac{a}{b} + \frac{b}{a} - ab = \frac{a^2 + b^2 - a^2b^2}{ab}.$$

We haven't used the given information that $ab = a - b$ yet. We do so by substituting $a^2b^2 = (ab)^2 = (a - b)^2 = a^2 - 2ab + b^2$ in for a^2b^2 in the numerator:

$$\frac{a^2 + b^2 - a^2b^2}{ab} = \frac{a^2 + b^2 - (a^2 - 2ab + b^2)}{ab} = \frac{a^2 + b^2 + 2ab - a^2 - b^2}{ab} = \frac{2ab}{ab} = \boxed{2}.$$

11.49

(a) The other factor must have an x^4 term and a y^4 term, so we start with $x^4 + y^4$:

$$(x + y)(x^4 + y^4) = x(x^4 + y^4) + y(x^4 + y^4) = x^5 + xy^4 + x^4y + y^5.$$

We have some extra terms. To eliminate the xy^4, we include a $-xy^3$ term in our second factor, since the product of this with y is $-xy^4$. Similarly, we include a $-x^3y$ term to eliminate the x^4y:

$$(x + y)(x^4 - x^3y - xy^3 + y^4) = x(x^4 - x^3y - xy^3 + y^4) + y(x^4 - x^3y - xy^3 + y^4)$$
$$= x^5 - x^4y - x^2y^3 + xy^4 + x^4y - x^3y^2 - xy^4 + y^5$$
$$= x^5 - x^2y^3 - x^3y^2 + y^5.$$

More extra terms. However, we see that we can eliminate them both by including $+x^2y^2$ in our second term:

$$\boxed{(x + y)(x^4 - x^3y + x^2y^2 - xy^3 + y^4)} = x(x^4 - x^3y + x^2y^2 - xy^3 + y^4) + y(x^4 - x^3y + x^2y^2 - xy^3 + y^4)$$
$$= x^5 - x^4y + x^3y^2 - x^2y^3 + xy^4 + x^4y - x^3y^2 + x^2y^3 - xy^4 + y^5$$
$$= x^5 + y^5.$$

We could also have used our $a^k - b^k$ factorization that we developed earlier. Specifically, we saw that

$$a^5 - b^5 = (a - b)(a^4 + a^3b + a^2b^2 + ab^3 + b^4).$$

If we let $a = x$ and $b = -y$, we have $x^5 - (-y)^5 = x^5 + y^5$ on the left. Making the substitution on the right gives us

$$x^5 + y^5 = [x - (-y)][x^4 + x^3(-y) + x^2(-y)^2 + x(-y)^3 + (-y)^4]$$
$$= (x + y)(x^4 - x^3y + x^2y^2 - xy^3 + y^4).$$

(b) We look at our sum of cubes and sum of fifth powers factorizations:

$$x^3 + y^3 = (x + y)(x^2 - xy + y^2),$$
$$x^5 + y^5 = (x + y)(x^4 - x^3y + x^2y^2 - xy^3 + y^4).$$

These give us a good guess for the second factor of $x^7 + y^7$:

$$\boxed{(x + y)(x^6 - x^5y + x^4y^2 - x^3y^3 + x^2y^4 - xy^5 + y^6)}$$
$$= x(x^6 - x^5y + x^4y^2 - x^3y^3 + x^2y^4 - xy^5 + y^6) + y(x^6 - x^5y + x^4y^2 - x^3y^3 + x^2y^4 - xy^5 + y^6)$$
$$= x^7 - x^6y + x^5y^2 - x^4y^3 + x^3y^4 - x^2y^5 + xy^6 + x^6y - x^5y^2 + x^4y^3 - x^3y^4 + x^2y^5 - xy^6 + y^7$$
$$= x^7 + y^7.$$

We also could have used the factorization of $a^7 - b^7$ to find the factorization of $x^7 + y^7$.

(c) Using our first two parts as a guideline, we once again have a very good guess at the factorization:

$$\boxed{(x + y)(x^{2n} - x^{2n-1}y + x^{2n-2}y^2 + \cdots + x^2y^{2n-2} - xy^{2n-1} + y^{2n})}$$
$$= x(x^{2n} - x^{2n-1}y + \cdots - xy^{2n-1} + y^{2n}) + y(x^{2n} - x^{2n-1}y + \cdots - xy^{2n-1} + y^{2n})$$
$$= x^{2n+1} - x^{2n}y + x^{2n-1}y^2 + \cdots - x^2y^{2n-1} + xy^{2n} + x^{2n}y - x^{2n-1}y^2 + \cdots + x^2y^{2n-1} - xy^{2n} + y^{2n+1}$$
$$= x^{2n+1} + y^{2n+1}.$$

(d) Our factorization in the previous part tells us that if k is odd, then we can write
$$x^k + y^k = (x + y)(x^{k-1} - x^{k-2}y + x^{k-3}y^2 + \cdots + x^2y^{k-3} - xy^{k-2} + y^{k-1}).$$

However, let's see if it's possible to have k be even in the factor
$$x^{k-1} - x^{k-2}y + x^{k-3}y^2 + \cdots + x^2y^{k-3} - xy^{k-2} + y^{k-1}.$$

There are k terms in this factor (one for each power of x from x^0 through x^{k-1}). So, when k is even, there is an even number of terms in this factor. In other words, the sign of the last term is negative, and this factor should read:
$$x^{k-1} - x^{k-2}y + x^{k-3}y^2 + \cdots - x^2y^{k-3} + xy^{k-2} - y^{k-1}.$$

So, the factorization
$$x^k + y^k = (x + y)(x^{k-1} - x^{k-2}y + x^{k-3}y^2 + \cdots + x^2y^{k-3} - xy^{k-2} + y^{k-1})$$

can't be true when k is even, because the second factor in the product on the right side is impossible when k is even. Maybe we think we can fix this by considering the product
$$(x + y)(x^{k-1} - x^{k-2}y + x^{k-3}y^2 + \cdots - x^2y^{k-3} + xy^{k-2} - y^{k-1}).$$

Unfortunately, this doesn't give us $x^k + y^k$; we can see this most quickly by noting that the y^k term in the expansion of this product is $-y^k$, not y^k.

Maybe we think we can fix the problem by considering the product
$$(x - y)(x^{k-1} - x^{k-2}y + x^{k-3}y^2 + \cdots - x^2y^{k-3} + xy^{k-2} - y^{k-1}).$$

This expansion will have a y^k term, but unfortunately, it will also have a $-2x^{k-1}y$ term (and a whole lot of other terms we don't want).

So, we conclude that we can't factor $x^k + y^k$ when k is even using the same tactics as we used in the first three parts.

11.50 The whole square has side length $a+b+c$, so its area is $(a+b+c)^2$. We can also find the area of each rectangle (or square) inside the big square as shown at right. When we add all the areas of these regions, we have $a^2 + b^2 + c^2 + 2ab + 2bc + 2ca$. Since these two methods of computing the area of the whole square must produce the same area, we have
$$(a + b + c)^2 = a^2 + b^2 + c^2 + 2ab + 2bc + 2ca.$$

c	ac	bc	c^2
b	ab	b^2	bc
a	a^2	ab	ac
	a	b	c

11.51 We can't factor either side as the equation is now, but if we subtract a^2 and ac from both sides, we'll have
$$bc - ac = b^2 - a^2 \quad \Rightarrow \quad c(b - a) = (b + a)(b - a).$$

Now, we can subtract $c(b - a)$ from both sides and factor more:
$$(b + a)(b - a) - c(b - a) = 0 \quad \Rightarrow \quad (b + a - c)(b - a) = 0.$$

Therefore, we must have either $b - a = 0$ or $b + a - c = 0$. From $b - a = 0$, we have $b = a$, and no restriction on c. Therefore, we have 5 choices for the value that a and b both have, and 5 choices for c. So, there are $5 \cdot 5 = 25$ solutions with $a = b$.

Turning to $b + a - c = 0$, we have $c = a + b$. We can count these solutions quickly with a little casework.

- $c = 2$. Then $a = b = 1$, and we have already counted this solution in the $a = b$ case.

- $c = 3$. We have two such solutions, $(a, b, c) = (2, 1, 3)$ and $(1, 2, 3)$.

- $c = 4$. We have two new solutions, $(a, b, c) = (3, 1, 4)$ and $(1, 3, 4)$. (We already counted $(2, 2, 4)$ in the $a = b$ case.)

- $c = 5$. We have four new solutions, $(a, b, c) = (4, 1, 5)$, $(3, 2, 5)$, $(2, 3, 5)$, and $(1, 4, 5)$.

Combining our solutions with $a = b$ with the new solutions we found with $a + b = c$, we have a total of $25 + 2 + 2 + 4 = \boxed{33}$ solutions.

11.52 First, we note that the side with length $1125(10^{2n+1}) + 8$ must be the longest side, since this number is larger than both $6(10^{n+2})$ and $1125(10^{2n+1}) - 8$ when n is a positive integer. So, if the three numbers given are the side lengths of a right triangle, then we must have

$$\left[6(10^{n+2})\right]^2 + \left[1125(10^{2n+1}) - 8\right]^2 = \left[1125(10^{2n+1}) + 8\right]^2.$$

That's a pretty intimidating equation! Rather than squaring the cumbersome binomials, we look for clever ways to simplify the equation. We can start by letting $k = 1125(10^{2n+1})$, so our equation is much less scary:

$$\left[6(10^{n+2})\right]^2 + (k - 8)^2 = (k + 8)^2.$$

We could square these binomials easily, and it's clear that the k^2 terms will cancel. We also can move the $(k - 8)^2$ term to the other side to create a difference of squares:

$$\left[6(10^{n+2})\right]^2 = (k + 8)^2 - (k - 8)^2.$$

If this equation is true, then the triangle is a right triangle. We can factor the right side as a difference of squares:

$$\left[6(10^{n+2})\right]^2 = [(k + 8) - (k - 8)][(k + 8) + (k - 8)] = 32k.$$

Now we're getting somewhere. Since

$$\left[6(10^{n+2})\right]^2 = 36(10^{2n+4}),$$

we must check if $32k$ also equals $36(10^{2n+4})$. Substituting our expression for k gives us

$$32k = 32(1125)(10^{2n+1}) = 16(2250)(10^{2n+1}) = 4(9000)(10^{2n+1}) = 36(10^3)(10^{2n+1}) = 36(10^{2n+4}).$$

Therefore, we have shown that

$$\left[6(10^{n+2})\right]^2 = \left[1125(10^{2n+1}) + 8\right]^2 - \left[1125(10^{2n+1}) - 8\right]^2,$$

so we can conclude that the triangle with the three given lengths as sides is a right triangle for all positive integers n.

11.53 We could just multiply everything out, but that doesn't look like much fun. Besides, finding the square root of large numbers usually isn't too easy. So, we look for an algebraic solution. We could let $x = 50$, and write our expression as

$$\sqrt{1 + 50 \cdot 51 \cdot 52 \cdot 53} = \sqrt{1 + x(x + 1)(x + 2)(x + 3)}.$$

However, expanding $x(x + 1)(x + 2)(x + 3)$ is a pain. We can make it a little easier by instead letting $z = 51$, so our expression is

$$\sqrt{1 + 50 \cdot 51 \cdot 52 \cdot 53} = \sqrt{1 + (z - 1)(z)(z + 1)(z + 2)}.$$

We recognize that we can multiply $z - 1$ and $z + 1$ to get $z^2 - 1$:

$$(z - 1)(z)(z + 1)(z + 2) = (z - 1)(z + 1)(z)(z + 2) = (z^2 - 1)(z)(z + 2) = (z^3 - z)(z + 2).$$

We expand $(z^3 - z)(z + 2)$ just as we do any product of binomials:

$$(z^3 - z)(z + 2) = z^3(z + 2) - z(z + 2) = z^4 + 2z^3 - z^2 - 2z.$$

So, our expression now is

$$\sqrt{1 + (z - 1)(z)(z + 1)(z + 2)} = \sqrt{z^4 + 2z^3 - z^2 - 2z + 1}.$$

So, if we can figure out what $z^4 + 2z^3 - z^2 - 2z + 1$ is the square of, we can find its square root. The z^4 suggests a z^2 in our square root, and the 1 at the end suggests either a 1 or a -1 at the end. However,

$$(z^2 + 1)^2 = (z^2)^2 + 2(z^2)(1) + 1^2 = z^4 + 2z^2 + 1.$$

Similarly, $(z^2 - 1)^2 = z^4 - 2z^2 + 1$. We need another term in our square root. But should we add this term to $z^2 + 1$ or $z^2 - 1$ when trying to find the square root?

We could try both in turn, but the fact that the coefficient of z^2 in $z^4 + 2z^3 - z^2 - 2z + 1$ is negative strongly suggests that if it has a square root, then that square root is of the form $z^2 + az - 1$, not $z^2 + az + 1$. The coefficient of z in the expansion of $(z^2 + az - 1)^2$ is $-2a$, so we must have $-2a = -2$. Therefore, we guess that $a = 1$. Testing this, we find:

$$(z^2 + z - 1)(z^2 + z - 1) = z^2(z^2 + z - 1) + z(z^2 + z - 1) - (z^2 + z - 1)$$
$$= z^4 + z^3 - z^2 + z^3 + z^2 - z - z^2 - z + 1$$
$$= z^4 + 2z^3 - z^2 - 2z + 1.$$

Phew. So, we have

$$\sqrt{z^4 + 2z^3 - z^2 - 2z + 1} = z^2 + z - 1.$$

Since $z = 51$, we have

$$z^2 + z - 1 = 51^2 + 51 - 1 = (50 + 1)^2 + 50 = 2500 + 100 + 1 + 50 = \boxed{2651}.$$

11.54 We know we can produce an x^4 term and a $4y^4$ term by squaring $x^2 + 2y^2$:

$$(x^2 + 2y^2)^2 = (x^2)^2 + 2(x^2)(2y^2) + (2y^2)^2 = x^4 + 4x^2y^2 + 4y^4.$$

So, we have

$$(x^2 + 2y^2)^2 = x^4 + 4x^2y^2 + 4y^4.$$

Close, but we have that extra $4x^2y^2$ term on the right. We'd like to get rid of it somehow. We notice that we have a perfect square on the left side, and that $4x^2y^2$ is itself a perfect square. So, if we subtract $4x^2y^2$ from both sides, we'll be able to factor:

$$x^4 + 4y^4 = (x^2 + 2y^2)^2 - 4x^2y^2 = (x^2 + 2y^2)^2 - (2xy)^2 = \boxed{(x^2 - 2xy + 2y^2)(x^2 + 2xy + 2y^2)}.$$

11.55 We borrow a strategy from the previous problem. We know that squaring $2^{11} + 1$ gives us 2^{22} and 1, plus an extra term:

$$(2^{11} + 1)^2 = (2^{11})^2 + 2(2^{11})(1) + 1^2 = 2^{22} + 2^{12} + 1.$$

Aha! Since 2^{12} is a perfect square, when we subtract it from both sides, we can factor:

$$2^{22} + 1 = (2^{11} + 1)^2 - 2^{12} = (2^{11} + 1)^2 - (2^6)^2$$
$$= (2^{11} + 1 + 2^6)(2^{11} + 1 - 2^6) = (2048 + 1 + 64)(2048 + 1 - 64) = \boxed{(2113)(1985)}.$$

11.56 We're not sure what to do about those scary exponents, but we do know what to do about $\sqrt{2} + 1$ in the denominator. We want to multiply that $\sqrt{2} + 1$ in the denominator by $\sqrt{2} - 1$ in order to rationalize it. However, to do so, we must multiply the denominator (and, of course, the numerator) by $(\sqrt{2} - 1)^{1+\sqrt{3}}$:

$$\frac{(\sqrt{2} - 1)^{1-\sqrt{3}}}{(\sqrt{2} + 1)^{1+\sqrt{3}}} = \frac{(\sqrt{2} - 1)^{1-\sqrt{3}}}{(\sqrt{2} + 1)^{1+\sqrt{3}}} \cdot \frac{(\sqrt{2} - 1)^{1+\sqrt{3}}}{(\sqrt{2} - 1)^{1+\sqrt{3}}} = \frac{(\sqrt{2} - 1)^{1-\sqrt{3}+1+\sqrt{3}}}{[(\sqrt{2} + 1)(\sqrt{2} - 1)]^{1+\sqrt{3}}}$$
$$= \frac{(\sqrt{2} - 1)^2}{(2 - 1)^{1+\sqrt{3}}} = \frac{(\sqrt{2})^2 + 2(\sqrt{2})(-1) + (-1)^2}{1^{1+\sqrt{3}}}$$
$$= \frac{2 - 2\sqrt{2} + 1}{1} = \boxed{3 - 2\sqrt{2}}.$$

11.57 Multiplying those fourth powers doesn't look like fun. Fourth powers are squares, however, so we can factor:

$$2003^4 - 1997^4 = (2003^2)^2 - (1997^2)^2 = (2003^2 - 1997^2)(2003^2 + 1997^2) = (2003 - 1997)(2003 + 1997)(2003^2 + 1997^2)$$

We could multiply out the squares now, but we notice that 1997 and 2003 are the same distance from 2000, so we have

$$2003^2 + 1997^2 = (2000 + 3)^2 + (2000 - 3)^2 = 2000^2 + 2(2000)(3) + 3^2 + 2000^2 - 2(2000)(3) + (-3)^2$$
$$= 2(2000^2) + 2(3^2) = 8000000 + 18 = 8000018.$$

Therefore, we have

$$2003^4 - 1997^4 = (2003 - 1997)(2003 + 1997)(2003^2 + 1997^2) = 6(4000)(8000018)$$
$$= 24000(8000018) = 192000432000.$$

The sum of the digits of this number is $1 + 9 + 2 + 4 + 3 + 2 = \boxed{21}$.

CHAPTER 12

Complex Numbers

Exercises for Section 12.2

12.2.1

$$(-8i)^2 = (-8)^2(i)^2 = (64)(-1) = \boxed{-64}$$
$$(i/2)^2 = (i^2)/(2^2) = (-1)/4 = \boxed{-1/4}$$
$$(5i^4) = 5^4 i^4 = (625)(1) = \boxed{625}$$

12.2.2

(a) We take the square root of both sides to find

$$x = \pm \sqrt{-36} = \pm \sqrt{36}\sqrt{-1} = \boxed{\pm 6i}.$$

(b) We isolate y^2 by dividing by 2 to get $y^2 = -20$. Taking the square root of both sides gives

$$y = \pm \sqrt{-20} = \pm \sqrt{20}\sqrt{-1} = \pm(2\sqrt{5})(i) = \boxed{\pm 2i\sqrt{5}}.$$

12.2.3 We have $i^6 = i^4 \cdot i^2 = 1 \cdot (-1) = -1$. We also have $i^{16} = (i^4)^4 = 1^4 = 1$, and $i^{-26} = 1/i^{26} = 1/(i^{24} \cdot i^2) = 1/[1 \cdot (-1)] = -1$. So, adding these three results gives $i^6 + i^{16} + i^{-26} = -1 + 1 - 1 = \boxed{-1}$.

12.2.4 $i^4 = i^8 = i^{12} = \cdots = i^{100} = 1$. $i^2 = i^6 = i^{10} = \cdots = i^{98} = -1$. So,

$$
\begin{aligned}
i^2 + i^4 + i^6 + \cdots + i^{98} + i^{100} &= -1 + 1 - 1 + 1 - 1 + 1 - \cdots - 1 + 1 \\
&= (-1 + 1) + (-1 + 1) + \cdots + (-1 + 1) \\
&= 0 + 0 + 0 + \cdots + 0 \\
&= \boxed{0}.
\end{aligned}
$$

12.2.5 If z is imaginary, then $z = bi$ for some real number b. Then, we have $zi = (bi)(i) = bi^2 = b(-1) = -b$. So, zi is $\boxed{\text{real}}$ if z is imaginary.

Exercises for Section 12.3

12.3.1

(a) $(1 + 9i) + (6 - 15i) = (1 + 6) + (9i - 15i) = \boxed{7 - 6i}$.

(b) $(3 - 2i) - (5 - 2i) = 3 - 2i - 5 + 2i = (3 - 5) + (-2i + 2i) = \boxed{-2}$.

(c) $2(3 - i) + i(2 + i) = 6 - 2i + 2i + i^2 = 6 - 2i + 2i - 1 = (6 - 1) + (-2i + 2i) = \boxed{5}$.

(d) $(2i)(3i)(2 - i) + (4 + 3i) = (6i^2)(2 - i) + 4 + 3i = (-6)(2 - i) + 4 + 3i = -12 + 6i + 4 + 3i = \boxed{-8 + 9i}$.

12.3.2

(a) $(2 + 3i)(1 - 2i) = 2(1) + 2(-2i) + 3i(1) + 3i(-2i) = 2 - 4i + 3i + 6 = \boxed{8 - i}$.

(b) $(3 - i)(6 + 2i) = 3(6) + 3(2i) - i(6) - i(2i) = 18 + 6i - 6i + 2 = \boxed{20}$.

(c) We know that $(1-2i)(1+2i)$ is a real number, so we rearrange our product first: $(1-2i)(3+7i)(1+2i) = (1 - 2i)(1 + 2i)(3 + 7i)$. Since $(1 - 2i)(1 + 2i) = 1^2 + 2^2 = 5$, we have $(1 - 2i)(1 + 2i)(3 + 7i) = 5(3 + 7i) = \boxed{15 + 35i}$.

(d) We first find $(5 - 3i)^2 = 5^2 + 2(5)(-3i) + (-3i)^2 = 25 - 30i + 9i^2 = 25 - 30i - 9 = 16 - 30i$. So, we have $(5 - 3i)^3 = (5 - 3i)^2(5 - 3i) = (16 - 30i)(5 - 3i) = 80 - 48i - 150i + 90i^2 = \boxed{-10 - 198i}$.

12.3.3 $\overline{-2 + 5i} = \boxed{-2 - 5i}$.

12.3.4 We know that the product of $5 - 6i$ and its conjugate is a real number, so one possible value of z is $\boxed{5 + 6i}$. We can find another possible value of z by multiplying $5 + 6i$ by any real number c besides 0 or 1 (we need a different value of z), because $[c(5 + 6i)](5 - 6i) = c(5 + 6i)(5 - 6i) = c(5^2 + 6^2) = 61c$.

12.3.5

(a) We multiply the numerator and denominator by the conjugate of the denominator, $3 + 4i$, to make the denominator real:

$$\frac{5 + 3i}{3 - 4i} = \frac{5 + 3i}{3 - 4i} \cdot \frac{3 + 4i}{3 + 4i} = \frac{(5 + 3i)(3 + 4i)}{3^2 + 4^2} = \frac{3 + 29i}{25} = \boxed{\frac{3}{25} + \frac{29}{25}i}.$$

(b) The conjugate of $7 - i$ is $7 + i$, so we have

$$\frac{-2i}{7 - i} = \frac{-2i}{7 - i} \cdot \frac{7 + i}{7 + i} = \frac{-2i(7 + i)}{7^2 + 1^2} = \frac{-2i(7 + i)}{50} = \frac{-i(7 + i)}{25} = \boxed{\frac{1}{25} - \frac{7}{25}i}.$$

(c) We start by noting $i^{577} = i^{576} \cdot i = (i^4)^{144} \cdot i = i$, so

$$\frac{10 - 9i}{2i^{577}} = \frac{10 - 9i}{2i} \cdot \frac{i}{i} = \frac{10i - 9i^2}{2i^2} = \frac{9 + 10i}{-2} = \boxed{-\frac{9}{2} - 5i}.$$

(d) The conjugate of $5 + 2i$ is $5 - 2i$ and we have:

$$\frac{-4 - 3i}{5 + 2i} = \frac{-4 - 3i}{5 + 2i} \cdot \frac{5 - 2i}{5 - 2i} = \frac{-20 - 15i + 8i + 6i^2}{5^2 + 2^2} = \boxed{-\frac{26}{29} - \frac{7}{29}i}.$$

Review Problems

12.12 We have $(2i)^2 = 2^2 i^2 = (4)(-1) = \boxed{-4}$, $(-3i)^3 = (-3)^3 i^3 = (-27)(-i) = \boxed{27i}$, and $(i/3)^4 = i^4/3^4 = \boxed{1/81}$.

12.13

(a) We isolate x^2 by subtracting 81 from both sides, which gives
$$x^2 = -81 \quad \Rightarrow \quad x = \pm\sqrt{-81} = \pm\sqrt{81}\sqrt{-1} = \boxed{\pm 9i}.$$

(b) $4z^2 + 9 = 0 \quad \Rightarrow \quad 4z^2 = -9 \quad \Rightarrow \quad z^2 = -9/4 \quad \Rightarrow \quad z = \pm\sqrt{-9/4} = \pm\sqrt{9/4}\sqrt{-1} = \boxed{\pm 3i/2}.$

12.14 2006 divided by 4 is 501 with a remainder of 2, so $i^{2006} = (i^4)^{501} i^2 = 1(-1) = \boxed{-1}$.

12.15 Each group of 4 consecutive powers of i adds to 0: $i + i^2 + i^3 + i^4 = i - 1 - i + 1 = 0$, $i^5 + i^6 + i^7 + i^8 = i^4(i + i^2 + i^3 + i^4) = 1(0) = 0$, and so on. Because 600 is divisible by 4, we know that if we start grouping the powers of i as suggested by our first two groups above, we won't have any 'extra' powers of i beyond i^{600}. We will, however, have the extra 1 before the i, so:
$$i^{600} + i^{599} + \cdots + i + 1 = (0) + (0) + \cdots + (0) + 1 = \boxed{1}.$$

12.16

(a) $(5 + 2i) - (-3 - 5i) = 5 + 2i + 3 + 5i = (5 + 3) + (2i + 5i) = \boxed{8 + 7i}$.

(b) $5(2 + 7i) + 3(4 - i) = 10 + 35i + 12 - 3i = (10 + 12) + (35i - 3i) = \boxed{22 + 32i}$.

12.17

(a) $(5 - 3i)(-4 + 3i) = 5(-4) + 5(3i) - 3i(-4) - 3i(3i) = -20 + 15i + 12i + 9 = \boxed{-11 + 27i}$.

(b) If we're very alert, we notice that $(-1 + 5i)(1 + 5i)$ equals an integer, so we rearrange the product as our first step:
$$(-1+5i)(2+8i)(1+5i) = [(-1+5i)(1+5i)](2+8i) = (-1^2 - 5i + 5i + 25i^2)(2+8i) = -26(2+8i) = \boxed{-52 - 208i}.$$

12.18 *Solution 1: Expand.* $(a + 6i)(5 - 3i) = 5a - 3ai + 30i + 18 = (5a + 18) + (30i - 3ai)$. Since this product is real, we have $30i - 3ai = 0$, so $a = \boxed{10}$.

Solution 2: Use the conjugate. We know the $(5 + 3i)(5 - 3i)$ equals a real number, but the coefficient of i in $5 + 3i$ is 3, not 6. However, if $(5 + 3i)(5 - 3i)$ is real, then we can multiply it by any real number to get another real number. Specifically, we can multiply it by 2 to make the coefficient of i equal to 6: $2(5 + 3i)(5 - 3i) = (10 + 6i)(5 - 3i)$. This product must be real, so $a = 10$.

12.19 In each part, we multiply the numerator and denominator by the conjugate of the denominator to make the denominator real.

(a) $\dfrac{6 - i}{4 + 2i} = \dfrac{6 - i}{4 + 2i} \cdot \dfrac{4 - 2i}{4 - 2i} = \dfrac{24 - 12i - 4i + 2i^2}{16 + 4} = \dfrac{22 - 16i}{20} = \boxed{\dfrac{11}{10} - \dfrac{4}{5}i}$.

Notice that the denominator of the original fraction is $4 + 2i = 2(2 + i)$, so we could have multiplied by $2 - i$ instead of $4 - 2i$.

(b) $\dfrac{11-13i}{13-11i} = \dfrac{11-13i}{13-11i} \cdot \dfrac{13+11i}{13+11i} = \dfrac{143-169i+121i-143i^2}{169+121} = \dfrac{286-48i}{290} = \boxed{\dfrac{143}{145} - \dfrac{24}{145}i}.$

12.20 Let z be the complex number we seek, so that $z(1+2i) = 13+i$. Dividing both sides by $1+2i$ gives

$$z = \frac{13+i}{1+2i} = \frac{13+i}{1+2i} \cdot \frac{1-2i}{1-2i} = \frac{13-26i+i-2i^2}{1+4} = \frac{15-25i}{5} = \boxed{3-5i}.$$

12.21 We have

$$\frac{1}{i-\dfrac{1}{i}} = \frac{1}{i-\dfrac{1}{i}\cdot\dfrac{i}{i}} = \frac{1}{i-\dfrac{i}{i^2}} = \frac{1}{i-\dfrac{i}{-1}} = \frac{1}{2i} = \frac{1}{2i}\cdot\frac{i}{i} = \boxed{-\dfrac{i}{2}}.$$

12.22 If $x = 0$, then both sides of the equation equal 0, so $x = 0$ is a solution to the equation. If x is not 0, then we can divide both sides of the equation by x^3 to get $x^2 = -1$. The solutions to this equation are $x = i$ and $x = -i$. Therefore, our solutions are $\boxed{0, i, \text{ and } -i}$.

12.23 We have $a = \dfrac{(2+i)^2}{3+i} = \dfrac{2^2 + 2(2)(i) + i^2}{3+i} = \dfrac{3+4i}{3+i}$, so

$$1 + \frac{1}{a} = 1 + \frac{1}{\dfrac{3+4i}{3+i}} = 1 + \frac{3+i}{3+4i}$$
$$= 1 + \frac{3+i}{3+4i} \cdot \frac{3-4i}{3-4i} = 1 + \frac{(3+i)(3-4i)}{(3+4i)(3-4i)}$$
$$= 1 + \frac{9-12i+3i-4i^2}{9-12i+12i-16i^2} = 1 + \frac{13-9i}{25}$$
$$= \frac{25}{25} + \frac{13-9i}{25} = \boxed{\dfrac{38}{25} - \dfrac{9}{25}i}.$$

Challenge Problems

12.24 First, we find x^2:

$$x^2 = \left(\frac{1-i\sqrt{3}}{2}\right)^2 = \frac{(1-i\sqrt{3})^2}{2^2} = \frac{1^2 + 2(1)(-i\sqrt{3}) + (-i\sqrt{3})^2}{4} = \frac{-2-2i\sqrt{3}}{4} = \frac{-1-i\sqrt{3}}{2}.$$

So,

$$\frac{1}{x^2 - x} = \frac{1}{\frac{-1-i\sqrt{3}}{2} - \frac{1-i\sqrt{3}}{2}} = \frac{1}{\frac{-1-i\sqrt{3}-1+i\sqrt{3}}{2}} = \frac{1}{\frac{-2}{2}} = \boxed{-1}.$$

12.25 Suppose $a + bi = \sqrt{-16+30i}$. Squaring both sides gives $(a+bi)^2 = -16 + 30i$. Expanding the left side gives

$$(a+bi)^2 = a^2 + 2(a)(bi) + (bi)^2 = a^2 - b^2 + 2abi.$$

So, we have $a^2 - b^2 + 2abi = -16 + 30i$. Matching up the real parts and the imaginary parts gives the system

$$a^2 - b^2 = -16,$$
$$2ab = 30.$$

Dividing the second equation by 2 gives $ab = 15$. Since $(1)(15) = (3)(5) = 15$, we try 1 and 15 for a and b, then 3 and 5. The latter works, since $3^2 - 5^2 = -16$. We must also remember that $(-3)(-5) = 15$, as well, so $a = -3$ and $b = -5$ is a solution in addition to $a = 3$ and $b = 5$. So, we have $\sqrt{-16 + 30i} = \boxed{\pm(3 + 5i)}$.

12.26 Let $w = a + bi$ and $z = c + di$, so that

$$w + z = a + bi + c + di = (a + c) + (b + d)i \qquad \text{and}$$
$$wz = (a + bi)(c + di) = ac + adi + bci + bdi^2 = (ac - bd) + (ad + bc)i.$$

We also have $\overline{w} = a - bi$ and $\overline{z} = c - di$. So, we have

$$\overline{w + z} = \overline{(a + c) + (b + d)i} = (a + c) - (b + d)i \qquad \text{and}$$
$$\overline{w} + \overline{z} = a - bi + c - di = a + c - bi - di = (a + c) - (b + d)i.$$

Therefore, $\overline{w + z} = \overline{w} + \overline{z}$.

We also have

$$\overline{wz} = \overline{(ac - bd) + (ad + bc)i} = (ac - bd) - (ad + bc)i \qquad \text{and}$$
$$\overline{w} \cdot \overline{z} = (a - bi)(c - di) = ac - adi - bci + bdi^2 = (ac - bd) - (ad + bc)i.$$

So, $\overline{wz} = \overline{w} \cdot \overline{z}$.

12.27

(a) We first find $\left(-\frac{1}{2} + \frac{\sqrt{3}}{2}i\right)^2$:

$$\left(-\frac{1}{2} + \frac{\sqrt{3}}{2}i\right)^2 = \left(-\frac{1}{2}\right)^2 + 2\left(-\frac{1}{2}\right)\left(\frac{\sqrt{3}}{2}i\right) + \left(\frac{\sqrt{3}}{2}i\right)^2 = \frac{1}{4} - \frac{i\sqrt{3}}{2} - \frac{3}{4} = \frac{-1 - i\sqrt{3}}{2}.$$

So, we have

$$\left(-\frac{1}{2} + \frac{\sqrt{3}}{2}i\right)^3 = \left(-\frac{1}{2} + \frac{\sqrt{3}}{2}i\right)^2 \left(-\frac{1}{2} + \frac{\sqrt{3}}{2}i\right) = \left(\frac{-1 - i\sqrt{3}}{2}\right)\left(-\frac{1}{2} + \frac{\sqrt{3}}{2}i\right)$$
$$= \frac{(-1 - i\sqrt{3})(-1 + i\sqrt{3})}{2 \cdot 2} = \frac{1 - i\sqrt{3} + i\sqrt{3} - i^2(\sqrt{3})^2}{4} = \frac{1 - (-1)(3)}{4} = 1.$$

(b) The expression we must cube in this part is the conjugate of the expression we cubed in part (a). So, we wish to compare $(a + bi)^3$ to $(a - bi)^3$. We have

$$(a + bi)^3 = (a + bi)^2(a + bi) = (a^2 - b^2 + 2abi)(a + bi)$$
$$= a^3 + a^2bi - ab^2 - b^3i + 2a^2bi + 2ab^2i^2 = (a^3 - 3ab^2) + (3a^2b - b^3)i,$$

and

$$(a - bi)^3 = (a - bi)^2(a - bi) = (a^2 - b^2 - 2abi)(a - bi)$$
$$= a^3 - a^2bi - ab^2 + b^3i - 2a^2bi + 2ab^2i^2 = (a^3 - 3ab^2) + (-3a^2b + b^3)i.$$

So, we see that $(a+bi)^3$ and $(a-bi)^3$ have the same real parts, but their imaginary parts are opposites: $-(3ab^2 - b^3) = -3ab^2 + b^3$. Therefore, $\left(-\frac{1}{2} - \frac{\sqrt{3}}{2}i\right)^3$ has the same real part as $\left(-\frac{1}{2} + \frac{\sqrt{3}}{2}i\right)^3$, but the opposite imaginary part. Since $\left(-\frac{1}{2} - \frac{\sqrt{3}}{2}i\right)^3$ has no imaginary part when expanded, we have

$$\left(-\frac{1}{2} - \frac{\sqrt{3}}{2}i\right)^3 = 1.$$

We also could have tackled this part using the result of the previous problem. Because $\overline{w \cdot z} = \overline{w} \cdot \overline{z}$, we can let $w = -\frac{1}{2} + \frac{\sqrt{3}}{2}i$ and find

$$(\overline{w})^3 = (\overline{w} \cdot \overline{w}) \cdot \overline{w} = \overline{w \cdot w} \cdot \overline{w} = \overline{w^2} \cdot \overline{w} = \overline{w^2 \cdot w} = \overline{w^3}.$$

Since $w^3 = 1$ from our first part, we therefore have

$$\left(-\frac{1}{2} - \frac{\sqrt{3}}{2}\right)^3 = (\overline{w})^3 = \overline{w^3} = \overline{1} = 1.$$

12.28 Because $\overline{a + bi} = a - bi$ and $(a + bi)^2 = a^2 - b^2 + 2abi$, we have $a - bi = a^2 - b^2 + 2abi$. Matching up the real parts and the imaginary parts, we have the system

$$a = a^2 - b^2,$$
$$-b = 2ab.$$

The second equation gives us $2ab + b = 0$, so $b(2a + 1) = 0$. Therefore, either $a = -1/2$ or $b = 0$. If $a = -1/2$, then the equation $a = a^2 - b^2$ becomes

$$-\frac{1}{2} = \frac{1}{4} - b^2 \quad \Rightarrow \quad b^2 = \frac{3}{4} \quad \Rightarrow \quad b = \pm\frac{\sqrt{3}}{2}.$$

So, we have the solutions $-\frac{1}{2} \pm \frac{\sqrt{3}}{2}i$.

If $b = 0$, then $a = a^2 - b^2$ becomes $a = a^2$, from which we have $a = 0$ or $a = 1$. These give us the solutions 0 and 1. Therefore, the four possible values of $a + bi$ are $\boxed{0, 1, -\frac{1}{2} + \frac{\sqrt{3}}{2}i, -\frac{1}{2} - \frac{\sqrt{3}}{2}i}$.

12.29 The only tool we know so far for solving a quadratic like this is factoring. So, we try to factor. We guess the quadratic will factor in the form

$$(5x + \_\_)(x + \_\_),$$

because this factorization will give us a $5x^2$ term when expanded. Because the coefficient of x in the quadratic is imaginary, we guess that we must put imaginary numbers in the blanks. So, we are looking for two imaginary numbers whose product is -9. We first try i and $9i$, and we find

$$(5x + i)(x + 9i) = 5x^2 + 45ix + ix + 9i^2 = 5x^2 + 46ix - 9, \qquad \text{and}$$
$$(5x + 9i)(x + i) = 5x^2 + 5ix + 9ix + 9i^2 = 5x^2 + 14ix - 9.$$

That didn't work, so we try $3i$ and $3i$:

$$(5x + 3i)(x + 3i) = 5x^2 + 15ix + 3ix + 9i^2 = 5x^2 + 18ix - 9.$$

Success! Now, we have $(5x + 3i)(x + 3i) = 0$, so we must have $5x + 3i = 0$ or $x + 3i = 0$, which give us the solutions $\boxed{x = -3i/5 \text{ and } x = -3i}$.

12.30 We seek the complex number $a + bi$ such that $(a + bi)^2 = i$. Expanding the left side gives $a^2 - b^2 + 2abi = i$. The real part of the left side must equal the real part of the right, so we have $a^2 - b^2 = 0$. Therefore, $a^2 = b^2$, which means $a = \pm b$. Similarly, the imaginary part of $a^2 - b^2 + 2abi$ must match the imaginary part of i, so we have $2ab = 1$. So, if $a = b$, we have $2a^2 = 1$, which gives $a = \pm\sqrt{1/2} = \pm\sqrt{2}/2$. Since $a = b$, this gives us the two solutions $\frac{\sqrt{2}}{2} + \frac{\sqrt{2}}{2}i$ and $-\frac{\sqrt{2}}{2} - \frac{\sqrt{2}}{2}i$.

The other case we must consider is $a = -b$. Then, $2ab = 1$ gives us $-2a^2 = 1$, from which we have $a^2 = -1/2$. However, a and b must be real, so there are no new solutions in this case.

So, the two complex numbers whose squares equal i are $\boxed{\dfrac{\sqrt{2}}{2} + \dfrac{\sqrt{2}}{2}i \text{ and } -\dfrac{\sqrt{2}}{2} - \dfrac{\sqrt{2}}{2}i}$.

CHAPTER 13

Quadratic Equations – Part 2

Exercises for Section 13.1

13.1.1

(a) $(r+7)^2 + 9 = 0 \quad \Rightarrow \quad (r+7)^2 = -9 \quad \Rightarrow \quad r+7 = \pm\sqrt{-9} \quad \Rightarrow \quad r+7 = \pm 3i \quad \Rightarrow \quad r = \boxed{-7 \pm 3i}$.

(b) $(2x-7)^2 + 16 = 0 \quad \Rightarrow \quad (2x-7)^2 = -16 \quad \Rightarrow \quad 2x-7 = \pm\sqrt{-16} \quad \Rightarrow \quad 2x-7 = \pm 4i$

$\Rightarrow \quad 2x = 7 \pm 4i \quad \Rightarrow \quad x = \boxed{\dfrac{7}{2} \pm 2i}$.

(c) $(y-7)^2 - 8 = 0 \quad \Rightarrow \quad (y-7)^2 = 8 \quad \Rightarrow \quad y-7 = \pm\sqrt{8} \quad \Rightarrow \quad y = \boxed{7 \pm 2\sqrt{2}}$.

(d) $(x-9)^2 + 18 = 0 \quad \Rightarrow \quad (x-9)^2 = -18 \quad \Rightarrow \quad x-9 = \pm\sqrt{-18} \quad \Rightarrow \quad x-9 = \pm 3i\sqrt{2}$

$\Rightarrow \quad x = \boxed{9 \pm 3i\sqrt{2}}$.

13.1.2

(a) Because $(x+a)^2 = x^2 + 2ax + a^2$, we have $x^2 + 2ax + a^2 = x^2 + 6x + c$. Equating coefficients of x gives us $2a = 6$, so $a = \boxed{3}$. Equating constants gives $c = a^2 = \boxed{9}$.

(b) Because $(y+a)^2 = y^2 + 2ay + a^2$, we have $y^2 + 2ay + a^2 = y^2 - 12y + c$. Equating coefficients of y gives $2a = -12$, so $a = \boxed{-6}$. Equating the constant terms gives $c = a^2 = \boxed{36}$.

(c) We have $x^2 + 2ax + a^2 = x^2 - \dfrac{x}{8} + c$. Equating coefficients of x gives us $2a = -\dfrac{1}{8}$, so $a = \boxed{-\dfrac{1}{16}}$.

Equating the constants on both sides of the equation gives $c = a^2 = \boxed{\dfrac{1}{256}}$.

(d) We have $y^2 + 2ay + a^2 = y^2 + \frac{y}{3} + c$. Equating coefficients of y gives $2a = 1/3$, so $a = \boxed{1/6}$. Equating the constant terms of the two quadratics then gives $c = a^2 = \boxed{1/36}$.

13.1.3 Because $(x+a)^2 = x^2 + 2ax + a^2$, in each part, we find the corresponding value of a by taking the positive square root of the constant term. Then we multiply this value by 2 to get c, because $2a$ is the coefficient of the linear term in the expansion of $(x+a)^2$.

(a) Because $\sqrt{25} = 5$, we must find the value of c such that $(x+5)^2 = x^2 + cx + 25$. Therefore,

$c = 2(5) = \boxed{10}$. (Remember, the problem asks for the positive value of c.)

(b) Because $\sqrt{400} = 20$, we must find the value of c that such that $(x + 20)^2 = x^2 + cx + 400$, which gives us $c = 2(20) = \boxed{40}$.

(c) Because $\sqrt{9/4} = 3/2$, we want the value of c such that $\left(x + \dfrac{3}{2}\right)^2 = x^2 + cx + \dfrac{9}{4}$. Therefore, $c = 2(3/2) = \boxed{3}$.

(d) Because $\sqrt{96} = \sqrt{16 \cdot 6} = 4\sqrt{6}$, we seek the value of c such that $(x + 4\sqrt{6})^2 = x^2 + cx + 96$. So, $c = 2(4\sqrt{6}) = \boxed{8\sqrt{6}}$.

13.1.4 We follow the process used in the text. We know how to solve an equation of the form $(x+a)^2 + c = 0$, so we try to manipulate the given equation into this form. We have the two terms $x^2 + 4x$. We need to add the appropriate constant to turn this into the square of a binomial. We know that $(x+2)^2 = x^2 + 4x + 4$, so we add 4 to both sides of our equation to give $(x^2 + 4x + 4) - 7 = 4$. Therefore, we have

$$(x + 2)^2 - 7 = 4 \quad \Rightarrow \quad (x+2)^2 = 11 \quad \Rightarrow \quad x + 2 = \pm\sqrt{11} \quad \Rightarrow \quad x = \boxed{-2 \pm \sqrt{11}}.$$

13.1.5 We have $(ax + b)^2 = (ax)^2 + 2(ax)(b) + b^2 = a^2x^2 + 2abx + b^2$.

(a) Suppose the quadratic is the square of the binomial $ax + b$. So, we have

$$a^2x^2 + 2abx + b^2 = 4x^2 + cx + 16.$$

Equating the coefficients of x^2 gives $a^2 = 4$, so $a = \pm 2$. Equating the constants gives $b^2 = 16$, so $b = \pm 4$. From the coefficients of x, we have $c = 2ab$. We want the positive value of c, which occurs when a and b are both positive or both negative. In either case, we have $c = 2(2)(4) = \boxed{16}$. Note: $(2x + 4)^2 = (2x)^2 + 2(2x)(4) + 4^2 = 4x^2 + 16x + 16$.

(b) Suppose the quadratic is the square of the binomial $ax + b$. So, we have

$$a^2x^2 + 2abx + b^2 = 25x^2 + cx + 81.$$

Equating the coefficients of x^2 gives $a^2 = 25$, so $a = \pm 5$. Equating the constants gives $b^2 = 81$, so $b = \pm 9$. From the coefficients of x, we have $c = 2ab$. We want the positive value of c, which occurs when a and b are both positive or both negative. In either case, we have $c = 2(5)(9) = \boxed{90}$. Note: $(5x + 9)^2 = (5x)^2 + 2(5x)(9) + 9^2 = 25x^2 + 90x + 81$.

13.1.6 We've seen a couple examples already in which we tackled a quadratic whose quadratic term has coefficient 1, so we start by dividing the equation by 3 to make the coefficient of r^2 equal to 1. This gives us $r^2 + 2r - \frac{7}{3} = 0$. Now, we're on familiar ground! We recognize that $(r + 1)^2 = r^2 + 2r + 1$, so we add 1 to both sides to complete the square, which gives

$$(r^2+2r+1) - \frac{7}{3} = 1 \quad \Rightarrow \quad (r+1)^2 - \frac{7}{3} = 1 \quad \Rightarrow \quad (r+1)^2 = \frac{10}{3} \quad \Rightarrow \quad r+1 = \pm\sqrt{\frac{10}{3}} = \pm\frac{\sqrt{10}}{\sqrt{3}} \cdot \frac{\sqrt{3}}{\sqrt{3}} = \pm\frac{\sqrt{30}}{3}.$$

Subtracting 1 from both sides gives $r = \boxed{-1 \pm \dfrac{\sqrt{30}}{3}}$.

Exercises for Section 13.2

13.2.1

(a) We first isolate the variable terms to give $x^2 - 2x = 15$. To complete the square we add $(-2/2)^2 = 1$ to both sides, which gives

$$x^2 - 2x + 1 = 15 + 1 \quad \Rightarrow \quad (x-1)^2 = 16 \quad \Rightarrow \quad x - 1 = \pm\sqrt{16} = \pm 4 \quad \Rightarrow \quad x = 1 \pm 4.$$

Therefore, our solutions are $x = 1+4 = \boxed{5}$ and $x = 1-4 = \boxed{-3}$. Notice that $x^2 - 2x - 15 = (x-5)(x+3)$, which we can use to find the solutions above.

(b) We isolate the variable terms to find $x^2 + x = -1$. To complete the square, we add $(1/2)^2 = 1/4$ to both sides, which gives

$$x^2 + x + \frac{1}{4} = -1 + \frac{1}{4} \quad \Rightarrow \quad \left(x+\frac{1}{2}\right)^2 = -\frac{3}{4} \quad \Rightarrow \quad x + \frac{1}{2} = \pm\sqrt{-\frac{3}{4}} = \pm\frac{\sqrt{3}}{2}i \quad \Rightarrow \quad x = \boxed{-\frac{1}{2} \pm \frac{\sqrt{3}}{2}i}.$$

(c) We isolate the variable terms, then divide by 12 to make the coefficient of r^2 equal 1. This gives $r^2 - 3r = 12$. We add $(-3/2)^2 = 9/4$ to both sides to complete the square, which gives

$$r^2 - 3r + \frac{9}{4} = 12 + \frac{9}{4} \quad \Rightarrow \quad \left(r - \frac{3}{2}\right)^2 = \frac{57}{4} \quad \Rightarrow \quad r - \frac{3}{2} = \pm\sqrt{\frac{57}{4}} = \pm\frac{\sqrt{57}}{2} \quad \Rightarrow \quad r = \boxed{\frac{3}{2} \pm \frac{\sqrt{57}}{2}}.$$

(d) We isolate the variable terms to get $3x^2 + 7x = 20$. Dividing this equation by 3 gives $x^2 + \frac{7}{3}x = \frac{20}{3}$. To complete the square, we add $[(7/3)/2]^2 = (7/6)^2 = 49/36$ to both sides, which gives

$$x^2 + \frac{7}{3}x + \frac{49}{36} = \frac{20}{3} + \frac{49}{36} \quad \Rightarrow \quad \left(x + \frac{7}{6}\right)^2 = \frac{289}{36} \quad \Rightarrow \quad x + \frac{7}{6} = \pm\frac{17}{6}.$$

Solving $x + \frac{7}{6} = \frac{17}{6}$ gives $x = \frac{5}{3}$ and solving $x + \frac{7}{6} = -\frac{17}{6}$ gives $x = -4$, so our solutions are $\boxed{x = \frac{5}{3} \text{ and } x = -4}$. Seeing that these solutions are rational makes us realize that we could have solved the equation by factoring. We have $3x^2 + 7x - 20 = (3x - 5)(x + 4) = 0$, so our solutions are $x = 5/3$ and $x = -4$.

(e) We isolate the variable terms and divide by 4 to make the coefficient of x^2 equal to 1. This gives $x^2 + 3x = -\frac{9}{4}$. To complete the square, we add $(3/2)^2 = 9/4$ to both sides, which gives

$$x^2 + 3x + \frac{9}{4} = -\frac{9}{4} + \frac{9}{4} \quad \Rightarrow \quad \left(x + \frac{3}{2}\right)^2 = 0 \quad \Rightarrow \quad x + \frac{3}{2} = 0 \quad \Rightarrow \quad x = \boxed{-\frac{3}{2}}.$$

We only found one solution to the quadratic; did we make a mistake?

No! Factoring the quadratic gives us $(2x + 3)^2 = 0$; our square was already "complete."

(f) We isolate the variable terms, then divide by 6, which gives $x^2 - \frac{1}{2}x = -\frac{1}{6}$. We complete the square by adding $[(-1/2)/2]^2 = (1/4)^2 = 1/16$ to both sides, which gives

$$x^2 - \frac{1}{2}x + \frac{1}{16} = -\frac{1}{6} + \frac{1}{16} \quad \Rightarrow \quad \left(x - \frac{1}{4}\right)^2 = -\frac{5}{48} \quad \Rightarrow \quad x - \frac{1}{4} = \pm\sqrt{-\frac{5}{48}} = \pm\frac{\sqrt{5}}{\sqrt{48}}i.$$

We have to do a little work to rationalize the denominator of the right side:

$$x - \frac{1}{4} = \pm \frac{\sqrt{5}}{\sqrt{48}} i = \pm \frac{\sqrt{5}}{4\sqrt{3}} i = \pm \frac{\sqrt{5}}{4\sqrt{3}} \cdot \frac{\sqrt{3}}{\sqrt{3}} i = \pm \frac{\sqrt{15}}{12} i.$$

Adding $\frac{1}{4}$ to both ends gives $x = \boxed{\frac{1}{4} \pm \frac{\sqrt{15}}{12} i}$.

13.2.2

(a) If x is a real number, then the smallest possible value of x^2 is 0, which occurs when $x = 0$. Therefore, the smallest possible value of $x^2 + 8$ is $0 + 8 = \boxed{8}$.

(b) In our first part, we saw that we can easily find the smallest possible value of an expression of the form $x^2 + c$, for some constant c. So, we'd like to manipulate $x^2 + 10x - 7$ into this form. This suggests completing the square, which will allow us to write the quadratic as the square of a binomial plus a constant. To complete the square, we must add $(10/2)^2 = 25$. However, there is no "other side" to add it to. Therefore, we both add and subtract 25 to the quadratic $x^2 + 10x - 7$ to give

$$x^2 + 10x - 7 = (x^2 + 10x + 25) - 7 - 25 = (x + 5)^2 - 32.$$

Because $(x + 5)^2$ is the square of a real number, the smallest it can be is 0. Therefore, the smallest that $(x + 5)^2 - 32$ can be is $\boxed{-32}$.

Exercises for Section 13.3

13.3.1

(a) We let $a = 1$, $b = 5$, and $c = 4$ in the quadratic formula to find

$$x = \frac{-5 \pm \sqrt{5^2 - 4(1)(4)}}{2(1)} = \frac{-5 \pm \sqrt{9}}{2} = \frac{-5 \pm 3}{2}.$$

So, our solutions are $x = (-5 + 3)/2 = \boxed{-1}$ and $x = (-5 - 3)/2 = \boxed{-4}$. We could also have found these solutions by factoring the quadratic to give $(x + 1)(x + 4) = 0$.

(b) We first rearrange the equation to give $r^2 - 3r - 7 = 0$. Letting $a = 1$, $b = -3$, and $c = -7$ in the quadratic formula, we have

$$r = \frac{-(-3) \pm \sqrt{(-3)^2 - 4(1)(-7)}}{2(1)} = \frac{3 \pm \sqrt{9 + 28}}{2} = \boxed{\frac{3 \pm \sqrt{37}}{2}}.$$

(c) We subtract $2t - 1$ from both sides to give $-t^2 + 3t + 1 = 0$. Letting $a = -1$, $b = 3$, and $c = 1$ in the quadratic formula gives

$$t = \frac{-3 \pm \sqrt{(3)^2 - 4(-1)(1)}}{2(-1)} = \frac{-3 \pm \sqrt{9 + 4}}{-2} = \boxed{\frac{3 \pm \sqrt{13}}{2}}.$$

Make sure you see why $-(-3 \pm \sqrt{13}) = 3 \pm \sqrt{13}$.

(d) Adding 2 to both sides gives $2z^2 - 3z + 2 = 0$. Letting $a = 2$, $b = -3$, and $c = 2$ in the quadratic formula gives

$$z = \frac{-(-3) \pm \sqrt{(-3)^2 - 4(2)(2)}}{2(2)} = \frac{3 \pm \sqrt{-7}}{4} = \boxed{\frac{3 \pm i\sqrt{7}}{4}}.$$

(e) First, we move all the terms to the left, which gives $4x^2 - 16x + 15 = 0$. Letting $a = 4$, $b = -16$, and $c = 15$ in the quadratic formula gives

$$x = \frac{-(-16) \pm \sqrt{(-16)^2 - 4(4)(15)}}{2(4)} = \frac{16 \pm \sqrt{256 - 240}}{8} = \frac{16 \pm \sqrt{16}}{8} = \frac{16 \pm 4}{8}.$$

So, our solutions are $x = (16 + 4)/8 = \boxed{5/2}$ and $x = (16 - 4)/8 = \boxed{3/2}$. Our solutions are rational, so we realize that we could have factored the quadratic $4x^2 - 16x + 15$. We have $4x^2 - 16x + 15 = (2x - 3)(2x - 5) = 0$, so our solutions are $x = 3/2$ and $x = 5/2$.

(f) Subtracting $36a$ from both sides gives $18a^2 - 36a + 81 = 0$. We could plow ahead with the quadratic formula, but we first notice that all the coefficients are divisible by 9. So, we divide by 9 to get $2a^2 - 4a + 9 = 0$. Finally, we use the quadratic formula to find

$$a = \frac{-(-4) \pm \sqrt{(-4)^2 - 4(2)(9)}}{2(2)} = \frac{4 \pm \sqrt{16 - 72}}{4} = \frac{4 \pm \sqrt{-56}}{4} = \frac{4 \pm 2i\sqrt{14}}{4} = \boxed{\frac{2 \pm i\sqrt{14}}{2}}.$$

13.3.2

(a) As we showed in the text, if the coefficients of a quadratic are real, but the roots are not, then the roots are complex conjugates. Therefore, if $2 - 3i$ is one root, then $\boxed{2 + 3i}$ is the other root.

(b) The quadratic formula tells us that the roots of the quadratic $ax^2 + bx + c = 0$ are

$$x = \frac{-b \pm \sqrt{b^2 - 4ac}}{2a} = -\frac{b}{2a} \pm \frac{\sqrt{b^2 - 4ac}}{2a}.$$

If the coefficients are rational, then $-b/(2a)$ is rational. Therefore, the only way there can be a square root as part of one of the roots of a quadratic with rational coefficients is from the second term in

$$x = -\frac{b}{2a} \pm \frac{\sqrt{b^2 - 4ac}}{2a}.$$

Because a, b, and c are rational, these two roots are of the form $p + q\sqrt{r}$ and $p - q\sqrt{r}$, where p, q, and r are rational. So, if $1 + \sqrt{2}$ is one root of a quadratic with rational coefficients, $\boxed{1 - \sqrt{2}}$ is the other root.

(c) We cannot determine anything about the other root of the quadratic unless we know that the coefficients are all real. Our solution to the next problem will explain why.

13.3.3 If some of the coefficients of the quadratic are imaginary, then it is possible for the discriminant to be positive, but for the roots to still not be real. For example, consider the quadratic equation

$$x^2 + ix - 2 = 0.$$

The discriminant of this equation is $b^2 - 4ac = i^2 - 4(1)(-2) = -1 + 8 = 7$. However, the quadratic formula tells us that the roots are

$$x = \frac{-i \pm \sqrt{7}}{2}.$$

Despite the fact that the discriminant is positive, the roots are still not real because b is not real. Therefore, in order to use the discriminant to determine whether or not the roots of a quadratic are real, we must also know that the coefficients of the quadratic are real.

13.3.4 The quadratic has real roots if its discriminant is nonnegative. The discriminant of $3x^2 + 4x + k$ is $4^2 - 4(3)(k) = 16 - 12k$. Therefore, we must have $16 - 12k \geq 0$. Solving this inequality gives $\boxed{k \leq 4/3}$.

13.3.5 Even though some of the coefficients are imaginary, we can still use the quadratic formula to find

$$x = \frac{-2 \pm \sqrt{2^2 - 4(2i)(9i)}}{2(2i)} = \frac{-2 \pm \sqrt{4 - 72i^2}}{4i} = \frac{-2 \pm \sqrt{76}}{4i} = \frac{-2 \pm 2\sqrt{19}}{4i} = \frac{-1 \pm \sqrt{19}}{2i}.$$

We can get the i out of the denominator by multiplying the numerator and denominator by i:

$$x = \frac{-1 \pm \sqrt{19}}{2i} \cdot \frac{i}{i} = \frac{(-1 \pm \sqrt{19})i}{2i^2} = -\frac{(-1 \pm \sqrt{19})i}{2} = \boxed{\left(\frac{1 \pm \sqrt{19}}{2}\right)i}.$$

Not all answers are pretty! (Make sure you see why $-(-1 \pm \sqrt{19}) = 1 \pm \sqrt{19}$.)

Exercises for Section 13.4

13.4.1 Because the coefficients of the quadratic are rational and $2 - \sqrt{3}$ is one root, we know that $2 + \sqrt{3}$ is the other root. We can use the sum and product of the roots to determine b and c, because $-b$ equals the sum of the roots and c equals the product of the roots. Therefore, we have $-b = (2 - \sqrt{3}) + (2 + \sqrt{3}) = 4$, so $\boxed{b = -4}$, and

$$c = (2 - \sqrt{3})(2 + \sqrt{3}) = 2^2 + 2\sqrt{3} - 2\sqrt{3} - (\sqrt{3})(\sqrt{3}) = 4 - 3 = \boxed{1}.$$

13.4.2

(a) Using the quadratic formula, we can write the sum of the squares of the roots as

$$\left(\frac{-b + \sqrt{b^2 - 4ac}}{2a}\right)^2 + \left(\frac{-b - \sqrt{b^2 - 4ac}}{2a}\right)^2$$

$$= \frac{(-b + \sqrt{b^2 - 4ac})^2}{(2a)^2} + \frac{(-b - \sqrt{b^2 - 4ac})^2}{(2a)^2}$$

$$= \frac{(-b)^2 + 2(-b)(\sqrt{b^2 - 4ac}) + (\sqrt{b^2 - 4ac})^2}{4a^2} + \frac{(-b)^2 - 2(-b)(\sqrt{b^2 - 4ac}) + (\sqrt{b^2 - 4ac})^2}{4a^2}$$

$$= \frac{b^2 - 2b\sqrt{b^2 - 4ac} + b^2 - 4ac + b^2 + 2b\sqrt{b^2 - 4ac} + b^2 - 4ac}{4a^2}$$

$$= \frac{4b^2 - 8ac}{4a^2} = \boxed{\frac{b^2 - 2ac}{a^2}}.$$

(b) Let r and s be the roots, so that $r + s = -\frac{b}{a}$ and $rs = \frac{c}{a}$. We seek $r^2 + s^2$, so we square the equation $r + s = -\frac{b}{a}$ to produce these two squares. Squaring this equation gives us $r^2 + 2rs + s^2 = \frac{b^2}{a^2}$. We're close to having $r^2 + s^2$, but we have that extra $2rs$ term. We take care of it by subtracting it from both sides and noting that $rs = \frac{c}{a}$, which gives

$$r^2 + s^2 = \frac{b^2}{a^2} - 2rs = \frac{b^2}{a^2} - 2\frac{c}{a} = \frac{b^2 - 2ac}{a^2}.$$

13.4.3 First, we simplify the huge fraction on the left side:

$$\frac{2}{3x + \frac{5}{x}} = \frac{2}{\frac{3x^2}{x} + \frac{5}{x}} = \frac{2}{\frac{3x^2 + 5}{x}} = \frac{2x}{3x^2 + 5}.$$

So, now we have $\frac{2x}{3x^2+5} = 1$. Multiplying both sides by $3x^2 + 5$ gives $2x = 3x^2 + 5$, so $3x^2 - 2x + 5 = 0$. Applying the quadratic formula gives us

$$x = \frac{-(-2) \pm \sqrt{(-2)^2 - 4(3)(5)}}{2(3)} = \frac{2 \pm \sqrt{-56}}{6} = \frac{2 \pm 2i\sqrt{14}}{6} = \boxed{\frac{1 \pm i\sqrt{14}}{3}}.$$

13.4.4 *Solution 1: Use sum and product of roots.* We use what we know about the sum and product of roots of a quadratic to write b and c in terms of r and s:

$$r + s = -b \qquad \text{and} \qquad rs = c.$$

Expanding the left side of $(r - 1)(s - 1) = 7$ gives us $rs - r - s + 1 = 7$, so $rs - (r + s) = 6$. Substituting $-b$ and c from above gives us $c - (-b) = 6$, so $b + c = \boxed{6}$.

Solution 2: Use factored form of the quadratic. Because r and s are the roots of $x^2 + bx + c$, we have

$$x^2 + bx + c = (x - r)(x - s).$$

This is true for any value of x. We know that $(r - 1)(s - 1) = 7$, so we let $x = 1$ in the equation above to find

$$1 + b + c = (1 - r)(1 - s) = [(-1)(r - 1)][(-1)(s - 1)] = (r - 1)(s - 1) = 7.$$

Therefore, $b + c = 7 - 1 = \boxed{6}$, as before.

13.4.5 *Solution 1: It's a quadratic with variable coefficients!* We can view the equation as a quadratic with a as the variable. The coefficient of a^2 is 1, the coefficient of a is b, and the "constant term" is b^2. Therefore, the quadratic formula tells us

$$a = \frac{-b \pm \sqrt{b^2 - 4(1)(b^2)}}{2(1)} = \frac{-b \pm \sqrt{-3b^2}}{2}.$$

The only way $-3b^2$ can be positive is if b is imaginary. If b is real and nonzero, then $-3b^2$ is negative, so a is imaginary. Therefore, there isn't a pair of real nonzero numbers a and b such that $a^2 + ab + b^2 = 0$.

Solution 2: Solve for a/b. We start by dividing the whole equation by b^2 (since $b \neq 0$):

$$\left(\frac{a}{b}\right)^2 + \frac{a}{b} + 1 = 0.$$

Viewing a/b as our variable, we now have a quadratic in which all three coefficients equal 1. We then apply the quadratic formula to get

$$\frac{a}{b} = \frac{-1 \pm \sqrt{1 - 4(1)(1)}}{2(1)} = \frac{-1 \pm i\sqrt{3}}{2}.$$

Since the ratio of a and b is not real, they cannot both be real.

13.4.6 Letting $x = 2 + \frac{\sqrt{2}}{2}$ in the left side of the original equation gives

$$\begin{aligned}
\frac{x}{x - 1} + \frac{1}{x - 2} &= \frac{2 + \frac{\sqrt{2}}{2}}{1 + \frac{\sqrt{2}}{2}} + \frac{1}{\frac{\sqrt{2}}{2}} = \frac{2 + \frac{\sqrt{2}}{2}}{1 + \frac{\sqrt{2}}{2}} \cdot \frac{2}{2} + \frac{2}{\sqrt{2}} \\
&= \frac{4 + \sqrt{2}}{2 + \sqrt{2}} + \frac{2}{\sqrt{2}} = \frac{4 + \sqrt{2}}{2 + \sqrt{2}} \cdot \frac{2 - \sqrt{2}}{2 - \sqrt{2}} + \frac{2}{\sqrt{2}} \cdot \frac{\sqrt{2}}{\sqrt{2}} \\
&= \frac{8 - 4\sqrt{2} + 2\sqrt{2} - 2}{2} + \frac{2\sqrt{2}}{2} = 3 - \sqrt{2} + \sqrt{2} = 3.
\end{aligned}$$

That one works; let's try the other one:

$$\begin{aligned}
\frac{x}{x - 1} + \frac{1}{x - 2} &= \frac{2 - \frac{\sqrt{2}}{2}}{1 - \frac{\sqrt{2}}{2}} + \frac{1}{-\frac{\sqrt{2}}{2}} = \frac{2 - \frac{\sqrt{2}}{2}}{1 - \frac{\sqrt{2}}{2}} \cdot \frac{2}{2} - \frac{2}{\sqrt{2}} \\
&= \frac{4 - \sqrt{2}}{2 - \sqrt{2}} - \frac{2}{\sqrt{2}} = \frac{4 - \sqrt{2}}{2 - \sqrt{2}} \cdot \frac{2 + \sqrt{2}}{2 + \sqrt{2}} - \frac{2}{\sqrt{2}} \cdot \frac{\sqrt{2}}{\sqrt{2}} \\
&= \frac{8 + 4\sqrt{2} - 2\sqrt{2} - 2}{2} - \frac{2\sqrt{2}}{2} = 3 + \sqrt{2} - \sqrt{2} = 3.
\end{aligned}$$

Yep, that one works, too!

Review Problems

13.17

(a) $(x+8)^2+81 = 0 \quad \Rightarrow \quad (x+8)^2 = -81 \quad \Rightarrow \quad x+8 = \pm\sqrt{-81} \quad \Rightarrow \quad x+8 = \pm 9i \quad \Rightarrow \quad x = \boxed{-8 \pm 9i}$.

(b) $2(3-2t)^2 - 12 = 0 \quad \Rightarrow \quad 2(3-2t)^2 = 12 \quad \Rightarrow \quad (3-2t)^2 = 6 \quad \Rightarrow \quad 3-2t = \pm\sqrt{6}$

$\Rightarrow \quad -2t = -3 \pm \sqrt{6} \quad \Rightarrow \quad t = \boxed{\dfrac{3}{2} \pm \dfrac{\sqrt{6}}{2}}$.

13.18

(a) Since $(x+a)^2 = x^2 + 2ax + a^2$, we have $x^2 + 2ax + a^2 = x^2 + cx + 121$. Comparing constants gives $a^2 = 121$, so $a = \pm 11$. Comparing the coefficients of x gives $c = 2a = \pm 22$. So, the positive value of c is $c = \boxed{22}$.

(b) Since $(x+a)^2 = x^2 + 2ax + a^2$, we have $x^2 + 2ax + a^2 = x^2 - 12x + c$. Comparing linear terms gives $2a = -12$, so $a = -6$. Comparing constants gives $c = a^2 = \boxed{36}$.

(c) Since $(y+a)^2 = y^2 + 2ay + a^2$, we have $y^2 + 2ay + a^2 = y^2 + 11y + c$. Comparing linear terms gives $2a = 11$, so $a = 11/2$, and comparing constants gives $c = a^2 = \boxed{121/4}$.

(d) Since $(az+b)^2 = a^2z^2 + 2abz + b^2$, we have $a^2z^2 + 2abz + b^2 = 4z^2 + cz + 9$. Comparing quadratic terms gives $a^2 = 4$, so $a = 2$. (We take the positive value of a because we seek the positive value of c.) Similarly, comparing constants gives $b^2 = 9$, so $b = 3$. Finally, comparing linear terms gives $c = 2ab = \boxed{12}$. (Notice that if we take $a = -2$ above, we still have $b^2 = 9$. Taking the negative solution $b = -3$ also gives us $c = 2ab = 12$.)

13.19

(a) We add $[(-3)/2]^2 = 9/4$ to both sides to complete the square:

$$x^2 - 3x + \frac{9}{4} = 5 + \frac{9}{4} \quad \Rightarrow \quad \left(x - \frac{3}{2}\right)^2 = \frac{29}{4} \quad \Rightarrow \quad x - \frac{3}{2} = \pm\sqrt{\frac{29}{4}} = \pm\frac{\sqrt{29}}{2} \quad \Rightarrow \quad x = \boxed{\frac{3}{2} \pm \frac{\sqrt{29}}{2}}.$$

(b) Adding $68r$ to both sides gives $10r^2 + 68r = 45$. We divide both sides by 10 to make the coefficient of r^2 equal to 1: $r^2 + \frac{34}{5}r = \frac{9}{2}$. We then add $[(34/5)/2]^2 = (17/5)^2 = 289/25$ to both sides to complete the square:

$$r^2 + \frac{34}{5}r + \frac{289}{25} = \frac{9}{2} + \frac{289}{25} \quad \Rightarrow \quad \left(r + \frac{17}{5}\right)^2 = \frac{225 + 578}{50} = \frac{803}{50}.$$

Taking the square root gives us

$$r + \frac{17}{5} = \pm\sqrt{\frac{803}{50}} = \pm\frac{\sqrt{803}}{5\sqrt{2}} \cdot \frac{\sqrt{2}}{\sqrt{2}} = \pm\frac{\sqrt{1606}}{10}.$$

Therefore, $r = \boxed{-\dfrac{17}{5} \pm \dfrac{\sqrt{1606}}{10}}$. And, we're really happy there's such a thing as the quadratic formula, so we don't have to wade through messes like that anymore.

13.20

(a) We use the quadratic formula, where $a = 1$, $b = -6$, and $c = 10$:

$$r = \frac{-(-6) \pm \sqrt{(-6)^2 - 4(1)(10)}}{2(1)} = \frac{6 \pm \sqrt{-4}}{2} = \frac{6 \pm 2i}{2} = \boxed{3 \pm i}.$$

(b) We rearrange the quadratic to get $x^2 - 3x + 9 = 0$. Then, we use the quadratic formula to find

$$x = \frac{-(-3) \pm \sqrt{(-3)^2 - 4(1)(9)}}{2(1)} = \frac{3 \pm \sqrt{-27}}{2} = \boxed{\frac{3 \pm 3i\sqrt{3}}{2}}.$$

(c) We notice that all the coefficients are even, so we simplify the equation by dividing both sides by 2, which gives $6y^2 - 11y + 3 = 0$. This quadratic factors! We have $6y^2 - 11y + 3 = (3y - 1)(2y - 3) = 0$, so our solutions are $y = \boxed{1/3}$ and $y = \boxed{3/2}$.

(d) Expanding the right and rearranging gives $3r^2 - 2r + 14 = 0$. The quadratic formula then gives

$$r = \frac{-(-2) \pm \sqrt{(-2)^2 - 4(3)(14)}}{2(3)} = \frac{2 \pm \sqrt{4 - 4(42)}}{6} = \frac{2 \pm \sqrt{4(-41)}}{6} = \frac{2 \pm 2i\sqrt{41}}{6} = \boxed{\frac{1 \pm i\sqrt{41}}{3}}.$$

Notice how we use factorization to avoid having to multiply out $4(3)(14)$.

(e) We multiply by 2 to get rid of the fraction, then rearrange, to get $x^2 - 6x - 16 = 0$. We factor this quadratic, to find $x^2 - 6x - 16 = (x - 8)(x + 2) = 0$, so our solutions are $x = \boxed{8}$ and $x = \boxed{-2}$.

(f) Expanding the right side, then rearranging, gives $3t^2 - 35t + 4 = 0$. We then use the quadratic formula to find

$$t = \frac{-(-35) \pm \sqrt{(-35)^2 - 4(3)(4)}}{2(3)} = \frac{35 \pm \sqrt{1225 - 48}}{6} = \boxed{\frac{35 \pm \sqrt{1177}}{6}}.$$

13.21 The quadratic equation $3x^2 + bx + 27 = 0$ has a double root if its discriminant is 0. Solving $b^2 - 4(3)(27) = 0$ gives us $b = \pm\sqrt{(4)(3)(27)} = \pm 2\sqrt{81} = \boxed{\pm 18}$.

13.22 We multiply both sides by $2 - z$ to find $3 = (2z + 7)(2 - z) = 4z - 2z^2 + 14 - 7z = -2z^2 - 3z + 14$. Moving all terms to the left gives $2z^2 + 3z - 11 = 0$. Applying the quadratic formula gives

$$z = \frac{-3 \pm \sqrt{3^2 - 4(2)(-11)}}{2(2)} = \frac{-3 \pm \sqrt{9 + 88}}{4} = \boxed{\frac{-3 \pm \sqrt{97}}{4}}.$$

13.23 We get rid of the fractions by multiplying both sides by $(x - 2)(x - 3)$. On the left we have

$$(x - 2)(x - 3)\left(\frac{2x}{x - 2} + \frac{3}{x - 3}\right) = (x - 3)(2x) + (x - 2)(3) = 2x^2 - 6x + 3x - 6 = 2x^2 - 3x - 6.$$

On the right, we have

$$1(x - 2)(x - 3) = x^2 - 5x + 6.$$

So, our equation is $2x^2 - 3x - 6 = x^2 - 5x + 6$. Moving all terms to the left gives us $x^2 + 2x - 12 = 0$. Applying the quadratic formula gives

$$x = \frac{-2 \pm \sqrt{2^2 - 4(1)(-12)}}{2(1)} = \frac{-2 \pm \sqrt{52}}{2} = \frac{-2 \pm 2\sqrt{13}}{2} = \boxed{-1 \pm \sqrt{13}}.$$

13.24

(a) Letting $a = 1$, $b = -i$, and $c = 2$ in the quadratic formula gives

$$x = \frac{-(-i) \pm \sqrt{(-i)^2 - 4(1)(2)}}{2(1)} = \frac{i \pm \sqrt{-1 - 8}}{2} = \frac{i \pm 3i}{2}.$$

Therefore, the two solutions are

$$x = \frac{i + 3i}{2} = \boxed{2i} \qquad \text{and} \qquad x = \frac{i - 3i}{2} = \boxed{-i}.$$

(b) The coefficients of the quadratic are not real, so the roots do not have to be complex conjugates. We see why by looking at the quadratic formula:

$$x = \frac{-b \pm \sqrt{b^2 - 4ac}}{2a}.$$

If the coefficients are real, the only way we can have a root that is not real is if $\sqrt{b^2 - 4ac}$ is imaginary. The $\pm$ before this square root ensures that these imaginary roots will be complex conjugates because the $-b$ in the numerator must be real. However, if both b and $\sqrt{b^2 - 4ac}$ are imaginary, then the two numbers

$$-b + \sqrt{b^2 - 4ac} \qquad \text{and} \qquad -b - \sqrt{b^2 - 4ac}$$

are two different imaginary numbers that are not opposites of each other. So, if a is real but b is not, then the two roots

$$x = \frac{-b + \sqrt{b^2 - 4ac}}{2a} \qquad \text{and} \qquad \frac{-b - \sqrt{b^2 - 4ac}}{2a}$$

will be numbers that are not real, but are not complex conjugates.

13.25 Let the common coefficient be a and the constant term be c, so our quadratic is $ax^2 + ax + c = 0$. Because the quadratic has a double root, the discriminant must be 0, so we have

$$a^2 - 4(a)(c) = 0 \quad \Rightarrow \quad a^2 = 4ac \quad \Rightarrow \quad a = 4c \quad \Rightarrow \quad \frac{a}{c} = \boxed{4}.$$

13.26 The discriminant of the quadratic is $(-4)^2 - 4(2)(-7) = 16 - (-56) = 72$. When we take the square root of this in the quadratic formula to find the roots, we have $\sqrt{72} = 6\sqrt{2}$. Because the coefficients of the quadratic are all rational, this tells us that the roots have $\sqrt{2}$ in them. Only choice (B) has $\sqrt{2}$ in it, so

$\boxed{(B)}$ is the only possible answer. See if you can also solve this problem by considering the product of the roots of the quadratic.

13.27 The product of the roots is $4/6 = 2/3$ and the sum of the roots is $-c/6$, so we must have

$$\frac{2}{3} = -\frac{c}{6} + 2.$$

Solving this equation gives $c = \boxed{8}$.

13.28 Because the quadratic has real coefficients, we know that if one root is not real, then the roots are complex conjugates. One root of the quadratic is $3 - i\sqrt{3}$, so the other root is $3 + i\sqrt{3}$. The sum of the roots of the quadratic is $-b$ and the product is c, so we have

$$-b = (3 - i\sqrt{3}) + (3 + i\sqrt{3}) = 6, \text{ and}$$
$$c = (3 - i\sqrt{3})(3 + i\sqrt{3}) = 9 - i^2(\sqrt{3})^2 = 9 + 3 = 12.$$

Therefore, $b = \boxed{-6}$ and $c = \boxed{12}$.

13.29 *Solution 1: Manipulate the original quadratic.* We want a quadratic with a constant term of 1 that has the same roots as the original quadratic. Dividing the original quadratic by 9 won't change the roots of the quadratic, and it gives

$$\frac{1}{9}x^2 + \frac{1}{3}x + 1 = 0.$$

Therefore, we have found a quadratic that has the same roots as $x^2 + 3x + 9 = 0$, but has a constant term of 1. From this quadratic, we see that $A = \boxed{1/9}$ and $B = \boxed{1/3}$.

Solution 2: Use the sum and product of the roots. The sum of the roots of our original quadratic is -3 and the product of its roots is 9. The sum of the roots of $Ax^2 + Bx + 1 = 0$ is $-B/A$ and the product of the roots is $1/A$. Since the roots of this equation are the same as those of $x^2 + 3x + 9 = 0$, we must have

$$-\frac{B}{A} = -3 \qquad \text{and} \qquad \frac{1}{A} = 9.$$

The second equation gives us $A = \boxed{1/9}$ and combining this with the first gives $B = \boxed{1/3}$.

13.30 *Solution 1: Use the quadratic formula.* The roots of the quadratic $ax^2 + bx + c = 0$ are

$$\frac{-b + \sqrt{b^2 - 4ac}}{2a} \qquad \text{and} \qquad \frac{-b - \sqrt{b^2 - 4ac}}{2a}.$$

Therefore, the sum of their reciprocals is

$$\frac{2a}{-b + \sqrt{b^2 - 4ac}} + \frac{2a}{-b - \sqrt{b^2 - 4ac}}.$$

We take care of the radical in the denominator of the first fraction in this sum by multiplying the numerator and denominator by $-b - \sqrt{b^2 - 4ac}$:

$$\frac{2a}{-b + \sqrt{b^2 - 4ac}} = \frac{2a}{-b + \sqrt{b^2 - 4ac}} \cdot \frac{-b - \sqrt{b^2 - 4ac}}{-b - \sqrt{b^2 - 4ac}} = \frac{-2ab - 2a\sqrt{b^2 - 4ac}}{(-b)^2 - (\sqrt{b^2 - 4ac})^2}$$

$$= \frac{-2ab - 2a\sqrt{b^2 - 4ac}}{b^2 - (b^2 - 4ac)} = \frac{-2ab - 2a\sqrt{b^2 - 4ac}}{4ac}$$

$$= \frac{2a(-b - \sqrt{b^2 - 4ac})}{(2a)(2c)} = \frac{-b - \sqrt{b^2 - 4ac}}{2c}.$$

We can similarly take care of the second fraction to find that

$$\frac{2a}{-b - \sqrt{b^2 - 4ac}} = \frac{-b + \sqrt{b^2 - 4ac}}{2c}.$$

Therefore,

$$\frac{2a}{-b + \sqrt{b^2 - 4ac}} + \frac{2a}{-b - \sqrt{b^2 - 4ac}} = \frac{-b - \sqrt{b^2 - 4ac}}{2c} + \frac{-b + \sqrt{b^2 - 4ac}}{2c} = \boxed{-\frac{b}{c}}.$$

Solution 2: Use the sum and product of roots. With such a simple final answer, there must be a simpler solution. Let r and s be the roots of the quadratic. We have

$$\frac{1}{r} + \frac{1}{s} = \frac{s}{rs} + \frac{r}{rs} = \frac{r + s}{rs}.$$

Because $r + s = -b/a$ and $rs = c/a$, we have

$$\frac{1}{r} + \frac{1}{s} = \frac{r + s}{rs} = \frac{-b/a}{c/a} = \boxed{-\frac{b}{c}}.$$

See if you can find another solution by letting $x = \frac{1}{y}$.

Challenge Problems

13.31 Solving $x - y = 3$ for y gives $y = x - 3$. We substitute this in the equation $x^2 - y = -1$ to find $x^2 - (x - 3) = -1$. Rearranging this equation gives $x^2 - x + 4 = 0$. The discriminant of $x^2 - x + 4$ is $(-1)^2 - 4(1)(4) = -15$. Because this discriminant is negative, the roots of the equation $x^2 - x + 4 = 0$ are not real. Therefore, there are $\boxed{\text{no real solutions}}$ to this system of equations.

13.32 If we let $y = z^2$, the equation becomes $y^2 - 4y + 3 = 0$, which we can factor as $(y - 3)(y - 1) = 0$. Therefore, $y = 3$ or $y = 1$. Since $y = z^2$, the solution $y = 3$ gives us $z^2 = 3$, so $z = \pm\sqrt{3}$. The solution $y = 1$ gives us $z^2 = 1$, so $z = \pm1$. Therefore, the four solutions are $\boxed{\pm\sqrt{3} \text{ and } \pm1}$.

13.33 If the equation has at least one real root, then its discriminant must be nonnegative. Therefore, we must have

$$(-5)^2 - 4(1)(k) \geq 0 \quad \Rightarrow \quad 25 - 4k \geq 0 \quad \Rightarrow \quad 25 \geq 4k \quad \Rightarrow \quad \boxed{\frac{25}{4}} \geq k.$$

When $k = 25/4$, our quadratic is $x^2 - 5x + \frac{25}{4} = \left(x - \frac{5}{2}\right)^2$. For any larger value of k, the quadratic will have a negative discriminant, and will have roots that are not real.

13.34 Let's get rid of the fractions. We do so by multiplying both sides by $ab(a + b)$. This gives

$$ab(a + b)\left(\frac{1}{a} + \frac{1}{b}\right) = ab(a + b)\left(\frac{2}{a + b}\right) \quad \Rightarrow \quad \frac{ab(a + b)}{a} + \frac{ab(a + b)}{b} = \frac{2ab(a + b)}{a + b}.$$

Simplifying all three fractions in our equation gives

$$b(a + b) + a(a + b) = 2ab \quad \Rightarrow \quad ab + b^2 + a^2 + ab = 2ab \quad \Rightarrow \quad a^2 + b^2 = 0.$$

If a and b are real and $a^2 + b^2 = 0$, then $a = b = 0$. However, $a = b = 0$ makes all the denominators of our original equation equal to zero, which is forbidden. Therefore, there are $\boxed{\text{no}}$ real numbers a and b that satisfy the equation.

13.35

(a) We could find the roots, then square them and add them, but that's a lot more work than necessary. Inspired by our work on an earlier example, we can use the sum and the product of the roots to find the sum of the squares of the roots. Suppose r and s are the roots of the quadratic, so we have

$$r + s = \frac{5}{3} \quad \text{and} \quad rs = \frac{8}{3}.$$

We seek $r^2 + s^2$, so we square the first equation to find $r^2 + 2rs + s^2 = \frac{25}{9}$. Since $rs = 8/3$, we subtract $2rs$ from both sides to find

$$r^2 + s^2 = \frac{25}{9} - 2rs = \frac{25}{9} - \frac{16}{3} = \boxed{-\frac{23}{9}}.$$

(b) We rewrite the equation as $x^2 + 2hx - 3 = 0$. Let r and s be the roots of this quadratic. We have $r + s = -2h$ and $rs = -3$. We use the first part as a guide. We square $r + s = -2h$ to find $r^2 + 2rs + s^2 = 4h^2$. Since $rs = -3$, we have

$$r^2 + s^2 = 4h^2 - 2rs = 4h^2 + 6.$$

We are told that $r^2 + s^2 = 10$, so we have $4h^2 + 6 = 10$. Solving for h^2 gives $h^2 = 1$, so $h = \boxed{\pm 1}$.

13.36 The roots of the quadratic equation $ax^2 + bx + c = 0$ are

$$x = \frac{-b \pm \sqrt{b^2 - 4ac}}{2a}.$$

If a, b, and c are rational, the only part of this expression that could result in an irrational number is the $\sqrt{b^2 - 4ac}$. If $b^2 - 4ac$ is the square of a rational number, then $\sqrt{b^2 - 4ac}$ will be rational, as will the roots. Otherwise, $\sqrt{b^2 - 4ac}$ will be irrational, as will the roots.

13.37 Solving the first equation for x in terms of y gives $x = 4 - 3y$. Substituting this into the second equation gives

$$(4-3y)^2 + y^2 - 4y = 12 \quad \Rightarrow \quad 16 - 24y + 9y^2 + y^2 - 4y = 12 \quad \Rightarrow \quad 10y^2 - 28y + 4 = 0 \quad \Rightarrow \quad 5y^2 - 14y + 2 = 0.$$

Applying the quadratic formula gives

$$y = \frac{-(-14) \pm \sqrt{(-14)^2 - 4(5)(2)}}{2(5)} = \frac{14 \pm \sqrt{156}}{10} = \frac{14 \pm 2\sqrt{39}}{10} = \frac{7 \pm \sqrt{39}}{5}.$$

When $y = \dfrac{7 + \sqrt{39}}{5}$, we have

$$x = 4 - 3y = 4 - 3\left(\frac{7 + \sqrt{39}}{5}\right) = 4 - \frac{21 + 3\sqrt{39}}{5} = \frac{20 - 21 - 3\sqrt{39}}{5} = \frac{-1 - 3\sqrt{39}}{5}.$$

When $y = \dfrac{7 - \sqrt{39}}{5}$, we have

$$x = 4 - 3y = 4 - 3\left(\frac{7 - \sqrt{39}}{5}\right) = 4 - \frac{21 - 3\sqrt{39}}{5} = \frac{20 - 21 + 3\sqrt{39}}{5} = \frac{-1 + 3\sqrt{39}}{5}.$$

Therefore, our solutions are $\boxed{(x, y) = \left(\dfrac{-1 + 3\sqrt{39}}{5}, \dfrac{7 - \sqrt{39}}{5}\right) \text{ and } (x, y) = \left(\dfrac{-1 - 3\sqrt{39}}{5}, \dfrac{7 + \sqrt{39}}{5}\right)}$.

13.38

(a) Factoring gives $(x - 12)(x + 12) = 0$, so our solutions are $x = \boxed{12 \text{ and } -12}$.

(b) Factoring $x^4 - 81$ as the difference of squares gives $(x^2 - 9)(x^2 + 9) = 0$. From $x^2 - 9 = 0$, we have the solutions $x = 3$ and $x = -3$. From $x^2 + 9 = 0$, we have the solutions $x = 3i$ and $x = -3i$. Therefore, our four solutions are $x = \boxed{3, -3, 3i, \text{ and } -3i}$.

(c) Factoring $x^3 - 27$ as a difference of cubes gives $(x - 3)(x^2 + 3x + 9) = 0$. From $x - 3 = 0$, we have the solution $x = 3$. We use the quadratic formula on $x^2 + 3x + 9 = 0$ to find the solutions $x = \dfrac{-3 \pm \sqrt{3^2 - 4(1)(9)}}{2} = \dfrac{-3 \pm 3i\sqrt{3}}{2}$. Therefore, our three solutions are $x = \boxed{3 \text{ and } \dfrac{-3 \pm 3i\sqrt{3}}{2}}$.

(d) Factoring $z^3 + 8$ as a sum of cubes gives $(z + 2)(z^2 - 2z + 4) = 0$. Therefore, we have the solutions $z = \boxed{-2}$ and $z = \dfrac{-(-2) \pm \sqrt{(-2)^2 - 4(1)(4)}}{2(1)} = \dfrac{2 \pm 2i\sqrt{3}}{2} = \boxed{1 \pm i\sqrt{3}}$.

(e) First, we factor $x^6 - 1$ as the difference of squares to get $(x^3 - 1)(x^3 + 1) = 0$. We then factor each of $x^3 - 1$ and $x^3 + 1$ to find

$$(x - 1)(x^2 + x + 1)(x + 1)(x^2 - x + 1) = 0.$$

We use the quadratic formula to determine the roots of the quadratic factors, and we have the six solutions

$$x = \boxed{1, -1, \frac{-1 \pm i\sqrt{3}}{2}, \text{ and } \frac{1 \pm i\sqrt{3}}{2}}.$$

13.39   We write the equation as $z^2 - z - 5 + 5i = 0$ so we can use the quadratic formula to find z:

$$z = \frac{-(-1) \pm \sqrt{(-1)^2 - 4(-5 + 5i)}}{2(1)} = \frac{1 \pm \sqrt{1 + 20 - 20i}}{2} = \frac{1 \pm \sqrt{21 - 20i}}{2}.$$

So, we must find $\sqrt{21 - 20i}$. If we let $a + bi = \sqrt{21 - 20i}$ and square both sides, we have $a^2 - b^2 + 2abi = 21 - 20i$. This gives us the system of equations

$$a^2 - b^2 = 21,$$
$$2ab = -20.$$

A little quick trial and error gives us $(a, b) = (5, -2)$ and $(a, b) = (-5, 2)$ as our possibilities. The $\pm$ before the $\sqrt{21 - 20i}$ means we don't have to worry about both possibilities, since $5 - 2i$ is the negative of $-5 + 2i$. So, our two solutions for z are

$$z = \frac{1 + (5 - 2i)}{2} = 3 - i \qquad \text{and} \qquad z = \frac{1 - (5 - 2i)}{2} = -2 + i.$$

The product of the real parts of these solutions is $(3)(-2) = \boxed{-6}$.

13.40 We let $\ell/w = x$ and try to write the given equation in terms of x. The right side of the equation is simply x. On the left, we have

$$\frac{\ell + w}{\ell} = \frac{\ell}{\ell} + \frac{w}{\ell} = 1 + \frac{1}{x}.$$

So, our equation is

$$1 + \frac{1}{x} = x \quad \Rightarrow \quad x\left(1 + \frac{1}{x}\right) = x(x) \quad \Rightarrow \quad x + 1 = x^2 \quad \Rightarrow \quad x^2 - x - 1 = 0.$$

Applying the quadratic formula gives

$$x = \frac{-(-1) \pm \sqrt{(-1)^2 - 4(1)(-1)}}{2(1)} = \frac{1 \pm \sqrt{5}}{2}.$$

Since the values of ℓ and w are positive (they are lengths), we must take the positive value of x, which is $\boxed{\dfrac{1 + \sqrt{5}}{2}}$.

13.41 *Solution 1: Use the sum and product of roots.* Because p and q are the roots of $x^2 + 2x + 2 = 0$, we have $p + q = -2$ and $pq = 2$. Suppose the quadratic $y^2 + by + c$ has roots $\frac{1}{p}$ and $\frac{1}{q}$. We have $\frac{1}{p} + \frac{1}{q} = -b$, so

$$b = -\left(\frac{1}{p} + \frac{1}{q}\right) = -\left(\frac{q}{pq} + \frac{p}{pq}\right) = -\left(\frac{p + q}{pq}\right) = -\frac{-2}{2} = 1.$$

Similarly,

$$c = \frac{1}{p} \cdot \frac{1}{q} = \frac{1}{pq} = \frac{1}{2}.$$

Therefore, the equation $\boxed{y^2 + y + \dfrac{1}{2} = 0}$ has roots $\dfrac{1}{p}$ and $\dfrac{1}{q}$.

   Solution 2: Clever substitution. If we let $x = \frac{1}{y}$, then $x = p$ when $y = \frac{1}{p}$ and $x = q$ when $y = \frac{1}{q}$. Since we know the equation $x^2 + 2x + 2 = 0$ is satisfied when $x = p$, and that $x = p$ when $y = \frac{1}{p}$, then we know that substituting $x = \frac{1}{y}$ into this equation gives us an equation that is satisfied when $y = \frac{1}{p}$:

$$\left(\frac{1}{y}\right)^2 + 2\left(\frac{1}{y}\right) + 2 = 0 \quad \Rightarrow \quad \frac{1}{y^2} + \frac{2}{y} + 2 = 0.$$

Multiplying this equation by y^2 gives $1 + 2y + 2y^2 = 0$. Just as this equation is satisfied by $y = \frac{1}{p}$, it is also satisfied by $y = \frac{1}{q}$. Therefore, $\boxed{2y^2 + 2y + 1 = 0}$ has roots $\frac{1}{p}$ and $\frac{1}{q}$. Notice that this equation is just 2 times the equation we found in the other solution.

13.42 The numerator of the left side is a difference of cubes, so we can factor it:

$$\frac{x^6 - 8}{x^2 - 2} = \frac{(x^2)^3 - 2^3}{x^2 - 2} = \frac{(x^2 - 2)[(x^2)^2 + (x^2)(2) + 2^2]}{x^2 - 2} = x^4 + 2x^2 + 4.$$

Therefore, our equation is $x^4 + 2x^2 + 4 = 12$, so $x^4 + 2x^2 - 8 = 0$. We can factor this equation just like we factor quadratics. (If you don't see why, let $y = x^2$ and rewrite the equation in terms of y.) Our equation then is $(x^2 + 4)(x^2 - 2) = 0$. We cannot have $x^2 - 2 = 0$ because this will make a denominator in our original equation equal to 0. From $x^2 + 4 = 0$, we have the solutions $x = \boxed{\pm 2i}$.

13.43 *Solution 1: Slick manipulation.* The given expression looks the quadratic formula turned upside-down. So, we wonder what happens if x is "turned upside-down". We let $x = 1/y$ in $ax^2 + bx + c = 0$, and we have

$$a\left(\frac{1}{y}\right)^2 + b\left(\frac{1}{y}\right) + c = 0 \quad \Rightarrow \quad \frac{a}{y^2} + \frac{b}{y} + c = 0 \quad \Rightarrow \quad a + by + cy^2 = 0.$$

Applying the quadratic formula to this equation (note that c is the coefficient of the quadratic term and a is the constant), we have

$$y = \frac{-b \pm \sqrt{b^2 - 4ac}}{2c}.$$

Since $x = 1/y$, we have

$$x = \frac{1}{y} = \frac{2c}{-b \pm \sqrt{b^2 - 4ac}}.$$

Solution 2: "Rationalize the denominator." We "rationalize the denominator" to find:

$$\frac{2c}{-b + \sqrt{b^2 - 4ac}} = \frac{2c}{-b + \sqrt{b^2 - 4ac}} \cdot \frac{-b - \sqrt{b^2 - 4ac}}{-b - \sqrt{b^2 - 4ac}} = \frac{-2bc - 2c\sqrt{b^2 - 4ac}}{(-b)^2 - (\sqrt{b^2 - 4ac})^2}$$

$$= \frac{-2bc - 2c\sqrt{b^2 - 4ac}}{b^2 - b^2 + 4ac} = \frac{-2bc - 2c\sqrt{b^2 - 4ac}}{4ac} = \frac{2c(-b - \sqrt{b^2 - 4ac})}{4ac} = \frac{-b - \sqrt{b^2 - 4ac}}{2a}.$$

Similarly, we can show that $\dfrac{2c}{-b - \sqrt{b^2 - 4ac}} = \dfrac{-b + \sqrt{b^2 - 4ac}}{2a}$, so the expression given for x in the problem is equivalent to the expression given by the quadratic formula.

13.44 Because r and s are the roots of $x^2 + bx + c$, we have $x^2 + bx + c = (x - r)(x - s)$. Similarly, we have $x^2 + ex + f = (x - r)(x - 3s)$. Therefore, we have

$$(x^2 + bx + c) + (x^2 + ex + f) = (x - r)(x - s) + (x - r)(x - 3s) = (x - r)[(x - s) + (x - 3s)] = (x - r)(2x - 4s).$$

So, we have $(x - r)(2x - 4s) = 0$, which gives us solutions $\boxed{x = r \text{ and } x = 2s}$.

13.45 We see $r^2 + 3r$ in both factors on the left, so we let $a = r^2 + 3r$ to have

$$a(a + 5) = 6 \quad \Rightarrow \quad a^2 + 5a - 6 = 0 \quad \Rightarrow \quad (a + 6)(a - 1) = 0.$$

Therefore, $a = -6$ or $a = 1$. If $a = -6$, we have

$$r^2 + 3r = -6 \quad \Rightarrow \quad r^2 + 3r + 6 = 0 \quad \Rightarrow \quad r = \frac{-3 \pm \sqrt{3^2 - 4(1)(6)}}{2(1)} = \boxed{\frac{-3 \pm i\sqrt{15}}{2}}.$$

If $a = 1$, we have

$$r^2 + 3r = 1 \quad \Rightarrow \quad r^2 + 3r - 1 = 0 \quad \Rightarrow \quad r = \frac{-3 \pm \sqrt{3^2 - 4(1)(-1)}}{2(1)} = \boxed{\frac{-3 \pm \sqrt{13}}{2}}.$$

13.46 If $x^2 + ax + b = 0$ has integer roots, then its discriminant is a perfect square, so $a^2 - 4b$ is a perfect square. Similarly, if $x^2 + ax + (b + 1)$ has integer roots, its discriminant is a perfect square, so $a^2 - 4(b+1) = a^2 - 4b - 4$ is a perfect square. Therefore, $a^2 - 4b$ and $a^2 - 4b - 4$ must both be perfect squares. Since $a^2 - 4b$ is 4 greater than $a^2 - 4b - 4$ and both must be perfect squares, we must have $a^2 - 4b = 4$ and $a^2 - 4b - 4 = 0$, since 4 and 0 are the only perfect squares that are 4 apart. (See if you can prove that 0 and 4 are the only perfect squares that differ by 4.) Therefore,

$$a^2 = 4b + 4 \quad \Rightarrow \quad a = \sqrt{4b + 4} = 2\sqrt{b + 1}.$$

(Remember, a and b nonnegative, so we can discard the negative value of a.) Since a must be an integer, $b + 1$ must be a square. Therefore, b must be 1 less than a perfect square.

To see that all such b will give us integer roots for both quadratics, we let $b = n^2 - 1$, so that

$$a = 2\sqrt{b + 1} = 2n.$$

So, our quadratics are $x^2 + 2nx + (n^2 - 1) = [x + (n-1)][x + (n+1)]$ and $x^2 + 2nx + n^2 = (x + n)^2$. Since b is one less than a positive perfect square and the largest positive perfect square less than 2002 is $44^2 = 1936$ ($45^2 = 2025$), there are $\boxed{44}$ possible values of b, each of which corresponds to a different value of a.

CHAPTER **14**

Graphing Quadratics

Exercises for Section 14.1

14.1.1   We start by completing the square to get $y = (x+2)^2 - 9$. The graph of this equation is a parabola with vertex $(-2, -9)$. We then choose a few more values of x and find corresponding values of y to graph the parabola as shown below.

x	y
-6	7
-4	-5
-3	-8
-2	-9
-1	-8
0	-5
2	7

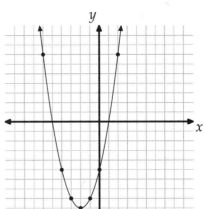

14.1.2   We first complete the square, as before. We have

$$y = -2\left(x^2 - 3x\right) - 7 \quad \Rightarrow \quad y - 2\left(\frac{3}{2}\right)^2 = -2\left[x^2 - 3x + \left(\frac{3}{2}\right)^2\right] - 7$$

$$\Rightarrow \quad y - \frac{9}{2} = -2\left(x - \frac{3}{2}\right)^2 - 7 \quad \Rightarrow \quad y = -2\left(x - \frac{3}{2}\right)^2 - \frac{5}{2}.$$

The graph of this equation is a parabola with vertex $(3/2, -5/2) = \boxed{\left(1\frac{1}{2}, -2\frac{1}{2}\right)}$. The graph's axis of

symmetry is $\boxed{x = 1\frac{1}{2}}$.   As shown below, we choose values of x on either side of this axis and find corresponding values of y. We combine these points with the vertex to graph the parabola.

x	y
0	-7
1	-3
$1\frac{1}{2}$	$-2\frac{1}{2}$
2	-3
3	-7

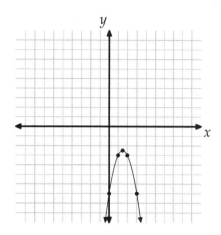

14.1.3 We start by completing the square:

$$x = 4y^2 + 16y + 9 \quad \Rightarrow \quad x = 4(y^2 + 4y) + 9 \quad \Rightarrow \quad x + 4(4) = 4(y^2 + 4y) + 4(4) + 9$$

$$\Rightarrow \quad x + 16 = 4(y^2 + 4y + 4) + 9 \quad \Rightarrow \quad x = 4(y + 2)^2 - 7.$$

The graph of this equation is a parabola with vertex $\boxed{(-7, -2)}$ and axis of symmetry $\boxed{y = -2}$. We find points on the parabola and graph it as shown below:

x	y
9	-4
-3	-3
-7	-2
-3	-1
9	0

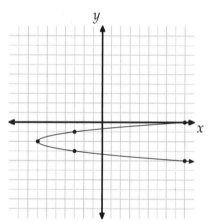

14.1.4 We find a few points on each graph, then graph the two equations as shown below.

x	y
2	-6
3	-4
4	-6

x	y
-6	2
-4	3
-6	4

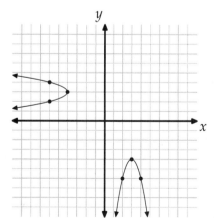

The vertex of the graph of the equation $y = -2(x - 3)^2 - 4$ is $(3, -4)$, and this graph is the downward-opening parabola above. The vertex of the graph of the equation $x = -2(y-3)^2 - 4$ is $(-4, 3)$, and its graph is the leftward-opening parabola above. Notice that the points on one parabola have coordinates that are the reverse of the coordinates on the other parabola. This is not a coincidence. We turn one equation into the other by swapping x and y, so if the point $(x, y) = (a, b)$ satisfies one equation, then $(x, y) = (b, a)$ satisfies the other.

As we'll investigate later in the text, this swapping of the coordinates causes one graph to be the mirror image of the other when reflected over the line $x = y$.

14.1.5

(a) Because the graph has a vertical axis of symmetry, we know that the equation is of the form $y = a(x-h)^2 + k$. Because the vertex is $(1, 2)$, we have $h = 1$ and $k = 2$. Therefore, our equation is now $y = a(x-1)^2 + 2$. Because the graph of this equation passes through $(3, 3)$, the values $(x, y) = (3, 3)$ must satisfy this equation, so $3 = a(3-1)^2 + 2$. Solving this equation gives $a = 1/4$, so an equation whose graph is the parabola described in the problem is $\boxed{y = \dfrac{1}{4}(x-1)^2 + 2}$.

(b) Because the graph has a horizontal axis of symmetry, we know that the equation is of the form $x = a(y-k)^2 + h$. Because the vertex is $(1, 2)$, we have $h = 1$ and $k = 2$. Therefore, our equation is now $x = a(y-2)^2 + 1$. Because the graph of this equation passes through $(3, 3)$, the values $(x, y) = (3, 3)$ must satisfy this equation, so $3 = a(3-2)^2 + 1$. Solving this equation gives $a = 2$, so an equation whose graph is the parabola described in the problem is $\boxed{x = 2(y-2)^2 + 1}$.

(c) Parabolas don't only open up or down or right or left. They can open in any direction, and their axes of symmetry can point in any direction. There are infinitely many other equations whose graphs are parabolas with vertex $(1, 2)$ such that the parabola passes through $(3, 3)$. However, the ones with a vertical or horizontal axis of symmetry are easiest to find.

14.1.6 Because the axis of symmetry of the parabola is horizontal, the parabola opens either left or right. Therefore, the general form of the equation that produces this graph is $x = ay^2 + by + c$. We create a system of equations involving a, b, and c by substituting the three points we are given on the graph into this equation:

$$1 = c,$$
$$2 = a + b + c,$$
$$2 = 4a - 2b + c.$$

Letting $c = 1$ in our second and third equations gives $a + b = 1$ and $4a - 2b = 1$. Solving this system of two equations gives $a = 1/2$ and $b = 1/2$, so our desired equation is $\boxed{x = \dfrac{1}{2}y^2 + \dfrac{1}{2}y + 1}$.

Exercises for Section 14.2

14.2.1 We complete the square for both the x terms and the y terms to find:

$$x^2 + y^2 - 10x + 4y = 20 \quad \Rightarrow \quad x^2 - 10x + y^2 + 4y = 20 \quad \Rightarrow$$

$$x^2 - 10x + 25 + y^2 + 4y + 4 = 20 + 25 + 4 \quad \Rightarrow \quad (x - 5)^2 + (y + 2)^2 = 49$$

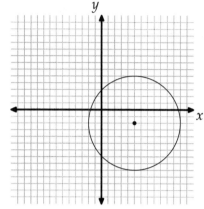

The graph of this equation is a circle with center $\boxed{(5, -2)}$ and radius $\sqrt{49} = \boxed{7}$. This graph is shown at right.

14.2.2 If the graphs of these two circles meet at a point (x, y), then that point must satisfy both equations. From our first equation, we have $y^2 = 4 - (x - 1)^2$. Substituting this into the second equation gives

$$(x + 3)^2 + 4 - (x - 1)^2 = 4 \quad \Rightarrow \quad x^2 + 6x + 9 - x^2 + 2x - 1 = 0 \quad \Rightarrow \quad x = -1.$$

Therefore, $y^2 = 4 - (x - 1)^2 = 0$, so $y = 0$. So, these two circles meet at $\boxed{(-1, 0)}$. See if you can find a geometric explanation for why the two graphs meet at this point.

14.2.3 We complete the square for both the x terms and the y terms to find

$$x^2 - 4x + y^2 + 6y = k \quad \Rightarrow \quad x^2 - 4x + 4 + y^2 + 6y + 9 = k + 4 + 9 \quad \Rightarrow \quad (x - 2)^2 + (y + 3)^2 = k + 13.$$

The graph of this equation is a circle if and only if $k + 13 > 0$. (The left side of the equation is nonnegative, so if $k + 13 < 0$, there are no real values of x and y that satisfy the equation. If $k + 13 = 0$, then only the point $(x, y) = (2, -3)$ satisfies the equation.) Therefore, we must have $\boxed{k > -13}$.

14.2.4 The center of a circle is the same distance from all points on the circle. Since the center is on the diameter and the endpoints of the diameter are on the circle, the center must be the midpoint of the diameter. Therefore, the center of the circle in this problem is $\left(\frac{-5+25}{2}, \frac{0+0}{2}\right) = (10, 0)$. Since the center of the circle, $(10, 0)$, is 15 units from $(25, 0)$, which is on the circle, the radius of the circle is 15. Therefore, an equation whose graph is the circle in the problem is

$$(x - 10)^2 + y^2 = 15^2.$$

So, if the point $(x, 15)$ is on the circle, we must have $(x - 10)^2 + 15^2 = 15^2$, so $(x - 10)^2 = 0$, which means $x = \boxed{10}$.

14.2.5 We seek the value of k such that there is exactly one solution (x, y) to the system of equations $x + ky = 4$, $x^2 + y^2 - 12x + 8y + 42 = 0$. Solving $x + ky = 4$ for x in terms of y gives $x = 4 - ky$. Substituting this into our other equation gives

$$(4 - ky)^2 + y^2 - 12(4 - ky) + 8y + 42 = 0 \quad \Rightarrow \quad 16 - 8ky + k^2y^2 + y^2 - 48 + 12ky + 8y + 42 = 0$$

$$\Rightarrow \quad (k^2 + 1)y^2 + (4k + 8)y + 10 = 0.$$

Since we want our system of equations to have exactly one solution (x, y), we want this quadratic to have exactly one root. Therefore, its discriminant must equal 0, so we must have

$$(4k + 8)^2 - 4(k^2 + 1)(10) = 0 \quad \Rightarrow \quad 16k^2 + 64k + 64 - 40k^2 - 40 = 0$$

$$\Rightarrow \quad 3k^2 - 8k - 3 = 0 \quad \Rightarrow \quad (k - 3)(3k + 1) = 0.$$

Therefore, the values of k for which our quadratic in y has exactly one solution are $\boxed{k = 3 \text{ and } k = -1/3}$. When $k = 3$, our quadratic above is $10y^2 + 20y + 10 = 0$, from which we have $(y + 1)^2 = 0$, so $y = -1$. Since $y = -1$, we have $x = 4 - ky = 4 - (3)(-1) = 7$. So, when $k = 3$, the two graphs meet only at $(7, -1)$. Similarly, when $k = -1/3$, we have $10y^2/9 + 20y/3 + 10 = 0$. Multiplying this equation by $9/10$ gives $y^2 + 6y + 9 = 0$, so we have $(y + 3)^2 = 0$. This gives us $y = -3$, so $x = 4 - ky = 3$ and the two graphs meet at $(3, -3)$.

Review Problems

14.15 Completing the square gives $y = (x + 1)^2 + 4$. Comparing this to the standard form $y = a(x - h)^2 + k$, we have $a = 1$, $h = -1$, and $k = 4$. Therefore, the vertex is $(h, k) = \boxed{(-1, 4)}$. The axis of symmetry is $x = h$, so the axis of the graph of this equation is $\boxed{x = -1}$.

14.16 We complete the square to find

$$x = 2(y^2 - 2y) + 4 \quad \Rightarrow \quad x + 2(1) = 2(y^2 - 2y + 1) + 4 \quad \Rightarrow \quad x = 2(y - 1)^2 + 2.$$

Therefore, the graph is a parabola with vertex $\boxed{(2, 1)}$. Since a is positive and y is squared, we expect the parabola to open rightward. We find some points on the graph, then graph it as shown below.

x	y
10	−1
4	0
2	1
4	2
10	3

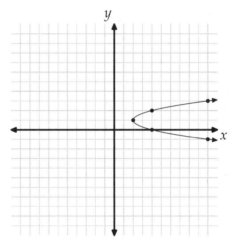

14.17

(a) The vertex of the parabola is $(3, -4)$, and the parabola opens upward, so the desired equation is of the form $y = a(x - 3)^2 - 4$. To find a, we identify another point on the parabola and substitute its coordinates in this equation. The graph passes through $(4, -2)$, so we have $-2 = a(4 - 3)^2 - 4$, which gives us $a = 2$. Therefore, our equation is $\boxed{y = 2(x - 3)^2 - 4}$.

(b) The vertex of the parabola is $(-6, -5)$ and the parabola opens rightward. Therefore, our desired equation is of the form $x = a(y + 5)^2 - 6$. As before, we find another point on the parabola and substitute its coordinates into this equation. The point $(-3, -2)$ is on the parabola, so we have $-3 = a(-2 + 5)^2 - 6$, which gives us $a = 1/3$. So, our equation is $\boxed{x = \dfrac{1}{3}(y + 5)^2 - 6}$.

14.18   We start by completing the squares in both x and y, which gives

$$x^2 + 2x + 1 + y^2 - 4y + 4 - 11 = 1 + 4 \quad \Rightarrow \quad (x+1)^2 + (y-2)^2 = 16.$$

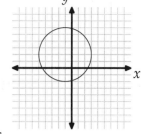

The graph of this equation is a circle with center $(-1, 2)$ and radius $\sqrt{16} = 4$, as shown at right.

14.19   Because a diameter has both its endpoints on the circle and passes through the center of the circle, its length is twice that of a radius. So, because a diameter of our circle is 6, its radius is 3. Therefore, our equation is $\boxed{(x-3)^2 + (y-5)^2 = 9}$.

14.20   We start by dividing everything by 4 to make the coefficients of x^2 and y^2 equal to 1. This gives

$$x^2 + y^2 + x - 4y = \frac{7}{4}.$$

Completing the square then gives us

$$x^2 + x + \frac{1}{4} + y^2 - 4y + 4 = \frac{7}{4} + \frac{1}{4} + 4 \quad \Rightarrow \quad \left(x + \frac{1}{2}\right)^2 + (y-2)^2 = 6.$$

The graph of this equation is circle with center $\boxed{(-1/2, 2)}$ and radius $\boxed{\sqrt{6}}$.

14.21   Expanding the squared binomial on the left gives $3x^2 + 6xy + 3y^2 = 6xy + 27$. Therefore, our equation is $3x^2 + 3y^2 = 27$, or $x^2 + y^2 = 9$. The graph of this equation is a circle centered at the origin with radius 3, as shown at right.

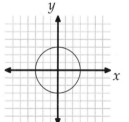

14.22   We complete the square to find

$$y = a\left(x^2 + \frac{b}{a}x\right) + c \quad \Rightarrow \quad y + a\left(\frac{b}{2a}\right)^2 = a\left[x^2 + \frac{b}{a}x + \left(\frac{b}{2a}\right)^2\right] + c$$

$$\Rightarrow \quad y = a\left(x + \frac{b}{2a}\right)^2 + c - \frac{b^2}{4a}.$$

Comparing this equation to our standard form $y = a(x-h)^2 + k$, we see that $h = -\frac{b}{2a}$ and $k = c - \frac{b^2}{4a}$, so our vertex is

$$(h, k) = \left(-\frac{b}{2a}, c - \frac{b^2}{4a}\right).$$

14.23   Because the axis of symmetry is parallel to the x-axis, it opens either leftward or rightward. Therefore, its general form is $x = ay^2 + by + c$. Substituting the three given points in this equation gives us the system

$$4 = c,$$
$$13 = a - b + c,$$
$$4 = 4a + 2b + c.$$

We have $c = 4$ from our first equation. Substituting this into our second and third equations gives $a - b = 9$ and $4a + 2b = 0$. Solving this system of equations gives $a = 3$ and $b = -6$. Therefore, our desired equation is $\boxed{x = 3y^2 - 6y + 4}$.

14.24 We seek the ordered pairs (x, y) that satisfy both equations. From our second equation, we have $y = 4x + 12$. Substituting this into the equation $y = 2x^2 + 5x - 3$, we have $4x + 12 = 2x^2 + 5x - 3$. Rearranging this gives $2x^2 + x - 15 = 0$, and factoring this equation gives $(2x - 5)(x + 3) = 0$. The solutions of this equation are $x = 5/2$ and $x = -3$. When $x = 5/2$, we have $y = 4x + 12 = 22$. When $x = -3$, we have $y = 4x + 12 = 0$. So, the graphs of the two given equations intersect at $\boxed{\left(2\frac{1}{2}, 22\right) \text{ and } (-3, 0)}$.

Challenge Problems

14.25 The center of the circle is $(2, -1)$ and its radius is $\sqrt{25} = 5$. Therefore, we seek lattice points that are 5 units from $(2, -1)$. There is one such point in each of the four vertical (up and down) or horizontal (left and right) directions. However, we also have $3^2 + 4^2 = 9 + 16 = 25 = 5^2$. So, for example, a point that is 3 units to the left of and 4 units above $(2, -1)$ is 5 units from $(2, -1)$. There are 4 points that are 3 units to the left or right of $(2, -1)$ and 4 units above or below $(2, -1)$, and there are 4 more that are 4 units horizontally and 3 units vertically from $(2, -1)$. Combining all these lattice points gives us $4 + 4 + 4 = \boxed{12}$ lattice points that are on the graph of the circle. These points are marked in the diagram at right.

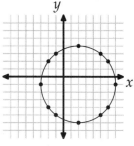

14.26 Finding the points where the graphs of these equations meet is the same as finding the ordered pairs (x, y) that satisfy both equations. Solving the linear equation for y in terms of x gives $y = 7 - 3x$. Substituting this into our other equation gives

$$x^2 + y^2 + 6x + 10y = 2 \quad \Rightarrow \quad x^2 + (7 - 3x)^2 + 6x + 10(7 - 3x) = 2$$
$$\Rightarrow \quad x^2 + 9x^2 - 42x + 49 + 6x + 70 - 30x = 2 \quad \Rightarrow \quad 10x^2 - 66x + 117 = 0.$$

The discriminant of this equation is $66^2 - 4(10)(117) = -324$, so there are no real solutions. So, there are no ordered pairs (x, y) that satisfy both equations, which means there are $\boxed{0}$ points at which the graphs of the two equations meet.

14.27 Substituting the two points that are given into the equation $y = ax^2 + bx + c$ gives

$$y_1 = a + b + c,$$
$$y_2 = a - b + c.$$

We seek b. We can eliminate a and c by subtracting the second equation from the first, which gives $y_1 - y_2 = 2b$. However, we are also given $y_1 - y_2 = -6$, so we must have $2b = -6$. Therefore, $b = \boxed{-3}$.

14.28

(a) The vertex of the initial parabola is $(5, 7)$. Because $(5, 7)$ is on the line $y = 7$, reflecting the vertex over the line $y = 7$ doesn't move the vertex at all. Therefore, the vertex of the new parabola is also $(5, 7)$.

 The original parabola opens upward. After reflecting the parabola over $y = 7$, the new parabola opens the opposite direction, downward. Therefore, the value of a in the standard form of the equation of the new parabola is negative. Reflecting the parabola over a line does not change the "width" of the parabola, and the value of a determines the width of the parabola. Because our new parabola opens in the opposite direction as our old parabola, but is just as "wide" (meaning it opens at the same rate as the old parabola), its value of a is just the negative of the corresponding coefficient of $(x - 5)^2$ in the original equation. Therefore, we have $a = -3$.

Because the vertex of the new parabola is $(5, 7)$ and the value of a in the desired equation whose graph produces this parabola is -3, the desired equation is $\boxed{y = -3(x - 5)^2 + 7}$.

(b) Again, the vertex of the original parabola is $(5, 7)$. This point is 2 units below the line $y = 9$. So, when the parabola is reflected over $y = 9$, the vertex of the new parabola is 2 units above the line $y = 9$, at $(5, 11)$. For the exact same reason as in the previous part, the value of a in the equation whose graph is the new parabola is -3. So, our desired equation is $\boxed{y = -3(x - 5)^2 + 11}$.

14.29

(a) The graph crosses the x-axis at the points where $y = 0$. Solving the equation $x^2 - 7x + 6 = 0$ gives us $x = 1$ and $x = 6$ as solutions, so the two points where the graph crosses the x-axis are $(1, 0)$ and $(6, 0)$. Completing the square for $y = x^2 - 7x + 6$ gives

$$y + \left(-\frac{7}{2}\right)^2 = x^2 - 7x + \left(-\frac{7}{2}\right)^2 + 6 \quad \Rightarrow \quad y = \left(x - \frac{7}{2}\right)^2 - \frac{25}{4}.$$

The x-coordinate of the vertex of this parabola is $3\frac{1}{2}$, which is the average of the x-coordinates of the two points where the parabola crosses the x-axis.

(b) The two points where the graph of $y = ax^2 + bx + c$ crosses the x-axis are the solutions to $ax^2 + bx + c = 0$. (Keep in mind, we are told that the parabola crosses the x-axis, so we are guaranteed that these points exist.) The solutions to this equation are given by the quadratic formula,

$$x = \frac{-b \pm \sqrt{b^2 - 4ac}}{2a}.$$

These are the x-coordinates of the points where the graph crosses the y-axis. The average of these is

$$\frac{\dfrac{-b + \sqrt{b^2 - 4ac}}{2a} + \dfrac{-b - \sqrt{b^2 - 4ac}}{2a}}{2} = -\frac{b}{2a}.$$

However, we saw in Problem 14.22 that the vertex of the parabola that is the graph of the equation $y = ax^2 + bx + c$ is also $-b/(2a)$. Therefore, the x-coordinate of the vertex of a parabola that meets the x-axis in two points and has a vertical axis is the average of the x-coordinates of the two points where the parabola meets the x-axis.

14.30 The expression on the left looks a lot like the square of the distance formula. If we take the square root of both sides, we have

$$\sqrt{(x-3)^2 + (y+2)^2} \le 6.$$

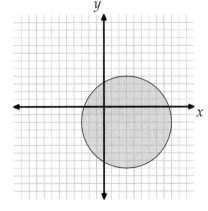

The left side of this expression represents the distance between the point (x, y) and the point $(3, -2)$. So, the points that satisfy this inequality are all those points that are no more than 6 units away from $(3, -2)$. The points that are exactly 6 units away from $(3, -2)$ form a circle with center $(3, -2)$ and radius 6. So, the graph of this inequality consists of this circle and all the points inside it, as shown at right. The area of this region is $\pi(6^2) = \boxed{36\pi}$.

14.31 First, because the vertex is below the x-axis and the parabola intersects the x-axis, we know that it opens upwards. Therefore, $\boxed{a \text{ must be positive}}$. To tackle b and c, we complete the square to get

$$y = ax^2 + bx + c \quad \Rightarrow \quad y = a\left(x^2 + \frac{b}{a}x\right) + c$$

$$\Rightarrow \quad y + a \cdot \frac{b^2}{4a^2} = a\left(x^2 + \frac{b}{a}x + \frac{b^2}{4a^2}\right) + c \quad \Rightarrow \quad y = a\left(x + \frac{b}{2a}\right)^2 + c - \frac{b^2}{4a}.$$

Because the vertex is $(4, -5)$, we must have

$$-\frac{b}{2a} = 4.$$

Therefore, $b = -8a$. Since a is positive, we conclude $\boxed{b \text{ must be negative}}$.

Finally, we turn to c. Our vertex only tells us that $c - \frac{b^2}{4a} = -5$. Since a is positive, this doesn't tell us anything about c except that it is larger than -5. Turning to the information about the x-intercepts, we first note that the x-coordinates of the x-intercepts of the graph are those points where $ax^2 + bx + c = 0$. In other words, these x-coordinates are the roots of the quadratic $ax^2 + bx + c = 0$. Since these x-intercepts are on opposite sides of the y-axis, one is negative and the other is positive. So, the product of these two roots is negative, which means c/a must be negative. Because a is positive and c/a is negative, we know that $\boxed{c \text{ is negative}}$.

14.32   The standard form of the equation whose graph is shown in the problem is $y = a(x-h)^2 + k$. From the vertex, we have $k = 2$ and $h = 1$, so the equation is $y = a(x-1)^2 + 2$. So, the equation the student is asked to graph is $x = a(y-1)^2 + 2$. Therefore, the graph should have had vertex $(2, 1)$ instead of $(1, 2)$. Because the student's graph opens downward, a is negative. Therefore, the graph the student should have drawn should open leftward. It should have the same shape as the shown graph, because the value of a is the same for both equations.

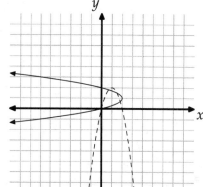

In the same way that swapping x and y reversed the coordinates of the vertex, it reverses the coordinates of every point on the graph. So, for example, since the shown graph goes through $(2, 0)$, the desired graph goes through $(0, 2)$, and so on. We can use this information to generate the desired graph, shown at right as a solid line (the original graph is dashed). As we will explore later in the text, reversing x and y is the same as reflecting the graph over the line $x = y$.

14.33   We seek the value of a such that the graphs of $y = x$ and $y = ax^2 + 6$ meet at exactly one point. In other words, we want the value of a such that there is only one pair (x, y) that satisfies both equations. Substituting $y = x$ from the linear equation into the quadratic gives $x = ax^2 + 6$, or $ax^2 - x + 6 = 0$. This equation has only one solution x if its discriminant equals 0. Therefore, we must have $(-1)^2 - 4a(6) = 0$, which gives us $a = \boxed{1/24}$.

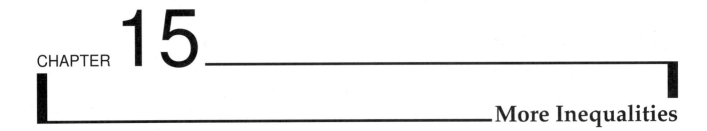

_____More Inequalities

Exercises for Section 15.1

15.1.1 Factoring the quadratic gives $(m + 1)(m + 5) \geq 0$. The product $(m + 1)(m + 5)$ is nonnegative when both $m + 1$ and $m + 5$ are nonnegative, or when they are both nonpositive. The factor $m + 1$ is nonnegative for $m \geq -1$ and $m + 5$ is nonnegative for $m \geq -5$, so they are both nonnegative when $m \geq -1$. Similarly, they are both nonpositive when $m \leq -5$. Putting these together, our solution is $\boxed{m \in (-\infty, -5] \cup [-1, \infty)}$. The graph of the solution set is shown below.

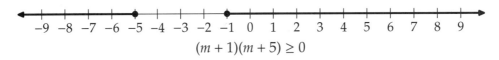

$$(m + 1)(m + 5) \geq 0$$

15.1.2 First, we bring all terms to the left side, which gives $2x^2 + 11x - 21 < 0$. Factoring the left side gives us $(2x - 3)(x + 7) < 0$. The product on the left side is only negative if one term is negative and the other is positive. If $2x - 3 > 0$, then $x > 3/2$, so $2x - 3$ is positive when $x > 3/2$. Similarly, $2x - 3$ is negative when $x < 3/2$. If $x + 7 > 0$, then $x > -7$. Similarly, $x + 7$ is negative when $x < -7$. So, we note that $2x - 3$ is negative and $x + 7$ is positive when $\boxed{-7 < x < 3/2}$. It is impossible for $2x - 3$ to be positive and $x + 7$ to be negative for the same value of x. We can also write our solution in interval notation as $\boxed{x \in (-7, 3/2)}$. The graph of our solutions is shown below.

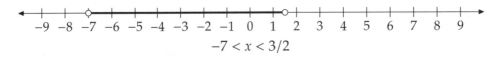

$$-7 < x < 3/2$$

15.1.3 First, we rearrange the equation to find $2r^2 - 3r + 7 > 0$. We can't factor the left side, so we use the quadratic formula to find the roots of $2r^2 - 3r + 7 = 0$:

$$r = \frac{3 \pm i\sqrt{47}}{4}.$$

As discussed in the text, the fact that the roots are not real tells us that the quadratic $2r^2 - 3r + 7$ is either greater than 0 for all real values of r or less than 0 for all real values of r. (Not convinced? Graph $y = 2x^2 - 3x + 7$. And review Problem 15.3 in the text.)

We can test the quadratic for one value of r. We find that $2r^2 - 3r + 7$ is positive when $r = 0$. So, $2r^2 - 3r + 7$ is positive $\boxed{\text{for all real values of } r}$.

15.1.4 First, we graph the parabola $x = -y^2 + 4$. We draw the graph with a dashed line to show that points on the parabola are not solutions to the strict inequality $x < -y^2 + 4$. Next, we test whether to shade inside the parabola or outside it. Because $(0,0)$ is inside the parabola, and it satisfies the inequality, we shade the inside of the parabola, as shown.

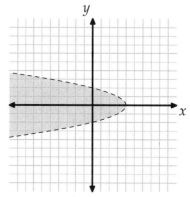

15.1.5

(a) The roots of a quadratic in x are the values of x for which the quadratic equals 0. If the quadratic is greater than 0 for all real values of x, then there are no real values of x for which the quadratic equals 0. Therefore, such a quadratic does not have real roots.

(b) A quadratic has real roots if its discriminant is nonnegative, and it does not have real roots if its discriminant is negative.

(c) We bring all terms to the left, which gives $x^2 - 6x + k - 17 > 0$. If the quadratic on the left does not equal 0 for any real value of x, then it cannot have any real roots. Therefore, its discriminant must be negative. So, we must have

$$(-6)^2 - 4(1)(k - 17) < 0 \quad \Rightarrow \quad 36 - 4k + 68 < 0 \quad \Rightarrow \quad \boxed{k > 26}.$$

Exercises for Section 15.2

15.2.1 We factor the quadratic to find $(2x - 3)(x + 3)(x + 5) \le 0$. The product on the left side is negative when all three factors are negative, and when one factor is negative and the other two are positive. We have $2x - 3 < 0$ when $x < 3/2$, $x + 3 < 0$ when $x < -3$, and $x + 5 < 0$ when $x < -5$. We can combine this information in a table:

	$2x - 3$	$x + 3$	$x + 5$	$(2x - 3)(x + 3)(x + 5)$
$x > 3/2$	$+$	$+$	$+$	$+$
$-3 < x < 3/2$	$-$	$+$	$+$	$-$
$-5 < x < -3$	$-$	$-$	$+$	$+$
$x < -5$	$-$	$-$	$-$	$-$

We see that one factor is negative when $-3 < x < 3/2$ and all three are negative when $x < -5$. Since the inequality is nonstrict, the values $x = 3/2$, $x = -3$, and $x = -5$ are also solutions, so the values of x that satisfy the inequality are $-3 \le x \le 3/2$ or $x \le -5$. In interval notation, this is $\boxed{x \in (-\infty, -5] \cup [-3, 3/2]}$.

15.2.2 We factor the quadratic as a difference of squares, and we find that we seek all values of r such that $(r - 3)^2(2r - 5)(2r + 5) > 0$. The factor $(r - 3)^2$ is positive for all values of r except $r = 3$. We have

$2r - 5 > 0$ for $r > 5/2$, and $2r + 5 > 0$ for $r > -5/2$. So, we see that all factors are positive for all values of r greater than $5/2$ except for $r = 3$. We also have two negative factors when $r < -5/2$, so the product is positive when $r < -5/2$, as well. So, the values of r that satisfy the inequality are

$$\boxed{\text{all } r > 5/2 \text{ except } r = 3, \text{ and all } r < -5/2}.$$

We can write this with intervals as $r \in (-\infty, -5/2) \cup (5/2, 3) \cup (3, \infty)$.

15.2.3 As our first step, we can change the $1 - x$ in the denominator of the first term to $x - 1$. (We don't have to do this, but it will help keep us from making mistakes later.) We have

$$-\frac{2x}{1-x} = \frac{2x}{-1(1-x)} = \frac{2x}{x-1}.$$

So, we can replace the $-\dfrac{2x}{1-x}$ on the left with $+\dfrac{2x}{x-1}$, and we have

$$\frac{1}{x-7} + \frac{2x}{x-1} \geq 2.$$

We bring all terms to the left and find a common denominator, which gives us:

$$\frac{1}{x-7} + \frac{2x}{x-1} - 2 \geq 0 \quad \Rightarrow \quad \frac{x-1}{(x-7)(x-1)} + \frac{2x(x-7)}{(x-7)(x-1)} - \frac{2(x-7)(x-1)}{(x-7)(x-1)} \geq 0$$

$$\Rightarrow \quad \frac{x-1}{(x-7)(x-1)} + \frac{2x^2 - 14x}{(x-7)(x-1)} - \frac{2x^2 - 16x + 14}{(x-7)(x-1)} \geq 0 \quad \Rightarrow \quad \frac{3x - 15}{(x-7)(x-1)} \geq 0.$$

We can factor a 3 out of $3x - 15$ to get $3(x - 5)$, then divide both sides by 3 to get

$$\frac{x-5}{(x-7)(x-1)} \geq 0.$$

We can make a table of the signs of our factors:

	$x-7$	$x-5$	$x-1$	$\frac{x-5}{(x-7)(x-1)}$
$x > 7$	+	+	+	+
$5 < x < 7$	−	+	+	−
$1 < x < 5$	−	−	+	+
$x < 1$	−	−	−	−

We see that our expression is positive for $x > 7$ and for $1 < x < 5$. Since the inequality is nonstrict, we must include the values of x that make the expression on the left side of the inequality equal to 0. Only $x = 5$ makes our expression 0; make sure you see why $x = 1$ and $x = 7$ do not. So, our solutions are all x such that $\boxed{x > 7 \text{ or } 1 < x \leq 5}$. In interval notation, this is $x \in (1, 5] \cup (7, \infty)$.

15.2.4 We bring all the terms to the left, which gives

$$\frac{x + 3 + \frac{1}{x+3}}{(x-3)(x+2)} - \frac{2}{(x-3)(x+2)} \geq 0 \quad \Rightarrow \quad \frac{x + 3 + \frac{1}{x+3} - 2}{(x-3)(x+2)} \geq 0.$$

To simplify this, we find a common denominator of all the terms in the numerator:

$$x + 3 + \frac{1}{x+3} - 2 = x + 1 + \frac{1}{x+3} = \frac{(x+1)(x+3)+1}{x+3} = \frac{x^2 + 4x + 4}{x+3}.$$

So, our inequality now is

$$\frac{\frac{x^2+4x+4}{x+3}}{(x-3)(x+2)} \geq 0 \quad \Rightarrow \quad \frac{(x+2)^2}{(x+3)(x-3)(x+2)} \geq 0 \quad \Rightarrow \quad \frac{x+2}{(x+3)(x-3)} \geq 0.$$

Even though $x = -2$ makes the left side of this final inequality equal to zero, we cannot include it among our solutions, because it makes a denominator equal to 0 in the original inequality. We turn to a table to determine the values of x that make the left side of our final inequality positive.

	$x-3$	$x+2$	$x+3$	$\frac{x+2}{(x+3)(x-3)}$
$x > 3$	+	+	+	+
$-2 < x < 3$	−	+	+	−
$-3 < x < -2$	−	−	+	+
$x < -3$	−	−	−	−

The expression $\frac{x+2}{(x+3)(x-3)}$ is positive for $x > 3$ (all three factors are positive) and for $-3 < x < -2$ (two negative factors, one positive). Even though the inequality is nonstrict, we can't have x equal to 3, -2, or -3, since each would cause division by 0 in the original inequality. Therefore, the values of x that satisfy the original inequality are all values of x such that $\boxed{-3 < x < -2 \text{ or } x > 3}$. In interval notation, this is $x \in (-3, -2) \cup (3, \infty)$.

Exercises for Section 15.3

15.3.1 We simply walk backwards through our solution. Let x be a positive real number. By the Trivial Inequality, we have $(x-1)^2 \geq 0$. Expanding the left side gives $x^2 - 2x + 1 \geq 0$. Adding $2x$ to both sides gives $x^2 + 1 \geq 2x$. Because x is positive, we can divide by x without changing the direction of the inequality signs. This gives us $\frac{x^2+1}{x} \geq 2$, or $x + \frac{1}{x} \geq 2$. Therefore, the sum of any positive number and its reciprocal is greater than or equal to 2.

15.3.2

(a) Since no square of a real number is negative, the sum of two squares is zero if and only if both squares equal 0. Therefore, we have $\boxed{x = y = 0}$.

(b) As in the first part, the sum of the squares of three real numbers equals 0 if and only if all three squares are 0. Therefore, we must have $a + 3 = b - 7 = c - a = 0$. From $a + 3 = 0$, we have $a = -3$. From $b - 7 = 0$, we have $b = 7$, and from $c - a = 0$, we have $c = a = -3$. Therefore, $a + b + c = -3 + 7 - 3 = \boxed{1}$.

15.3.3 *Solution 1: Algebra.* We wish to show that $\frac{x+y}{2} \geq \sqrt{xy}$. Working backwards, we multiply both sides by 2 to get $x + y \geq 2\sqrt{xy}$. We get rid of the square root by squaring both sides. We don't have to worry about this changing the direction of the inequality because x and y are nonnegative. So, we now have $(x + y)^2 \geq 4xy$. Expanding the left side, then rearranging gives

$$x^2 + 2xy + y^2 \geq 4xy \quad \Rightarrow \quad x^2 - 2xy + y^2 \geq 0 \quad \Rightarrow \quad (x - y)^2 \geq 0.$$

This final inequality is true by the Trivial Inequality.

We can now walk through our steps backwards to write our proof. By the Trivial Inequality, we have $(x - y)^2 \geq 0$. Expanding the left side gives $x^2 - 2xy + y^2 \geq 0$. Adding $4xy$ to both sides gives $x^2 + 2xy + y^2 \geq 4xy$, or $(x + y)^2 \geq 4xy$. Because x and y are positive, we can take the square root of both sides to give $x + y \geq 2\sqrt{xy}$. Dividing both sides by 2 gives the desired $(x + y)/2 \geq \sqrt{xy}$.

Solution 2: Use what we already proved. In the text, we proved that $(a^2 + b^2)/2 \geq ab$. If we let $a = \sqrt{x}$ and $b = \sqrt{y}$, we have $(x + y)/2 \geq \sqrt{xy}$, as desired.

Exercises for Section 15.4

15.4.1

(a) We complete the square to find

$$a^2 - 8a + 4 = a^2 - 8a + 16 - 16 + 4 = (a - 4)^2 - 12.$$

Because $(a-4)^2 \geq 0$ by the Trivial Inequality, we can subtract 12 from both sides to find $(a-4)^2 - 12 \geq -12$. Therefore, $a^2 - 8a + 4 \geq -12$, so $a^2 - 8a + 4$ cannot be less than -12. Moreover, $a^2 - 8a + 4 = -12$ when $a = 4$, so the smallest value $a^2 - 8a + 4$ can equal is $\boxed{-12}$.

We can also note that $(a-4)^2 - 12$ equals some nonnegative number minus 12. Therefore, $(a-4)^2 - 12$ cannot be less than -12. Since $(a - 4)^2 - 12$ equals -12 when $a = 4$, we see that the smallest value the quadratic can take on is -12.

(b) We factor -4 out of the two terms with variables, and we have

$$-4t^2 - 40t - 36 = -4(t^2 + 10t) - 36.$$

To complete the square inside the parentheses, we must add $(10/2)^2 = 25$ inside parentheses. So, we must add $-4(25)$ outside the parentheses. We must therefore also add $+4(25)$ to keep the expression equal to our original quadratic. So, we have

$$\begin{aligned}
-4t^2 - 40t - 36 &= -4(t^2 + 10t) - 36 \\
&= -4(t^2 + 10t) - 4(25) + 4(25) - 36 \\
&= -4(t^2 + 10t + 25) + 4(25) - 36 \\
&= -4(t + 5)^2 + 64.
\end{aligned}$$

Since $-4(t + 5)^2$ is nonpositive, the greatest possible value of the expression is $-4(0)^2 + 64 = \boxed{64}$, which occurs when $\boxed{t = -5}$.

(c) We factor $\frac{1}{3}$ out of the two terms with variables, which gives

$$\frac{x^2}{3} + 2x + 9 = \frac{1}{3}(x^2 + 6x) + 9.$$

To complete the square inside the parentheses, we must add $(6/2)^2 = 9$ inside parentheses. So, we must add $\frac{1}{3}(9)$ outside the parentheses. We must therefore also add $-\frac{1}{3}(9)$ to keep the expression equal to our original quadratic. So, we have

$$\begin{aligned}
\frac{x^2}{3} + 2x + 9 &= \frac{1}{3}(x^2 + 6x) + 9 \\
&= \frac{1}{3}(x^2 + 6x) + \frac{1}{3}(9) - \frac{1}{3}(9) + 9 \\
&= \frac{1}{3}(x^2 + 6x + 9) - \frac{1}{3}(9) + 9 \\
&= \frac{1}{3}(x + 3)^2 + 6.
\end{aligned}$$

Since $\frac{1}{3}(x+3)^2$ is nonnegative, the smallest possible value of the expression is $\frac{1}{3}(0)^2 + 6 = \boxed{6}$, which occurs when $\boxed{x = -3}$.

15.4.2

(a) If I speed up x miles per hour, my car will run for $10x$ less minutes. Each hour has 60 minutes, so $10x$ minutes equals $(10x)/60 = x/6$ hours. So, my speed is $60 + x$ and the amount of time the car will run is $5 - \frac{x}{6}$ hours.

(b) Therefore, on one tank of gas, I can drive

$$(60 + x)\left(5 - \frac{x}{6}\right) = 300 - 10x + 5x - \frac{x^2}{6} = -\frac{x^2}{6} - 5x + 300$$

miles.

(c) Completing the square then gives us

$$\begin{aligned}
-\frac{x^2}{6} - 5x + 300 &= -\frac{1}{6}(x^2 + 30x) + 300 \\
&= -\frac{1}{6}(x^2 + 30x + 15^2) + \frac{1}{6}(225) + 300 \\
&= -\frac{1}{6}(x + 15)^2 + \frac{675}{2}.
\end{aligned}$$

Because $(x + 15)^2$ must be nonnegative, the distance I can drive is $675/2 = 337.5$ miles minus some nonnegative number. Therefore, the longest distance I can drive is $\boxed{337.5}$ miles.

(d) In order to maximize the distance I drive, $-\frac{1}{6}(x + 15)^2$ must equal 0. Therefore, $x + 15 = 0$, which means $x = -15$. Therefore, I must drive $60 - 15 = \boxed{45}$ miles per hour to maximize the distance I can drive.

15.4.3

(a) The ball hits the ground when its height above the ground is 0. Therefore, we seek the value of t for which $-16t^2 + 48t + 45 = 0$. Multiplying this equation by -1 gives $16t^2 - 48t - 45 = 0$. Factoring gives $(4t - 15)(4t + 3) = 0$. We want the positive value of t, because we want the amount of time after Adam kicks the ball. Solving $4t - 15 = 0$ gives us $t = \boxed{15/4}$.

(b) Finding the maximum height of the ball means finding the maximum value of $-16t^2 + 48t + 45$. Since we want the maximum value of a quadratic, we complete the square:

$$-16t^2 + 48t + 45 = -16(t^2 - 3t) + 45 = -16\left(t^2 - 3t + \frac{9}{4}\right) + 16 \cdot \frac{9}{4} + 45$$

$$= -16\left(t - \frac{3}{2}\right)^2 + 81.$$

Since $-16(t - \frac{3}{2})^2$ is at most 0, the greatest height the ball achieves is $\boxed{81 \text{ feet}}$. The ball reaches this point 3/2 seconds after Adam kicks it.

15.4.4 Let the length in meters of the side of the chicken run that is opposite the barn be x and the length in meters of the other two fenced sides be y. Let the area of the chicken run be A, so we have $A = xy$. Because the farmer has 20 meters of fence, and one of the sides of the run is the barn, we must have $x + 2y = 20$. Therefore, we wish to maximize xy given that $x + 2y = 20$. Solving $x + 2y = 20$ for x in terms of y gives $x = 20 - 2y$, so we have $A = xy = (20 - 2y)(y) = 20y - 2y^2$. To maximize A, we complete the square of this quadratic:

$$A = -2(y^2 - 10y) = -2(y^2 - 10y + 25) + 2(25) = -2(y - 5)^2 + 50.$$

Therefore, A reaches its maximum when $y = 5$, which gives us $A = \boxed{50 \text{ square meters}}$.

Review Problems

15.18 Putting all terms on the same side gives $x^2 - 4x - 21 \le 0$. Factoring the left side gives $(x-7)(x+3) \le 0$. The product $(x - 7)(x + 3)$ is negative when one of $x - 7$ and $x + 3$ is negative and the other is positive. We have $x - 7 < 0$ when $x < 7$ and $x - 7 > 0$ when $x > 7$. Similarly, $x + 3$ is negative when $x < -3$ and positive when $x > -3$. We can include all this information in a table:

	$x - 7$	$x + 3$	$(x - 7)(x + 3)$
$x > 7$	$+$	$+$	$+$
$-3 < x < 7$	$-$	$+$	$-$
$-3 < x$	$-$	$-$	$+$

We therefore see that the product $(x - 7)(x + 3)$ is negative when $-3 < x < 7$. Because the inequality is nonstrict, we must also include the values of x for which $(x - 7)(x + 3)$ equals 0, which are $x = -3$ and $x = 7$. Therefore, the solutions to the inequality are $\boxed{-3 \le x \le 7}$. In interval notation, this is $\boxed{x \in [-3, 7]}$. The graph of these solutions on the number line is shown below.

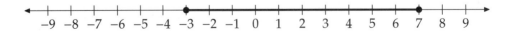

15.19 Putting all terms on the left gives $t^2 - 3t - 28 > 0$. Factoring then gives $(t-7)(t+4) > 0$. The product $(t-7)(t+4)$ is positive only if both $t-7$ and $t+4$ are positive or if both are negative. Therefore, our product is positive when $t > 7$ and when $t < -4$. So, our solutions are $\boxed{t < -4 \text{ or } t > 7}$. In interval notation, this is $\boxed{t \in (-\infty, -4) \cup (7, \infty)}$.

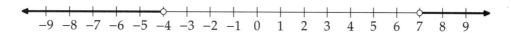

15.20 Factoring the left side gives $(r-4)^2 > 0$. Since all squares are nonnegative, the only value of r for which this inequality is not satisfied is $r = 4$, since then we have $(r-4)^2 = 0$. Therefore, all values of r except $r = 4$ satisfy the inequality. In interval notation, we could write this as $\boxed{r \in (-\infty, 4) \cup (4, \infty)}$.

15.21 The expression on the left side is negative when one of $5x + 4$, $x + 2$, and $x - 3$ is negative and the other two positive, or when all three are negative. The term $5x + 4$ is negative when $x < -4/5$ and positive when $x > -4/5$. The term $x + 2$ is negative when $x < -2$ and positive when $x > -2$, and $x - 3$ is negative when $x < 3$ and positive when $x > 3$. Putting these together in a table, we have

	$x - 3$	$5x + 4$	$x + 2$	$\frac{5x+4}{(x+2)(x-3)}$
$x > 3$	$+$	$+$	$+$	$+$
$-4/5 < x < 3$	$-$	$+$	$+$	$-$
$-2 < x < -4/5$	$-$	$-$	$+$	$+$
$x < -2$	$-$	$-$	$-$	$-$

Therefore, the expression $\dfrac{5x + 4}{(x + 2)(x - 3)}$ is negative when $-4/5 < x < 3$ or $x < -2$. Because the inequality is nonstrict, we also have as solutions the values of x for which the numerator of the left side equals zero. Since $5x + 4 = 0$ when $x = -4/5$, the solutions to the inequality $\dfrac{5x + 4}{(x + 2)(x - 3)} \le 0$ are $\boxed{-4/5 \le x < 3 \text{ or } x < -2}$. In interval notation, this is $\boxed{x \in (-\infty, -2) \cup [-4/5, 3)}$ Make sure you see why we do not include $x = -2$ or $x = 3$ in the solution.

15.22 We don't like dealing with quadratics that have a negative coefficient of the quadratic term, so we start by multiplying both sides of the inequality by -1. On the left, we have

$$-(2x + 4)(-3x^2 - 5x + 2) = (2x + 4)[(-1)(-3x^2 - 5x + 2)] = (2x + 4)(3x^2 + 5x - 2).$$

Because we multiply both sides by -1, we must reverse the direction of the inequality, which gives us

$$(2x + 4)(3x^2 + 5x - 2) > 0.$$

We factor 2 out of the linear term, and we factor the quadratic, to give

$$2(x + 2)(3x - 1)(x + 2) > 0.$$

Dividing by 2 gives us $(x + 2)(3x - 1)(x + 2) > 0$. Notice that two of our factors in the product are the same! So, we can combine them to give $(x + 2)^2(3x - 1) > 0$. The term $3x - 1$ is positive for $x > 1/3$ and the term $(x + 2)^2$ is always nonnegative, so the product of these two terms is positive for $\boxed{x > 1/3}$. In interval notation, this is $\boxed{x \in (1/3, \infty)}$.

15.23 Converting the words to math makes our problem $(t + 1)t > 2$. Expanding the left side and putting all terms on the left, we have $t^2 + t - 2 > 0$. Factoring gives us $(t + 2)(t - 1) > 0$. The product $(t + 2)(t - 1)$ is positive when $t > 1$ and when $t < -2$. So, the values of t that satisfy the problem are $\boxed{t < -2 \text{ or } t > 1}$. In interval notation, this is $\boxed{t \in (-\infty, -2) \cup (1, \infty)}$.

15.24

(a) The problem with this solution is in the cross-multiplication step. Cross-multiplying is the same as multiplying both sides by $(x - 3)(x - 5)$. However, this quantity is sometimes negative (in which case we switch the direction of the inequality sign) and sometimes positive (in which case we don't switch the direction of the inequality sign).

(b) Instead of worrying about when to reverse the inequality sign, we bring all terms to one side and find a common denominator. We have

$$\frac{6}{x - 5} - \frac{5}{x - 3} \geq 0 \quad \Rightarrow \quad \frac{6(x - 3)}{(x - 5)(x - 3)} - \frac{5(x - 5)}{(x - 3)(x - 5)} \geq 0$$

$$\Rightarrow \quad \frac{6(x - 3) - 5(x - 5)}{(x - 5)(x - 3)} \geq 0 \quad \Rightarrow \quad \frac{x + 7}{(x - 5)(x - 3)} \geq 0.$$

Now we have a problem we know how to handle. We seek those values of x for which either all three of $x + 7$, $x - 5$, and $x - 3$ are positive, or for which one term is positive and the others negative. As before, we build a table to find the appropriate values of x:

	$x - 5$	$x - 3$	$x + 7$	$\frac{x+7}{(x-5)(x-3)}$
$x > 5$	$+$	$+$	$+$	$+$
$3 < x < 5$	$-$	$+$	$+$	$-$
$-7 < x < 3$	$-$	$-$	$+$	$+$
$x < -7$	$-$	$-$	$-$	$-$

We see that $\dfrac{x + 7}{(x - 5)(x - 3)}$ is positive when $x > 5$ and also when $-7 < x < 3$. Because the inequality is nonstrict, we must also include as solutions to our inequality the values of x for which the numerator $x + 7$ equals 0. Therefore, we include $x = -7$ among our solutions, so the values of x that satisfy the inequality are $\boxed{-7 \leq x < 3 \text{ or } x > 5}$. In interval notation, this is $\boxed{x \in [-7, 3) \cup (5, \infty)}$.

15.25 We start by completing the square to give

$$x \geq -2(y^2 - 3y) + 1 \quad \Rightarrow \quad x - 2\left(\frac{-3}{2}\right)^2 \geq -2\left[y^2 - 3y + \left(\frac{-3}{2}\right)^2\right] + 1$$

$$\Rightarrow \quad x - \frac{9}{2} \geq -2\left(y - \frac{3}{2}\right)^2 + 1 \quad \Rightarrow \quad x \geq -2\left(y - \frac{3}{2}\right)^2 + \frac{11}{2}.$$

To graph this inequality, we first graph the parabola $x = -2\left(y - \frac{3}{2}\right)^2 + \frac{11}{2}$. Because the inequality is nonstrict, the points on this parabola are solutions to the inequality. Therefore, we graph this parabola with a solid line. Next, we must decide whether the points to the right or the points to the left of the parabola satisfy the inequality. Those points for which x is greater than $-2\left(y - \frac{3}{2}\right)^2 + \frac{11}{2}$ are those that are to the right of the parabola, i.e., those points (x, y) for which the x value is greater than the value of $-2\left(y - \frac{3}{2}\right)^2 + \frac{11}{2}$. (We could also have noted that the point $(0, 0)$ is to the left of the parabola, but it does not satisfy the inequality, so it is the region to the right of the parabola that must be shaded.)

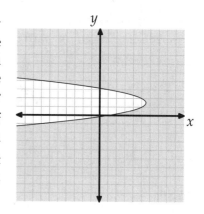

15.26

(a) $\boxed{\text{False}}$. Factoring the left side gives $(x + 5)^2 > 0$. This inequality is not true when $x = -5$, so it is not true for all real values of x.

(b) $\boxed{\text{True}}$. Factoring the left side gives $(x + 5)^2 \geq 0$. Since the square of a real number must be nonnegative, $(x + 5)^2$ is nonnegative for all real values of x. Therefore, we must have $(x + 5)^2 \geq 0$ for all real values of x.

(c) $\boxed{\text{True}}$. The graph of $y = ax^2 + bx + c$, when $a \neq 0$ is a parabola that either opens upward or downward. When the parabola opens upward (for $a > 0$), its vertex is the lowest point on the parabola. In other words, there are no points on the parabola with a y value less than that of the vertex. So, the y-coordinate of this vertex gives us the minimum value of $ax^2 + bx + c$ when the parabola opens upward. Similarly, if $a < 0$, the parabola opens downward, and the y-coordinate of the vertex gives us the greatest possible value of $y = ax^2 + bx + c$.

(d) $\boxed{\text{False}}$. We can see why by considering the graph of $y = ax^2 + bx + c$. Suppose the graph of $y = ax^2 + bx + c$ is entirely below the x-axis. This is possible, because the graph could be a downward-opening parabola with its vertex below the x-axis. The graph of $y = -x^2 - 1$ is such a graph. Because x^2 is nonnegative, the expression $-x^2$ must be nonpositive. So, the expression $-x^2 - 1$ equals a nonpositive number minus 1. Such a number can never be greater than or equal to 0 if x is real. Therefore, for all real values of x, we have $-x^2 - 1 < 0$.

15.27 We first work backwards. We bring all the terms to the left side and have $a^2 - 4a + 4 \geq 0$. Factoring the left side gives $(a - 2)^2 \geq 0$. The Trivial Inequality tells us that this inequality is true for all real values of a.

To write our solution forwards, we reverse our steps above. By the Trivial Inequality, we have $(a - 2)^2 \geq 0$. Expanding the left side gives $a^2 - 4a + 4 \geq 0$. Adding $4a$ to both sides gives $a^2 + 4 \geq 4a$.

Make sure you see the difference between working backwards and working forwards. In the first paragraph, we started from what we wanted to prove and worked backwards until we reached something that was true. However, when we write our final solution, we should start from something we know is true, then work forwards to reach the inequality we want to prove. We did this in the second paragraph, starting from $(a - 2)^2 \geq 0$, which the Trivial Inequality tells us is true, and ending with $a^2 + 4 \geq 4a$, which is what we wanted to prove.

Notice that we proved earlier in the text that $(a^2 + b^2)/2 \geq ab$. If we let $b = 2$ in this inequality, we have $(a^2 + 4)/2 \geq 2a$. Multiplying both sides of this inequality by 2 gives us the desired $a^2 + 4 \geq 4a$.

15.28 Because all squares of real numbers are nonnegative, the sum of the squares of two real numbers equals 0 if and only if both of the numbers equals 0. Therefore, we must have $a - 2b = 0$ and $b - 2 = 0$. From $b - 2 = 0$, we have $b = 2$. From $a - 2b = 0$, we have $a = 2b = 2(2) = 4$. So, we have $ab = (2)(4) = \boxed{8}$.

15.29 Completing the square gives us

$$P = (x^2 - 6x + 9) - 9 - 13 = (x - 3)^2 - 22.$$

Because $(x - 3)^2 \geq 0$ for all real x, we have $(x - 3)^2 - 22 \geq -22$ for all real x. When $\boxed{x = 3}$, we have $(x - 3)^2 - 22 = -22$. So, the smallest possible value for P is $P = \boxed{-22}$.

15.30

(a) Let $A = -2t^2 + 12t - 8$, so that our goal is to find the maximum possible value of A. Completing the square gives

$$A = -2(t^2 - 6t) - 8 \quad \Rightarrow \quad A - 2(9) = -2(t^2 - 6t + 9) - 8 \quad \Rightarrow \quad A = -2(t - 3)^2 + 10.$$

Because $(t - 3)^2$ must be nonnegative, we see that A equals 10 minus some nonnegative number. So, the largest possible value A can be is $\boxed{10}$, which occurs when $t = 3$.

(b) Again, we let $A = -2t^2 + 12t - 8$. In the previous part, we saw that $A = -2(t - 3)^2 + 10$. From this expression, we see that if t is very large, then A equals 10 minus a very large number. So, A itself will be very negative. If we then increase t even more, A will get even more negative. There's no limit to how negative A can get, because we can keep increasing t to make A more and more negative.

15.31 We complete the square for both the x terms and the y terms to find

$$\begin{aligned}
3x^2 + 6x + y^2 - 4y + 11 &= 3(x^2 + 2x) + (y^2 - 4y) + 11 \\
&= 3(x + 2x + 1) - 3(1) + (y^2 - 4y + 4) - 4 + 11 \\
&= 3(x + 1)^2 + (y - 2)^2 + 4.
\end{aligned}$$

Since the terms $(x + 1)^2$ and $(y - 2)^2$ are both squares of real numbers, both $3(x + 1)^2$ and $(y - 2)^2$ must be nonnegative. Therefore, the smallest possible value of our expression is $0 + 0 + 4 = \boxed{4}$, which occurs when $x = -1$ and $y = 2$ (since these values make both of our squares equal to 0).

15.32 Suppose there are x groups of 3 items added to the ballot, so that there are $30 + 3x$ items on the ballot. Since there are 10 fewer voters for every 3 items added, there are $240 - 10x$ voters if $3x$ items are added to the ballot. Let V be the total number of votes cast. Since there are $240 - 10x$ voters and each casts a vote on $30 + 3x$ items, the number of total votes in terms of x is

$$(240 - 10x)(30 + 3x) = 7200 + 720x - 300x - 30x^2 = -30x^2 + 420x + 7200.$$

Completing the square gives us

$$\begin{aligned}
-30x^2 + 420x + 7200 &= -30(x^2 - 14x) + 7200 = -30(x^2 - 14x + 49) + 30(49) + 7200 \\
&= -30(x - 7)^2 + 8670.
\end{aligned}$$

Therefore, the number of total votes equals 8670 minus some nonnegative number. So, the largest possible number of votes is 8670, which occurs when $(x - 7)^2 = 0$. So, to maximize the total number of votes, we must have $x = 7$. This means there should be 7 groups of 3 more items added to the ballot, for a total of $30 + 3 \cdot 7 = \boxed{51}$ items on the ballot.

15.33 Suppose the shop increases the price of bracelets by x dollars. Then, the price of the each bracelet is $10 + x$ dollars. Since they sell 2 fewer bracelets for each dollar they raise the price, the shop will sell $50 - 2x$ bracelets. Therefore, their revenue is $(10 + x)(50 - 2x)$.

(a) Let the shop's revenue be R. Above, we found that $R = (10 + x)(50 - 2x)$. We find the maximum possible value of the revenue by completing the square:

$$R = 500 - 20x + 50x - 2x^2 \quad \Rightarrow \quad R = -2x^2 + 30x + 500 \quad \Rightarrow \quad R = -2\left(x^2 - 15x\right) + 500$$

$$\Rightarrow \quad R - 2\left(\frac{15}{2}\right)^2 = -2\left[x^2 - 15x + \left(\frac{15}{2}\right)^2\right] + 500 \quad \Rightarrow \quad R = -2\left(x - \frac{15}{2}\right)^2 + \frac{1225}{2}.$$

Therefore, R equals $1225/2$ minus some nonnegative number. So, R is maximized when that nonnegative number equals 0. This occurs when $x - \frac{15}{2} = 0$, or $x = 15/2$. So, the shop should sell its bracelets for $10 + \frac{15}{2} = 10 + 7.5 = \boxed{17.5 \text{ dollars}}$.

(b) Above, we found that $R = -2x^2 + 30x + 500$, where R is the shop's revenue. In this part we are asked to find those values of x for which $R \geq 600$, so we have

$$-2x^2 + 30x + 500 \geq 600 \quad \Rightarrow \quad -2x^2 + 30x - 100 \geq 0.$$

We can multiply both sides by -1 to make the coefficient of x^2 positive. We must remember to change the direction of the inequality when we do so. We then have

$$2x^2 - 30x + 100 \leq 0 \quad \Rightarrow \quad x^2 - 15x + 50 \leq 0 \quad \Rightarrow \quad (x - 10)(x - 5) \leq 0.$$

The product $(x - 10)(x - 5)$ is negative only if exactly one of $x - 10$ and $x - 5$ is negative. This happens when $5 < x < 10$ (when $x > 10$, both are positive, and when $x < 5$, both are negative). We also must include $x = 5$ and $x = 10$ as solutions for x, because these values of x make $(x - 10)(x - 5)$ equal to zero. So, we have $5 \leq x \leq 10$. Therefore, if the shop increases the price of bracelets by anywhere from 5 to 10 dollars, its revenue will be at least \$600. The initial price of the bracelets is \$10, so the shop will have a revenue of at least \$600 if it charges any price from $\boxed{\$15 \text{ to } \$20}$ (including both \$15 and \$20).

Notice that our answer to part (a) is the midpoint of the range of possibilities we found in part (b). Is this a coincidence? Why or why not? (Hint: We showed that the revenue in terms of x equals $-2x^2 + 30x + 500$. Think about the graph of $y = -2x^2 + 30x + 500$.)

Challenge Problems

15.34 Since x, y and z are integers, the smallest nonzero possible value of each of the squares in our sum is 1. Therefore, if the sum of the three squares is 1, then one of the squares is 1 and the other two equal 0. If $(x + 5)^2 = 1$, then x is -6 or -4. Meanwhile, if $(x + 5)^2 = 1$, then $(y - 2)^2$ and $(z + 3)^2$ must be zero, which gives $y = 2$ and $z = -3$. So, we have the two solutions $(x, y, z) = (-6, 2, -3)$ and $(-4, 2, -3)$. Similarly, if $(y - 2)^2 = 1$, we have $y = 3$ or $y = 1$, while $x = -5$ and $z = -3$ make the other two squares equal to zero. So, we have the two solutions $(x, y, z) = (-5, 3, -3)$ and $(-5, 1, -3)$. Finally, if $(z + 3)^2 = 1$, we have $z = -4$ or $z = -2$, while $x = -5$ and $y = 2$ make the other squares 0. This gives us our final two solutions, $(x, y, z) = (-5, 2, -4)$ and $(-5, 2, -2)$.

15.35 Since we have a sum of squares that equals zero, each of the squares must equal 0. Therefore, we have

$$x - y - 3 = 0,$$
$$x + z + 2 = 0.$$

We wish to find $y + z$, so we want to eliminate x. Subtracting the second equation from the first does the job, and leaves us $-y - 3 - z - 2 = 0$, from which we find $y + z = \boxed{-5}$.

15.36 Bringing all the terms to one side gives $x^2 - x - 1 < 0$. Unfortunately, we can't factor the left side. However, we can figure out the solutions to the inequality by thinking about the graph of $y = x^2 - x - 1$. As shown at right, the graph of this equation is a parabola that opens upward. We want to find the values of x such that if (x, y) is on the parabola, then y is negative. In other words, we want the values of x for which the point (x, y) on the parabola is below the x-axis. Since this parabola opens upward, the point (x, y) on the parabola is below the x-axis if and only if x between the x-coordinates of the two points where the parabola crosses the x-axis. The parabola crosses the x-axis at the two points where $y = 0$. These are the solutions to the equation $0 = x^2 - x - 1$. Using the quadratic formula, we find

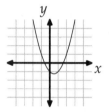

$$x = \frac{1 \pm \sqrt{5}}{2}.$$

Since the values of x that satisfy our inequality are those values of x between these two roots, we have

$$\boxed{\frac{1 - \sqrt{5}}{2} < x < \frac{1 + \sqrt{5}}{2}}.$$

15.37 First, we solve the inequality to determine what the possible values of x are. We factor the quadratic to find $(x - 2)(x - 3) < 0$. The product $(x - 2)(x - 3)$ is negative only when one factor is negative and the other is positive, which occurs when $2 < x < 3$.

Next, we find the possible values of P when x is in this range. When $x = 2$, we have $P = 2^2 + 5 \cdot 2 + 6 = 20$. When $x = 3$, we have $P = 9 + 15 + 6 = 30$. To see that this information tells us that $\boxed{20 < P < 30}$, we consider the graph of $y = x^2 + 5x + 6$. This graph is an upward-opening parabola whose vertex has x-coordinate $-5/2$. Therefore, the portion of this graph from $x = 2$ to $x = 3$ is entirely on the right of the vertex. So, the graph goes upward the whole time from $(2, 20)$ to $(3, 30)$. So, for any value of y between 20 and 30, there is some value of x between 2 and 3 such that $y = x^2 + 5x + 6$. Finally, we exclude 20 and 30 from the possibilities for P because we cannot have $x = 2$ or $x = 3$.

15.38 Suppose $k > 5$. Then, all three factors in the product are positive for $x > k$, and one is nonnegative while the other two are negative for $2 \le x \le 5$. So, there are two intervals in the graph if $k > 5$. Similarly, there are two intervals if $k < 2$ or if $2 < k < 5$. But what if $k = 5$? Then, our inequality is $(x - 2)(x - 5)^2 \ge 0$. The expression $(x - 5)^2$ is nonnegative for all real values of x, so our inequality is satisfied only when $x - 2 \ge 0$. Therefore, we have $x \ge 2$, a single interval.

We also must investigate the case $k = 2$, which gives us $(x - 2)^2(x - 5) \geq 0$. Here, $x = 2$ and $x \geq 5$ are the solutions because $(x - 2)^2$ is nonnegative for all values of x, and $x - 5$ is nonnegative only if $x \geq 5$. Since $x = 2$ together with $x \geq 5$ is more than a single interval, we only have a single interval for our solution if $\boxed{k = 5}$.

15.39 We first simplify the expression inside the parentheses:

$$\left(\frac{5x + 2}{x + 1} - 5\right)^2 + 10 = \left(\frac{5x + 2}{x + 1} - \frac{5x + 5}{x + 1}\right)^2 + 10 = \left(\frac{-3}{x + 1}\right)^2 + 10.$$

Since the square of a real number is nonnegative, this expression is minimized when the square equals 0. However, $-3/(x + 1)$ cannot equal 0! We can get as close to 0 as we like by choosing a very, very large number (or a very, very negative number) for x. So, we can get our square very close to zero, but we can't make it equal to zero. Therefore, while the expression cannot be less than 10, it cannot equal 10, either. Moreover, we can make it equal to any value larger than 10.

So, the expression $\boxed{\text{does not have a minimum value}}$ – for any value just above 10 that the expression equals, we can find another value a little closer to 10 that the expression equals. But, the expression can never equal 10. This tells us that there is not a single number the expression equals that is lower than all other possible values of the expression.

15.40 The equation is a quadratic in y. The roots of this quadratic are real if the discriminant of the quadratic is nonnegative. The discriminant of this quadratic is

$$(4x)^2 - 4(4)(x + 6) = 16x^2 - 16x - 96.$$

So, we must have $16x^2 - 16x - 96 \geq 0$. Dividing by 16 gives $x^2 - x - 6 \geq 0$, and factoring gives $(x - 3)(x + 2) \geq 0$. Solving this inequality gives $\boxed{x \in (-\infty, -2] \cup [3, \infty)}$.

15.41

(a) First, we use our equations $a + b = c + d = 10$ to reduce the number of variables we have to deal with. We have $b = 10 - a$ and $d = 10 - c$, so $ab = a(10 - a) = 10a - a^2$ and $cd = c(10 - c) = 10c - c^2$. We can compare the values of $10a - a^2$ and $10c - c^2$ by considering the graph $y = 10x - x^2$ for $x = a$ and $x = c$. Whichever produces the larger y-coordinate tells us which of $10a - a^2$ and $10c - c^2$ is larger. We can learn more about the graph of $y = 10x - x^2$ by completing the square:

$$y = 10x - x^2 \quad \Rightarrow \quad y = -(x^2 - 10x + 25) + 25 \quad \Rightarrow \quad y = -(x - 5)^2 + 25.$$

Therefore, the graph is a downward-opening parabola with vertex $(5, 25)$. Because $y = -(x-5)^2+25$, the closer that x is to 5, the greater y is.

Furthermore, since $a + b = 10$, we see that a and b are the same distance from 5 (for example, if a is 2 more than 5, then b is 2 less than 5, etc.). The same holds for c and d. Because c and d are closer to each other than a and b are, and 5 is both directly between c and d and directly between a and b, we see that c is closer to 5 than a is. Because c is closer to 5 than a is, we know that the point on $y = 10x - x^2 = -(x - 5)^2 + 25$ with $x = c$ has a greater y-coordinate than the point with $x = a$. In other words, we have $10c - c^2 > 10a - a^2$, so we know that $cd > ab$.

Another way we could have approached this problem is to note that $a + b = c + d = 10$ means that there is some m such that $a = 5 + m$ and $b = 5 - m$, and some n such that $c = 5 + n$ and $5 - n$.

We assume $a > b$ and $c > d$ so that m and n are positive. (We can handle the other possible cases essentially in the same way we handle this one.) When we compare ab to cd, we are comparing $ab = (5 + m)(5 - m) = 25 - m^2$ to $cd = (5 + n)(5 - n) = 25 - n^2$. If a and b are farther apart than c and d, we must have $m > n$. Since $m > n$ and both m and n are positive, we have $m^2 > n^2$, so $25 - m^2 < 25 - n^2$. This means that $ab < cd$, as before.

(b) There's nothing special about 10 in the previous part. We can replace it with any positive number and the result still holds. If $a + b = c + d = S$ and a, b, c, d are positive numbers such that c and d are closer to each other than a and b are, then $cd > ab$.

(c) As explained in the previous parts, to minimize the product, we must make the two numbers as far apart as possible. Therefore, we let the two integers be 1 and 364, which have a product of $(1)(364) = \boxed{364}$.

(d) We have $54321 + 54322 = 54320 + 54323$. Since 54321 and 54322 are closer together than 54320 and 54323 and they have the same sum, we have $\boxed{54321 \cdot 54322} > 54320 \cdot 54323$.

15.42

(a) We work backwards. Squaring both sides gives

$$\frac{x^2 + y^2}{2} \geq \frac{x^2 + 2xy + y^2}{4} \quad \Rightarrow \quad 2(x^2 + y^2) \geq x^2 + 2xy + y^2 \quad \Rightarrow \quad x^2 - 2xy + y^2 \geq 0 \quad \Rightarrow \quad (x - y)^2 \geq 0.$$

The last inequality is true by the Trivial Inequality.

We reverse our steps above to write our proof. By the Trivial Inequality, we have $(x - y)^2 \geq 0$. Expanding the left side gives $x^2 - 2xy + y^2 \geq 0$, so $x^2 + y^2 \geq 2xy$. Adding $x^2 + y^2$ to both sides gives $2(x^2 + y^2) \geq x^2 + 2xy + y^2$. Dividing both sides by 4 gives

$$\frac{x^2 + y^2}{2} \geq \frac{x^2 + 2xy + y^2}{4} \quad \Rightarrow \quad \frac{x^2 + y^2}{2} \geq \frac{(x + y)^2}{4}.$$

Taking the square root of both sides of this gives the desired $\sqrt{\dfrac{x^2 + y^2}{2}} \geq \dfrac{x + y}{2}$.

(b) We look through our steps in our proof above to see how equality can occur. In our first step, we had $(x - y)^2 \geq 0$ by the Trivial Inequality. Equality only holds here if and only if $x = y$. If we let $x = y$ in both sides of our inequality, we see that both sides simplify to x, so the two sides are equal if and only if $x = y$.

15.43 The sum of squares suggests using the inequality we proved in the previous problem. Letting $x = 12344$ and $y = 12346$, we have

$$\sqrt{\frac{12344^2 + 12346^2}{2}} \geq \frac{12344 + 12346}{2}.$$

Squaring both sides, then multiplying by 2, gives

$$12344^2 + 12346^2 \geq \frac{(12344 + 12346)^2}{2}.$$

Finally, we note that $12344 + 12346 = 2(12345)$, so we have

$$\frac{(12344 + 12346)^2}{2} = \frac{(2 \cdot 12345)^2}{2} = \frac{4 \cdot 12345^2}{2} = 2 \cdot 12345^2.$$

Therefore, we have $12344^2 + 12346^2 \geq 2(12345)^2$. We can quickly show that the two sides are not equal, either by noting that we do not have $x = y$, or by noting that the last digits of the two sides are different. So, we have $12344^2 + 12346^2 > 2(12345)^2$.

We can also show that $12344^2 + 12346^2 > 2(12345)^2$ by letting $x = 12345$ and writing both sides in terms of x.

15.44 We have information about the minimum of an expression that consists of two quadratics, so we complete the square:

$$2x^2 + ax + 2y^2 - ay + a^2 = 2\left(x^2 + \frac{a}{2}x\right) + 2\left(y^2 - \frac{a}{2}\right) + a^2$$

$$= 2\left(x^2 + \frac{a}{2}x + \frac{a^2}{16}\right) - 2 \cdot \frac{a^2}{16} + 2\left(y^2 - \frac{a}{2} + \frac{a^2}{16}\right) - 2 \cdot \frac{a^2}{16} + a^2$$

$$= 2\left(x + \frac{a}{4}\right)^2 + 2\left(y - \frac{a}{4}\right)^2 + \frac{3a^2}{4}.$$

This expression is minimized when the two squares equal zero. The resulting minimal value of the expression is $3a^2/4$. We are told that this equals 72, so we have $3a^2/4 = 72$. Solving for a^2 gives $a^2 = 96$. Since we are told a is positive, we have $a = \sqrt{96} = \boxed{4\sqrt{6}}$.

15.45 We start by completing the square to see what information that will give us:

$$x^2 - 10x + y^2 - 8y = 8 \quad \Rightarrow \quad x^2 - 10x + 25 + y^2 - 8y + 16 = 8 + 25 + 16 \quad \Rightarrow \quad (x-5)^2 + (y-4)^2 = 49.$$

We recognize this equation. Its graph is a circle with center $(5, 4)$ and radius 7. Since we want the largest possible value of x, we need to find the rightmost point on the graph. The rightmost point on the circle is directly to the right of the center. Because the radius is 7 and the center is $(5, 4)$, this point is $(5 + 7, 4) = (12, 4)$. So, the largest possible value of x is $\boxed{12}$.

We also could have solved this problem by noting that to maximize x, we must maximize $(x-5)^2$. Therefore, we want $(y-4)^2$ to be 0, so that $(x-5)^2$ can be as large as possible. This gives us $(x-5)^2 = 49$, so $x = 12$ or $x = -2$. We want the largest value of x, not the smallest, so our answer is $x = \boxed{12}$.

15.46 We don't know much about dealing with three variables in an inequality, but we have proved inequalities that include some of these terms. For example, in the text we showed that

$$\frac{x^2 + y^2}{2} \geq xy.$$

This has squares on the larger side and a product on the smaller side, just like the inequality we want to prove. This looks promising, so we write this inequality for each of the three possible pairs of variables:

$$\frac{x^2 + y^2}{2} \geq xy,$$

$$\frac{y^2 + z^2}{2} \geq yz,$$

$$\frac{z^2 + x^2}{2} \geq zx.$$

Aha! We see that adding all three of these inequalities gives us the desired

$$x^2 + y^2 + z^2 \geq xy + yz + zx.$$

CHAPTER **16**

_____Functions

Exercises for Section 16.1

16.1.1 We have $g(0) = 3 \cdot 0 - 4 = 0 - 4 = \boxed{-4}$. Since $g(a) = 3a - 4$, the equation $g(a) = 0$ means $3a - 4 = 0$. Solving this equation gives $a = \boxed{4/3}$.

16.1.2

(a) Since each person has exactly one date of birth, this describes a function.

(b) Some houses have more than one person who lives there. So, for some inputs there is more than one possible output. Therefore, this does not describe a function.

(c) For any positive integer, there is only one possible value of the sum of its digits. Therefore, this does describe a function.

16.1.3

(a) Because $g(x) = 3$, no matter what we input to g, the output is 3. So, $g(2) = \boxed{3}$.

(b) As we saw in the first part, no matter what we input to g, the output is 3. So, $g(-7) = \boxed{3}$.

(c) There are no restrictions on what we can input to g, so the domain is $\boxed{\text{all real numbers}}$.

(d) No matter what goes in, only 3 comes out. Therefore, the range is only $\boxed{\text{the number 3}}$.

16.1.4

(a) We have $f(3) = 2(3^2) - 4 \cdot 3 + 9 = 18 - 12 + 9 = 15$ and $f(-3) = 2(-3)^2 - 4(-3) + 9 = 18 + 12 + 9 = 39$. So, we have $2f(3) + 3f(-3) = 2(15) + 3(39) = 30 + 117 = \boxed{147}$.

(b) We can input any real number to f, so the domain of f is $\boxed{\text{all real numbers}}$.

(c) We find the minimum possible value of a quadratic expression by completing the square. Completing the square gives

$$f(x) = 2x^2 - 4x + 9 = 2(x^2 - 2x) + 9 = 2(x^2 - 2x + 1) - 2(1) + 9 = 2(x - 1)^2 + 7.$$

Because $2(x - 1)^2$ is nonnegative, we see that $f(x)$ equals a nonnegative number plus 7. Therefore, $f(x)$ cannot be less than 7. When $x = 1$, we have $f(1) = 2(1 - 1)^2 + 7 = 2 \cdot 0 + 7 = 7$. So, the smallest possible value of $f(x)$ is $\boxed{7}$.

(d) In the previous part, we found that $f(x)$ cannot be less than 7. However, the output can be any number greater than or equal to 7. To see why, suppose y is a number greater than or equal to 7. We find the input to f that produces y as the output by solving the equation $f(x) = y$. Therefore, we have

$$2(x-1)^2 + 7 = y \quad \Rightarrow \quad 2(x-1)^2 = y - 7 \quad \Rightarrow \quad (x-1)^2 = \frac{y-7}{2}.$$

Because $y \geq 7$, we know that $y - 7 \geq 0$, so we can take the square root of both sides of the last equation above to give $x - 1 = \pm\sqrt{\dfrac{y-7}{2}}$. Solving this equation for x gives us $x = 1 \pm \sqrt{\dfrac{y-7}{2}}$. So, for any value of y that is greater than or equal to 7, we can use the equation

$$x = 1 \pm \sqrt{\frac{y-7}{2}}$$

to find a value of x we can input to f in order to produce y as the output. So, the range is

$$\boxed{\text{all numbers greater than or equal to 7.}}$$

16.1.5

(a) We have $f(2) = \dfrac{2^2 - 1}{2+1} = \boxed{1}$ and $g(2) = 2 - 1 = \boxed{1}$.

(b) We have $f(7) = \dfrac{7^2 - 1}{7+1} = \boxed{6}$ and $g(7) = 7 - 1 = \boxed{6}$.

(c) If x is not equal to -1, then $(x^2 - 1)/(x+1) = (x-1)(x+1)/(x+1) = x - 1$. So, if $x \neq -1$, we see that $f(x)$ and $g(x)$ are the same expression. The value -1 is not in the domain of f, and for all other values of x, $f(x)$ and $g(x)$ are equal. Therefore, there are no values of x in the domains of both f and g such that $f(x)$ and $g(x)$ are unequal.

(d) $\boxed{\text{No}}$. They are not the same because -1 is in the domain of g but not in the domain of f. While the two functions are the same for every other input, the two functions are still different because their domains are different. (Their ranges are also different, because -2 is not in the range of f.)

16.1.6 Since $h(1) = 5$, we have $a \cdot 1 + b = 5$, so $a + b = 5$. Since $h(-1) = 1$, we have $a \cdot (-1) + b = 1$, so $-a + b = 1$. Adding these two equations gives $2b = 6$, so $b = 3$. From $a + b = 5$, we find $a = 2$. Therefore, $h(x) = 2x + 3$, so $h(6) = 2 \cdot 6 + 3 = \boxed{15}$.

16.1.7

(a) First, we find $f(2, -3) = 2^2 + 4 \cdot 2 + (-3)^2 + 5 = 26$ and $f(0, 5) = 0^2 + 4 \cdot 0 + 5^2 + 5 = 30$. So, $f(2, -3) + f(0, 5) = \boxed{56}$.

(b) We complete the square in the terms with x to find

$$f(x, y) = x^2 + 4x + y^2 + 5 = x^2 + 4x + 4 - 4 + y^2 + 5 = (x+2)^2 + y^2 + 1.$$

So, for any values of x and y, we see that $f(x, y)$ is the sum of two squares and of 1. Therefore, $f(x, y)$ can never be less than 1. When $x = -2$ and $y = 0$ (which makes both our squares equal to 0), we have $f(-2, 0) = 1$, so the smallest possible value of $f(x, y)$ is $\boxed{1}$.

16.1.8

(a) In our list, $x = 5$ is paired with -3, so $f(5) = \boxed{-3}$.

(b) The only possible values of x in our list of $(x, f(x))$ pairs are $\boxed{2, 4, 5, \text{ and } 6}$, so our domain consists of these values.

(c) The only possible values of $f(x)$ in our list of $(x, f(x))$ pairs are $\boxed{-3, -2, 2, \text{ and } 3}$, so our range consists of these values.

Exercises for Section 16.2

16.2.1

(a) We have $s(x) = f(x) + 2g(x) = x^2 - 3 + 2(x + 3) = x^2 + 2x + 3$. Clearly, whatever input we give to s will produce exactly one output. So, $\boxed{\text{yes}}$, s is a function.

(b) Suppose we input the value a to r. In order to show that r is a function, we must show that there is at most one possible output, $r(a)$ for each such input a.

  We have $r(a) = p(a) + 2q(a)$. Suppose both $p(a)$ and $q(a)$ are defined. (If either is not defined, then $r(a)$ is not defined for that value of a.) Because p is a function, there's only one possible value of $p(a)$: whatever the result is when we input a to the function p. Similarly, there's only one possible value of $q(a)$. Since there is only one possible $q(a)$ and one possible $q(a)$ for each a, there's only one possible value of $p(a) + 2q(a)$. So, there's only one possible value of $r(a)$ for each input a that is in the domains of both p and q.

  There's nothing special about a. The argument above holds for any valid input to r. Because each valid input to r produces only one output, $\boxed{r \text{ is a function}}$.

16.2.2

(a) The domain of f is all nonnegative numbers, and the domain of g is all numbers greater than or equal to 4. The domain of s consists of the values that are in the domains of both f and g, so the domain of s is $\boxed{\text{all numbers greater than or equal to 4}}$.

(b) When $x = -4$, we have $\sqrt{x^2 - 4x} = \sqrt{16 + 16} = 4\sqrt{2}$. So, $\sqrt{x^2 - 4x}$ is indeed defined when $x = -4$. However, -4 is not in the domain of either f or g. Because $p = f \cdot g$ and -4 is not in the domain of f or g, the value -4 is not in the domain of p.

(c) As with s in the first part, the domain of p consists of all numbers that are in the domains of f and g. Therefore, the domain of p is $\boxed{\text{all numbers greater than or equal to 4}}$.

Exercises for Section 16.3

16.3.1 We have $g(4) = 3 \cdot 4 + 5 = 17$, so $f(g(4)) = f(17) = 17 + 3 = 20$. We also have $f(4) = 4 + 3 = 7$, so $g(f(4)) = g(7) = 3 \cdot 7 + 5 = 26$. Therefore, $f(g(4)) - g(f(4)) = 20 - 26 = \boxed{-6}$.

16.3.2 We substitute $f(x)$ in for x in our definition of $f(x)$ to find

$$f(f(x)) = 5[f(x)]^2 - 5 = 5(5x^2 - 5)^2 - 5$$
$$= 5[(5x^2)^2 + 2(5x^2)(-5) + (-5)^2] - 5 = 5(25x^4 - 50x^2 + 25) - 5$$
$$= \boxed{125x^4 - 250x^2 + 120}.$$

16.3.3 Since $f(x) = x+6$ and $g(x) = ax+b$, we have $f(g(x)) = ax+b+6$. Since we are given $f(g(x)) = 3x+3$, we have $ax + b + 6 = 3x + 3$. Equating the coefficients of x gives $a = 3$, and equating the constants gives $b + 6 = 3$, so $b = -3$. Therefore, $a + b = \boxed{0}$.

16.3.4 We don't know $g(x)$, so we don't have an expression we can simply stick 4 in to get an answer. We do, however, know that $g(f(x)) = 5 - 4x$. So, if we can figure out what to put into $f(x)$ such that 4 is output, we can use our expression for $g(f(x))$ to find $g(4)$.

Since $f(x) = 2x - 3$, the value of x such that $f(x) = 4$ is the solution to the equation $2x - 3 = 4$, which is $x = 7/2$. So, we have $f(7/2) = 4$. Therefore, if we let $x = 7/2$ in $g(f(x)) = 5 - 4x$, we have

$$g(f(7/2)) = 5 - 4 \cdot \frac{7}{2} \quad \Rightarrow \quad g(4) = 5 - 14 = \boxed{-9}.$$

Exercises for Section 16.4

16.4.1 Since f returns 5 when 3 is input, the inverse of f reverses f and returns $\boxed{3}$ when 5 is input.

16.4.2

(a) Let g be the inverse of f, so $f(g(x)) = g(f(x)) = x$. Letting $g(x) = y$, we have $f(g(x)) = f(y) = 3y + 2$. Since we also have $f(g(x)) = x$, we have $3y + 2 = x$. Solving for y, we find $y = (x - 2)/3$. So, we have $\boxed{f^{-1}(x) = (x - 2)/3}$.

(b) Suppose g is the inverse of f. What is $g(13)$? Since $f(x)$ equals 13 for all values of x, it's impossible to reverse f to find the single value of x such that $f(x) = 13$. Therefore, this function has $\boxed{\text{no inverse}}$.

(c) We seek the function g such that $f(g(x)) = x$, so we must solve

$$\frac{4g(x) - 5}{g(x) - 4} = x.$$

We let $y = g(x)$, so that our equation is $\dfrac{4y - 5}{y - 4} = x$. We wish to solve this equation for y. We start by multiplying both sides by y to get $4y - 5 = x(y - 4) = xy - 4x$. We move the y terms to one side and the other terms to the other side, then isolate y to find:

$$4y - xy = 5 - 4x \quad \Rightarrow \quad y(4 - x) = 5 - 4x \quad \Rightarrow \quad y = \frac{5 - 4x}{4 - x} = \frac{4x - 5}{x - 4}.$$

Therefore, we have $\boxed{f^{-1}(x) = \dfrac{4x - 5}{x - 4}}$. Notice that f^{-1} and f are the same. This means that f is its own inverse!

(d) Notice that both $f(1)$ and $f(-1)$ equal 5. Since f has the same output for two different inputs, f does not have an inverse .

(e) Suppose g is the inverse of f. We must have $f(g(x)) = x$, which gives $[g(x)]^3 = x$. Taking the cube root of both sides gives $g(x) = \sqrt[3]{x}$.

(f) Suppose g is the inverse of f. We must have $f(g(x)) = x$, which gives

$$\frac{1}{2g(x)} = x.$$

Multiplying both sides by $g(x)$ gives $1 = 2x[g(x)]$. Dividing by $2x$ gives

$$g(x) = \frac{1}{2x}.$$

Because f and g are the same, f is its own inverse.

16.4.3 If f is its own inverse, then $f(f(x)) = x$ for all x in the domain of f. So, we have

$$\frac{f(x)}{f(x) - a} = x \quad \Rightarrow \quad \frac{\frac{x}{x-a}}{\frac{x}{x-a} - a} = x \quad \Rightarrow \quad \frac{\frac{x}{x-a}}{\frac{x}{x-a} - \frac{a(x-a)}{x-a}} = x.$$

Therefore,

$$\frac{\frac{x}{x-a}}{\frac{x-ax+a^2}{x-a}} = x \quad \Rightarrow \quad \frac{x}{x - ax + a^2} = x.$$

Multiplying both sides by $x - ax + a^2$ gives $x = x(x - ax + a^2) = x^2 - ax^2 + a^2x$. If the equation

$$x = x^2 - ax^2 + a^2x$$

is true for all x, then the two x^2 terms must cancel out. This can only happen if $a = \boxed{1}$, which does give us $x = x^2 - ax^2 + a^2x = x$.

16.4.4 When finding the inverse of g, we solved the equation $g(f(x)) = x$ for $f(x)$. We then let $y = f(x)$, so that our equation became $g(y) = x$. From this equation, we see that x is the output of g. Therefore, any x in this equation is in the range of g. So, we consider the range of g. The definition of $g(x)$ was

$$g(x) = \frac{2x - 3}{x + 5}.$$

In Problem 16.4 of the text, we found that the range of this function is all values *except* 2. So, 2 is not in the range of g. Therefore, when we solve the equation $g(y) = x$ for y, it's OK to divide by $x - 2$, because x, which is in the range of g, cannot equal 2.

Exercises for Section 16.5

16.5.1 Since we know what $g(2x + 5)$ is, to determine $g(-3)$, we must determine what value of x makes $2x + 5$ equal to -3. Solving $2x + 5 = -3$ gives us $x = -4$. Letting $x = -4$ in $g(2x + 5) = 4x^2 - 3x + 2$ gives $g(-3) = 4(-4)^2 - 3(-4) + 2 = 4 \cdot 16 + 12 + 2 = \boxed{78}$.

16.5.2

(a) Let $A(t)$ be the number of meters Alice has run after Bob has run for t seconds. In the 20 seconds before Bob starts running, Alice runs $3(20) = 60$ meters. After that Alice runs 3 meters per second for t seconds, so she covers $3t$ more meters during these t seconds. Therefore $A(t) = 3t + 60$.

Let $B(t)$ be the number of meters Bob runs in t seconds. Since Bob runs 5 m/s, he runs $5t$ meters in t seconds. Therefore, $B(t) = 5t$.

(b) Since Bob has run $B(t)$ meters after t seconds, and Alice has run $A(t)$ meters, we seek the time t such that $B(t) = 1.5A(t)$. Using our expressions for $A(t)$ and $B(t)$ from the previous part, we have $5t = 1.5(3t + 60)$. Solving this equation gives $t = \boxed{180}$ seconds.

16.5.3 Since $f(2x) = 2/(2 + x)$, we let $x = y/2$ to find $f(y)$:

$$f\left(2 \cdot \frac{y}{2}\right) = \frac{2}{2 + \frac{y}{2}} \quad \Rightarrow \quad f(y) = \frac{2}{\frac{4+y}{2}} = \frac{4}{y + 4}.$$

Therefore, $f(x) = 4/(x + 4)$, so $2f(x) = \boxed{8/(x + 4)}$.

16.5.4 We let $N = 10a + b$, so $P(N) = ab$ and $S(N) = a + b$. Therefore, $N = P(N) + S(N)$ gives us $10a + b = ab + a + b$. Grouping all terms on the left side gives us $9a - ab = 0$. Since $a \neq 0$ because N is a two-digit number, we can divide this equation by a to give $9 - b = 0$. So, $b = \boxed{9}$.

16.5.5 We can make our equation have $f(1, 0)$ on the right side by letting $x = 2$ and $y = 2$:

$$f(2, 2) = 2 + f(2 - 1, 2 - 2) = 2 + f(1, 0) = 7.$$

Similarly, we can take another step by noting that we have $f(2, 2)$ on the right side by letting $x = 3$ and $y = 1$:

$$f(3, 1) = 3 + f(3 - 1, 3 - 1) = 3 + f(2, 2) = 10.$$

Letting $x = 4$, $y = 3$ produces $f(3, 1)$ on the right side:

$$f(4, 3) = 4 + f(4 - 1, 4 - 3) = 4 + f(3, 1) = 14.$$

Aha! Letting $x = 5$, $y = 2$ produces $f(4, 3)$ on the right side:

$$f(5, 2) = 5 + f(5 - 1, 5 - 2) = 5 + f(4, 3) = \boxed{19}.$$

Exercises for Section 16.6

16.6.1 We have $5 \,\&\, 8 = \dfrac{5}{8} + \dfrac{8}{5} = \dfrac{5^2 + 8^2}{5 \cdot 8} = \boxed{\dfrac{89}{40}}$.

16.6.2 We have $a \star b = \dfrac{1}{a} + \dfrac{1}{b} = \dfrac{b + a}{ab}$. We have $a + b = 9$ and $ab = 20$, so $a \star b = \dfrac{a + b}{ab} = \boxed{\dfrac{9}{20}}$.

16.6.3 Since $3 \odot y = 4 \cdot 3 - 3y + 3 \cdot y = 12$, we see that $3 \odot y = 12$ for $\boxed{\text{all values of } y}$.

16.6.4 Since $h \otimes h = h^3 - h$, we have $h \otimes (h \otimes h) = h \otimes (h^3 - h) = h^3 - (h^3 - h) = \boxed{h}$.

16.6.5 We have

$$a@1 = \frac{a^3 - 1}{a - 1} = \frac{(a - 1)(a^2 + a + 1)}{a - 1} = a^2 + a + 1,$$

so we seek the number of real solutions to $a^2 + a + 1 = 0$. Because the discriminant of $a^2 + a + 1$ is $1^2 - 4(1)(1) = -3$, there are $\boxed{\text{no real values}}$ that satisfy the equation.

Review Problems

16.25

(a) A given magazine can only have one possible number of pages, so this describes a function.

(b) Soccer teams have more than one player, so there is more than one possible output for a given input. Therefore, this does not describe a function.

16.26 Since $r(a) = a^2 - a + 14$, we seek the values of a such that $a^2 - a + 14 = 20$. Collecting all terms on the left side gives $a^2 - a - 6 = 0$, from which we have $(a - 3)(a + 2) = 0$. Therefore, the desired values of a are $\boxed{a = 3 \text{ and } a = -2}$.

16.27

(a) $f(8) = 3\sqrt{2 \cdot 8 - 7} - 8 = 3\sqrt{9} - 8 = \boxed{1}$.

(b) The expression inside the square root sign must be nonnegative, so we must have $2x - 7 \geq 0$. Solving this inequality gives $x \geq 7/2$. This is the only restriction on the input to our function, so our domain is all values of x such that $\boxed{x \geq 7/2}$.

(c) Since $\sqrt{2x - 7}$ must be nonnegative, we have $f(x) = 3\sqrt{2x - 7} - 8 \geq 3 \cdot 0 - 8 = -8$. So, we see that $f(x)$ must be greater than or equal to -8. All such values are in the range, so the range of the function f is $\boxed{\text{all numbers greater than or equal to } -8}$.

To see that all numbers greater than -8 are in the range, we solve the equation $3\sqrt{2x - 7} - 8 = y$ for x, and find

$$x = \frac{(y + 8)^2}{18} + \frac{7}{2}.$$

From this equation, we can find an input x to f that produces y as an output for any value of y greater than or equal to -8.

16.28

(a) We have $f(3) = \dfrac{3 + 2}{3 - 2} = \boxed{5}$.

(b) We cannot divide by zero, so we cannot have $x - 2 = 0$. There are no other restrictions on inputs to f, so the domain of f is $\boxed{\text{all numbers except 2}}$.

(c) Suppose y is the output of f. Then, we have $f(x) = y$, or

$$\frac{x + 2}{x - 2} = y \quad \Rightarrow \quad x + 2 = y(x - 2) \quad \Rightarrow \quad x + 2 = xy - 2y \quad \Rightarrow \quad x - xy = -2y - 2 \quad \Rightarrow \quad x = \frac{-2y - 2}{1 - y}.$$

We can substitute any desired output y besides $y = 1$ into this equation to find the input, x, to f that produces this desired output. However, we can't let $y = 1$ in this equation because we can't divide by 0. Therefore, the range of f is $\boxed{\text{all numbers except 1}}$.

16.29 Because $g(-6) = 0$ and $h = f/g$, we will have division by 0 if we input -6 to h. So, -6 is $\boxed{\text{not in the domain}}$ of h.

16.30 We have $f(1) = 2 \cdot 1 = 2$, so $g(f(g(f(1)))) = g(f(g(2)))$. Since $g(2) = 2 + 1 = 3$, we have $g(f(g(2))) = g(f(3))$. Since $f(3) = 2 \cdot 3 = 6$, we have $g(f(3)) = g(6) = 6 + 1 = \boxed{7}$.

16.31 Since $q(4) = 3 \cdot 4 - b = 12 - b$, we can write $p(q(4)) = 7$ as $p(12 - b) = 7$. Since $p(x) = 2x - 7$, we have $p(12 - b) = 2(12 - b) - 7 = 17 - 2b$. Substituting this into $p(12 - b) = 7$ gives $17 - 2b = 7$, from which we have $b = \boxed{5}$.

16.32 First, make sure you see why the answer is *not* $[f(8)]^{19} = 13^{19}$. The expression $f^{19}(8)$ does not mean raise $f(8)$ to the nineteenth power. Instead it means we compose 19 f's: $f(f(f(\cdots f(f(8))\cdots)))$.

We find $f^{19}(8)$ from the inside out. We have

$$f(8) = 8 + 5 = 13,$$
$$f^2(8) = f(f(8)) = f(13) = 13 + 5 = 18,$$
$$f^3(8) = f(f(f(8))) = f(18) = 18 + 5 = 23,$$
$$f^4(8) = f(f^3(8)) = f(23) = 23 + 5 = 28\ldots,$$

We see a pattern; $f^2(x)$ is 5 more than $f(x)$, $f^3(x)$ is 5 more than $f^2(x)$, $f^4(x)$ is 5 more than $f^3(x)$, and so on. Similarly, $f^n(x)$ is always 5 more than $f^{n-1}(x)$ for any $n > 1$:

$$f^n(8) = f(f^{n-1}(8)) = f^{n-1}(8) + 5.$$

So, each time we include another f in the composition of f's, it is the same as adding another 5. Therefore $f^{19}(8)$ is the same as adding nineteen 5's to the input, 8, which gives

$$f^{19}(8) = 8 + 19 \cdot 5 = \boxed{103}.$$

16.33 We don't know $g(x)$, so we don't have an expression we can simply stick -5 in to get an answer. We do, however, know that $g(f(x)) = 2x^2 + 5x - 3$. So, we if we can figure out what to put into $f(x)$ such that -5 is output, we can use our expression for $g(f(x))$ to find $g(-5)$.

If $f(x) = -5$, we have $3x - 8 = -5$, so $x = 1$. Therefore, letting $x = 1$ in $g(f(x)) = 2x^2 + 5x - 3$ gives

$$g(f(1)) = 2 \cdot 1^2 + 5 \cdot 1 - 3 \quad \Rightarrow \quad g(-5) = \boxed{4}.$$

16.34 If g is the inverse of f, then $f(g(x)) = x$. Since $f(x) = 4 - 5x$, we have $f(g(x)) = 4 - 5g(x)$. So, we must have $4 - 5g(x) = x$. Solving for $g(x)$ gives $g(x) = \boxed{(4 - x)/5}$.

16.35 Because $f(1) = 8$ and $f(-1) = 8$, there are two different inputs to f such that 8 is the output. Since f returns the same output for two different inputs, it $\boxed{\text{does not have an inverse}}$.

16.36 *Solution 1: Use the definition of inverse.* When we put 3 into the inverse of f, it outputs whatever input x makes $f(x) = 3$. So, $f^{-1}(3)$ equals the value of x such that $f(x) = 3$. Since $f(x) = 3x - 8$ and we seek the value of x such that $f(x) = 3$, we must have $3x - 8 = 3$. Solving this equation gives $x = \boxed{11/3}$. Or, in short, let $x = f^{-1}(3)$. Then, we must have

$$f(x) = f(f^{-1}(3)) = 3 \quad \Rightarrow \quad 3x - 8 = 3 \quad \Rightarrow \quad x = \frac{11}{3}.$$

Solution 2: Find $f^{-1}(x)$. Since f^{-1} is the inverse of f, we must have $f(f^{-1}(x)) = x$. Therefore, we must have $3f^{-1}(x) - 8 = x$. Solving for $f^{-1}(x)$, we have $f^{-1}(x) = (x+8)/3$. Therefore, $f^{-1}(3) = (3+8)/3 = \boxed{11/3}$.

16.37

(a) We have $f(-5) = 3(-5) + 10 = -5$. So, $f(f(-5)) = f(-5) = -5$. And, we have $f(f(f(-5))) = f(f(-5)) = f(-5) = \boxed{-5}$ and $f(f(f(f(-5)))) = f(-5) = \boxed{-5}$. Similarly, $f^n(-5) = -5$ for all positive integers n. So, if we put -5 into f, then take the output and put it back into f, then take the output of that and input it back into f, and so on, then at each step, the output is always -5. It's "fixed."

(b) If x is a fixed point of g, then $g(x) = x$. Since $g(x) = x^2 - 6$, we have $x^2 - 6 = x$ for any fixed point, x, of g. Solving this equation gives $x = -2$ and $x = 3$ as solutions. We can test these by evaluating $g(x)$ for each of these values of x. Since $g(-2) = (-2)^2 - 6 = -2$ and $g(3) = 3^2 - 6 = 3$, we see that $\boxed{-2 \text{ and } 3}$ are, indeed, fixed points of g.

(c) We seek those values of x such that $f(x) = x$. Therefore, we must solve the equation $ax + b = x$ for x in terms of a and b. Putting the x terms on one side and the b on the other, we have

$$ax - x = -b \quad \Rightarrow \quad x(a-1) = -b \quad \Rightarrow \quad x = -\frac{b}{a-1}.$$

So, the fixed point of f is $\boxed{-b/(a-1)}$. To test this answer, we input $-b/(a-1)$ to f:

$$f\left(-\frac{b}{a-1}\right) = a \cdot \frac{-b}{a-1} + b = \frac{-ab}{a-1} + \frac{b(a-1)}{a-1} = \frac{-ab + ab - b}{a-1} = -\frac{b}{a-1}.$$

We put $-b/(a-1)$ in, and we get $-b/(a-1)$ out; it's a fixed point. (The $a - 1$ in the denominator is why we must have $a \neq 1$. If $a = 1$ and $b = 0$, then all real numbers are fixed points of f. If $a = 1$ and $b \neq 0$, then there are no fixed points of f.)

16.38 The domain of g is all real numbers while the domain of f is all numbers except -1. Since these two functions have different domains, they are not the same function.

16.39 We have $3 \oplus 1 = 3 \cdot 3 + 4 \cdot 1 = \boxed{13}$.

16.40 We have $2t \star t = \sqrt{(2t)^2 + t^2} = \sqrt{4t^2 + t^2} = \sqrt{5t^2}$. Therefore, we want those values of t for which $\sqrt{5t^2} = 15$. Squaring both sides of this equation gives $5t^2 = 225$. Solving this equation gives $t = \boxed{\pm 3\sqrt{5}}$.

16.41 We can turn $f(x - 9)$ into $f(x)$ by choosing the proper value for x. Specifically, if we let $z = x - 9$, we have $x = z + 9$. Substituting this into $f(x - 9) = 2x + 7$, we get

$$f(z) = 2(z + 9) + 7 = 2z + 25.$$

The z is just a dummy variable, so we can simply change it to whatever letter we want, like x:

$$f(x) = \boxed{2x + 25}.$$

16.42

(a) We don't yet have an expression for $h(z)$ to plug -9 into to find $h(-9)$. Instead, we have an expression for $h(2z + 3)$. We must find the value of z that makes $2z + 3 = -9$ in order to use our expression for $h(2z + 3)$. Solving $2z + 3 = -9$ gives $z = -6$. Substituting this into our expression for $h(2z + 3)$ gives

$$h(2 \cdot (-6) + 3) = \frac{5 - 2(-6)}{4 + (-6)} \quad \Rightarrow \quad h(-9) = \boxed{-\frac{17}{2}}.$$

(b) In the previous problem we turned $f(x - 3)$ into $f(z)$ by letting $z = x - 3$. We do similarly here by letting $y = 2z + 3$ to turn $h(2z + 3)$ into $h(y)$. Solving $y = 2z + 3$ for z in terms of y gives $z = (y - 3)/2$. Substituting this into our expression for $h(2z + 3)$ gives

$$h(y) = \frac{5 - 2 \cdot \frac{y-3}{2}}{4 + \frac{y-3}{2}} = \frac{5 - (y - 3)}{\frac{8+y-3}{2}} = \frac{8 - y}{\frac{5+y}{2}} = \frac{16 - 2y}{5 + y}.$$

As before, y is a dummy variable, so we can simply change it to whatever letter we want, like z:

$$h(z) = \boxed{\frac{16 - 2z}{5 + z}}.$$

(c) We could find $h^{-1}(z)$ as the first step, but as we saw in the text, there is a slicker approach. Since h^{-1} is the inverse of h, $h^{-1}(4)$ equals the value of x such that $h(x) = 4$. We found in the previous part that $h(x) = (16 - 2x)/(5 + x)$, so we seek the solution to

$$\frac{16 - 2x}{5 + x} = 4.$$

Multiplying both sides by $5 + x$ gives

$$16 - 2x = 4(5 + x) \quad \Rightarrow \quad 16 - 2x = 20 + 4x \quad \Rightarrow \quad -4 = 6x \quad \Rightarrow \quad x = \boxed{-2/3}.$$

16.43 We know $f(0)$ and seek $f(1998)$. If we let $x = 1998$ and $y = 0$, we get an equation involving $f(1998)$ and $f(0)$:

$$f(1998 + 0) = 1998 + f(0) \quad \Rightarrow \quad f(1998) = 1998 + 2 = \boxed{2000}.$$

Challenge Problems

16.44 Because $f(-3) = 2$, we have $81a - 9b - 3 + 5 = 2$, so $81a - 9b = 0$. We also have $f(3) = 81a - 9b + 3 + 5 = 81a - 9b + 8$. We already found $81a - 9b = 0$ so $f(3) = 81a - 9b + 8 = \boxed{8}$.

16.45   We have $f(-4) = 3(-4)^2 - 7 = 41$, so we seek $g(f(-4)) = g(41)$. But what's $g(41)$? So, we turn to the other information we are given, $g(f(4)) = 9$. Since $f(4) = 3(4)^2 - 7 = 41$, this equation gives us $g(41) = \boxed{9}$.

16.46   Since $g(a) = 2a - \frac{6}{a}$, we seek all solutions to the equation

$$2a - \frac{6}{a} = -4.$$

Multiplying this equation by a gives $2a^2 - 6 = -4a$. Dividing by two gives $a^2 - 3 = -2a$, rearranging gives $a^2 + 2a - 3 = 0$, and factoring gives $(a + 3)(a - 1) = 0$. Therefore, the values of a such that $g(a) = -4$ are $\boxed{a = 1 \text{ and } a = -3}$.

16.47   We have $f(\sqrt{2}) = a(\sqrt{2})^2 - \sqrt{2} = 2a - \sqrt{2}$. Therefore, we have

$$f(f(\sqrt{2})) = f(2a - \sqrt{2}) = a(2a - \sqrt{2})^2 - \sqrt{2}.$$

We must have $f(f(\sqrt{2})) = -\sqrt{2}$, so we have the equation

$$a(2a - \sqrt{2})^2 - \sqrt{2} = -\sqrt{2}.$$

Therefore, we have $a(2a - \sqrt{2})^2 = 0$. We want the positive solution for a, so we disregard the solution $a = 0$ to this equation. From $2a - \sqrt{2} = 0$, we find $a = \boxed{\dfrac{\sqrt{2}}{2}}$.

16.48   First, we note $f(x) = 1 - \dfrac{1}{x} = \dfrac{x - 1}{x}$.

(a)   We have $f(f(x)) = 1 - \dfrac{1}{f(x)} = 1 - \dfrac{1}{\frac{x-1}{x}} = 1 - \dfrac{x}{x - 1} = \dfrac{x - 1}{x - 1} - \dfrac{x}{x - 1} = -\dfrac{1}{x - 1}$.

(b)   Using our expression from the last part for $f(f(x))$, we have

$$f(f(f(x))) = 1 - \dfrac{1}{f(f(x))} = 1 - \dfrac{1}{-\frac{1}{x-1}} = 1 + x - 1 = x.$$

(c)   We use our expression for $f(f(f(x)))$ to find $f(f(f(f(x))))$:

$$f(f(f(f(x)))) = 1 - \dfrac{1}{f(f(f(x)))} = 1 - \dfrac{1}{x}.$$

(d)   In the previous part, we learned that $f^4(x) = f(x)$, and in part (c), we saw that $f^3(x) = x$. So, we have $f^7(x) = f^4(f^3(x)) = f^4(x) = f(x)$. In the same way, we have $f^{10}(x) = f(x)$, $f^{13}(x) = f(x)$, and so on. We therefore find that if n is one more than a multiple of 3, then $f^n(x) = f(x)$. So,

$$f^{34}(5) = f(5) = 1 - \dfrac{1}{5} = \boxed{\dfrac{4}{5}}.$$

16.49

(a)   We have $x \star y = \sqrt{x^2 + y^2}$ and $y \star x = \sqrt{y^2 + x^2}$. Because $\sqrt{x^2 + y^2} = \sqrt{y^2 + x^2}$ for any x and y, we have $x \star y = y \star x$. Therefore, $\star$ is commutative.

(b) We have $x \odot y = x^2 - y^2$ and $y \odot x = y^2 - x^2$. Since it is not true that $x^2 - y^2 = y^2 - x^2$ for all x and y (try $x = 3$, $y = 0$), we see that $\odot$ is not commutative.

(c) We have $x \spadesuit y = xy + x + y$ and $y \spadesuit x = yx + y + x$. Since $xy + x + y = yx + y + x$ for all x and y, the operation $\spadesuit$ is commutative.

(d) We have $x \diamond y = (x \clubsuit y) \heartsuit (y \clubsuit x)$ and $y \diamond x = (y \clubsuit x) \heartsuit (x \clubsuit y)$ Because $\clubsuit$ is commutative, we have $y \clubsuit x = x \clubsuit y$, so

$$y \diamond x = (y \clubsuit x) \heartsuit (x \clubsuit y) = (x \clubsuit y) \heartsuit (y \clubsuit x) = x \diamond y.$$

So, $\diamond$ is commutative. Notice that we didn't even need the fact that $\heartsuit$ is commutative. Extra challenge: What if $\clubsuit$ is not commutative but $\heartsuit$ is? Is $\diamond$ still commutative?

16.50 If we find two values of x that provide the same value of $f(x)$, then we know that f does not have an inverse. We could just use trial and error to find these values, but completing the square helps us find such values more easily:

$$f(x) = 3x^2 + 6x + 5 = 3(x^2 + 2x) + 5 = 3(x^2 + 2x + 1) - 3 \cdot 1 + 5 = 3(x + 1)^2 + 2.$$

So, we have $f(0) = 3(0 + 1)^2 + 2 = 3(1) + 2 = 5$ and $f(-2) = 3(-2 + 1)^2 + 2 = 3(1) + 2 = 5$. Therefore, $f(0) = f(-2)$, so f does not have an inverse.

16.51

(a) $f(f(x)) = \dfrac{f(x)}{1 - f(x)} = \dfrac{\dfrac{x}{1-x}}{1 - \dfrac{x}{1-x}} = \dfrac{\dfrac{x}{1-x}}{\dfrac{1 - x - x}{1 - x}} = \boxed{\dfrac{x}{1 - 2x}}.$

(b) We evaluate $f(f(f(x)))$ by putting $f(f(x))$ into f.

$$f(f(f(x))) = \frac{f(f(x))}{1 - f(f(x))} = \frac{\dfrac{x}{1 - 2x}}{1 - \dfrac{x}{1 - 2x}} = \frac{\dfrac{x}{1 - 2x}}{\dfrac{1 - 2x - x}{1 - 2x}} = \boxed{\frac{x}{1 - 3x}}.$$

(c) We evaluate $f(f(f(f(x))))$ by putting $f(f(f(x)))$ into f:

$$f(f(f(f(x)))) = \frac{f(f(f(x)))}{1 - f(f(f(x)))} = \frac{\dfrac{x}{1 - 3x}}{1 - \dfrac{x}{1 - 3x}} = \frac{\dfrac{x}{1 - 3x}}{\dfrac{1 - 3x - x}{1 - 3x}} = \boxed{\frac{x}{1 - 4x}}.$$

(d) Our first three parts suggest that $f^n(x) = \dfrac{x}{1 - nx}$. To show this, we show that if $f^k(x) = \dfrac{x}{1 - kx}$, then $f^{k+1}(x) = \dfrac{x}{1 - (k + 1)x}$. Since $f^{k+1}(x) = f(f^k(x))$, then $f^k(x) = \dfrac{x}{1 - kx}$ gives us

$$f^{k+1}(x) = \frac{f^k(x)}{1 - f^k(x)} = \frac{\dfrac{x}{1 - kx}}{1 - \dfrac{x}{1 - kx}} = \frac{\dfrac{x}{1 - kx}}{\dfrac{1 - kx - x}{1 - kx}} = \frac{x}{1 - (k + 1)x}.$$

So, we've shown that if $f^k(x) = \dfrac{x}{1 - kx}$, then $f^{k+1}(x) = \dfrac{x}{1 - (k + 1)x}$. In other words, if $f^n(x)$ fits our pattern for $n = k$, then it fits it for $n = k + 1$, too. In our first three parts, we showed that $f^n(x)$ fits our pattern for $n = 2, 3$, and 4. Since it works for 4, it works for 5. Since it works for 5, it works for 6. Similarly, it works for all positive integers.

The reasoning we used in this proof is called mathematical induction. We'll be exploring induction in much more detail in later books in this series.

16.52   We try to learn more about F by substituting values of x into

$$F(x + 1) = F(x) + F(x - 1).$$

We choose values of x that let us use the information we are given about $F(1)$ and $F(4)$. For example, letting $x = 2$ gives

$$F(3) = F(2) + F(1) = F(2) + 1.$$

Letting $x = 3$ gives

$$F(4) = F(3) + F(2) \quad \Rightarrow \quad 1 = F(3) + F(2).$$

We now have a system of two linear equations with two variables:

$$F(3) = F(2) + 1,$$
$$1 = F(2) + F(3).$$

Subtracting the second equation from the first gives $F(3) - 1 = 1 - F(3)$, from which we find $F(3) = 1$. Using this, we find $F(2) = 0$. Now, we can find all the values of $F(x)$ that we want. Specifically, $F(5) = F(4) + F(3) = 2$, $F(6) = F(5) + F(4) = 3$, $F(7) = F(6) + F(5) = 5$, $F(8) = F(7) + F(6) = 8$, $F(9) = F(8) + F(7) = 13$, and $F(10) = F(9) + F(8) = \boxed{21}$.

16.53   Letting $x = 100$ and $y = 6$ gives $f(600) = f(100)/6$. So, if we find $f(100)$, we can find $f(600)$. Letting $x = 100$ and $y = 5$ gives $f(500) = f(100)/5$, so $f(100) = 5f(500)$. Since $f(500) = 3$, we have $f(100) = 5f(500) = 15$. Therefore, $f(600) = f(100)/6 = \boxed{5/2}$.

16.54   *Solution 1: Let $f(x) = y$.* If we let $y = f(x)$, then we have $g(y) = 2x^2 - 3x + 5$. Therefore, if we can find x in terms of y, then we can find $g(y)$. Because $y = f(x)$, we have $y = 3 - 2x$. Solving for x in terms of y gives $x = (3 - y)/2$. Substituting this into $g(y) = 2x^2 - 3x + 5$ gives us

$$g(y) = 2\left(\frac{3-y}{2}\right)^2 - 3\left(\frac{3-y}{2}\right) + 5 = \frac{y^2}{2} - \frac{3y}{2} + 5.$$

Because $g(y) = \dfrac{y^2}{2} - \dfrac{3y}{2} + 5$, we have $g(x) = \boxed{\dfrac{x^2}{2} - \dfrac{3x}{2} + 5}$.

Solution 2: Educated guess. We guess that $g(x)$ is quadratic, because we know that if $g(x)$ is quadratic, then $g(3 - 2x)$ will be quadratic, as well. So, we let $g(x) = ax^2 + bx + c$, for some constants a, b, and c. Therefore,

$$g(f(x)) = g(3-2x) = a(3-2x)^2 + b(3-2x) + c = a(9-12x+4x^2) + 3b - 2bx + c = 4ax^2 + (-2b-12a)x + 9a + 3b + c.$$

Equating the coefficients of this to the coefficients of $g(f(x)) = 2x^2 - 3x + 5$, we have $4a = 2$, $-2b - 12a = -3$, and $9a + 3b + c = 5$. The first equation gives $a = 1/2$. Combining this with the second equation gives $b = -3/2$. Letting $a = 1/2$ and $b = -3/2$ in the third equation gives $c = 5 - 9a - 3b = 5$. Therefore,

$$g(x) = \boxed{\frac{x^2}{2} - \frac{3x}{2} + 5}.$$

16.55   We seek a function g such that $g(f(x,y)) = (x,y)$. Since $f(x,y) = (x+y, x-y)$, we need the function g such that $g(x+y, x-y) = (x,y)$. So, we seek a function g that combines $x+y$ and $x-y$ to get x as the first number in its ordered pair output, and that combines $x+y$ and $x-y$ to give y as the second number in its ordered pair output. We note that $[(x+y) + (x-y)]/2 = 2x/2 = x$ and $[(x+y) - (x-y)]/2 = 2y/2 = y$, so if $g(x,y) = \left(\frac{x+y}{2}, \frac{x-y}{2}\right)$, then we have

$$g(x+y, x-y) = \left(\frac{x+y+x-y}{2}, \frac{x+y-(x-y)}{2}\right) = (x,y),$$

as desired. Therefore, the inverse of f is $\boxed{g(x,y) = \left(\dfrac{x+y}{2}, \dfrac{x-y}{2}\right)}$.

16.56   Let $h = f \circ g$ and $j = g^{-1} \circ f^{-1}$. We must show that $h(j(x)) = x$ and $j(h(x)) = x$. When we put j into h, we have
$$h(j(x)) = f(g(g^{-1}(f^{-1}(x)))).$$
Because g and g^{-1} are inverses, we have $g(g^{-1}(x)) = x$ for any input x. So, we have $g(g^{-1}(f^{-1}(x))) = f^{-1}(x)$. Therefore,
$$h(j(x)) = f(g(g^{-1}(f^{-1}(x)))) = f(f^{-1}(x)) = x.$$
Similarly, we have
$$g^{-1}(f^{-1}(f(g(x)))) = g^{-1}(g(x)) = x.$$
So, $g^{-1} \circ f^{-1}$ is the inverse of $f \circ g$.

16.57   If we could find $f(x)$, we could use it to find the desired values of a. But how do we find $f(x)$? That $f(1/x)$ term seems very hard to deal with. We think to try to get rid of it by letting $z = 1/x$, since we know x cannot be 0. This will get rid of the $f(1/x)$, since then $f(1/x) = f(z)$. But when we make this substitution elsewhere in our equation, we have

$$f(1/z) + 2f(z) = 3/z.$$

Uh-oh. We still have $f(1/z)$. However, this equation isn't exactly the same as our first one. If we write both equations with the same variable, we have the system

$$f(y) + 2f\left(\frac{1}{y}\right) = 3y,$$
$$2f(y) + f\left(\frac{1}{y}\right) = \frac{3}{y}.$$

Now we can get rid of the nasty $f(1/y)$ term! We subtract 2 times the second equation from the first to give
$$-3f(y) = 3y - \frac{6}{y} \quad \Rightarrow \quad f(y) = -y + \frac{2}{y}.$$

We wish to find those values of a such that $f(a) = f(-a)$, or $-a + \frac{2}{a} = a - \frac{2}{a}$. Rearranging this gives $2a = \frac{4}{a}$, so $2a^2 = 4$. Therefore, our desired values of a are $\boxed{\pm \sqrt{2}}$.

CHAPTER **17**

Graphing Functions

Exercises for Section 17.1

17.1.1

(a)
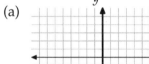

(d)
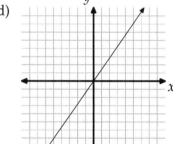

(b)

(e)
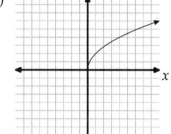

(c)
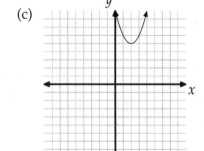

(f)
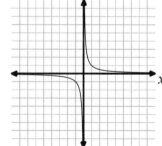

17.1.2

(a) No two points on the graph are on the same vertical line. This graph could be the graph of a function. The domain of the function is all real numbers from −8 to 8, including −8 and 8. We can't tell exactly what the range is from our graph because we can't tell the exact y value for the highest and lowest points. We can estimate and say that the range is all real numbers between approximately −6.3 and 6.3.

(b) There are many vertical lines that pass through two points. This is not the graph of a function.

(c) No two points on the graph are on the same vertical line. The possible x values are the integers from −8 through 8, so this is the domain of our function. The y-coordinates of the points give us our range, so our range consists of the values 6, 4.5, 3, 1.5, 0, −1.5, −3, −4.5, and −6.

(d) There are several vertical lines that pass through two points, so this is not the graph of a function.

17.1.3

(a) The point $(2, 1)$ is on the graph of $y = f(x)$, so $f(2) = \boxed{1}$.

(b) The point $(-1, 1)$ is on the graph of $y = g(x)$, so $g(-1) = \boxed{1}$.

(c) In part (a), we found that $f(2) = 1$, so $f(g(f(2))) = f(g(1))$. Because the point $(1, 1)$ is on the graph of $y = g(x)$, we have $g(1) = 1$, so $f(g(1)) = f(1)$. We can't easily read $f(1)$ from the graph of $y = f(x)$, but we can see that the portion of this graph between $x = -1$ and $x = 2$ is a line segment connecting $(-1, -1)$ to $(2, 1)$. The slope of this line is $2/3$, so an equation for this line is $y - 1 = (2/3)(x - 2)$. So, when $x = 1$, we have $y - 1 = (2/3)(1 - 2) = -2/3$, which gives us $y = 1/3$. Therefore, $f(1) = 1/3$, so $f(g(f(2))) = f(g(1)) = f(1) = \boxed{1/3}$.

17.1.4   For each value of x in the domain of h, we find the y-coordinate on the graph of $y = f(x) + g(x)$ by adding the y-coordinates of the graphs of $y = f(x)$ and $y = g(x)$. The domain of h consists of all values in the domains of f and g, which are all values of x such that $-4 \le x \le 4$. The resulting graph is shown at right.

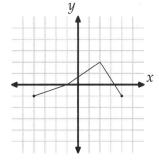

17.1.5   Let $x = f^{-1}(-1)$. Because f^{-1} is the inverse of f, we must have $f(x) = -1$. To determine x, we find the point on the graph of $y = f(x)$ such that $y = -1$. Since $(2, -1)$ is on the graph, we have $f(2) = -1$. Therefore, $f^{-1}(-1) = \boxed{2}$.

Exercises for Section 17.2

17.2.1   As discussed in the text, the graph of $y = f(2x)$ is a horizontal scaling of the graph of $y = f(x)$ by a factor of $1/2$, and the graph of $y = f(x + 3)$ is a leftward shift of the graph of $y = f(x)$ by 3. Once we have the graph of $y = f(2x)$, we can produce the graph of $y = 2f(x + 3)$ by scaling the graph of $y = f(x + 3)$ vertically away from the y-axis by a factor of 2, and we can produce the graph of $y = f(2x) + 3$ by shifting the graph of $y = f(2x)$ upward by 3. The resulting four graphs are shown below.

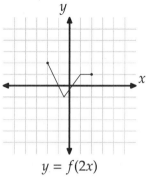

$y = f(2x)$

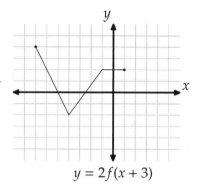

$y = 2f(x + 3)$

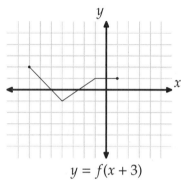

$y = f(x + 3)$

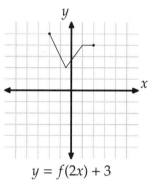

$y = f(2x) + 3$

17.2.2

(a) Since $g(5) = 3$, if we let $x = 2.5$ in the equation $y = g(2x)$, we have $y = g(2 \cdot 2.5) = g(5) = 3$. Therefore, the point $\boxed{(2.5, 3)}$ is on the graph of $y = g(2x)$. We could also note that the graph of $y = g(2x)$ is a horizontal scaling by $1/2$ of the graph of $y = g(x)$. Since we are told that $(5, 3)$ is on the graph of $y = g(x)$, we know that scaling this point horizontally by $1/2$, which gives $(5/2, 3)$, gives us a point on the graph of $y = g(2x)$.

(b) Since $g(5) = 3$, if we let $x = 12$ in $y = g(x - 7)$, we have $y = g(12 - 7) = g(5) = 3$, so the point $\boxed{(12, 3)}$ is on the graph of $y = g(x - 7)$. (Also, we could think of the graph of $y = g(x - 7)$ as a 7-unit rightward shift of the graph of $y = g(x)$, thus shifting the point $(5, 3)$ to $(12, 3)$.)

(c) Letting $x = 5$ in $y = g(x) + 5$ gives us $y = g(5) + 5 = 3 + 5 = 8$, so $\boxed{(5, 8)}$ is on the graph of $y = g(x) + 5$. We can also note that the graph of $y = g(x) + 5$ is a 5-unit upward shift of the graph of $y = g(x)$.

(d) We let $x = 10$ in $y = g(x/2) + 3$ to give $y = g(10/2) + 3 = g(5) + 3 = 3 + 3 = 6$. When $x = 10$, we have $y = 6$, so $\boxed{(10, 6)}$ is on the graph of $y = g(x/2) + 3$.

(e) We let $x = 5$ in $y = 3g(x) + 5$ to find $y = 3g(5) + 5 = 9 + 5 = 14$. So, the point $\boxed{(5, 14)}$ is on the graph of $y = 3g(x) + 5$.

(f) We know what happens when we put 5 into g. So, we must find the value of x that makes $3x - 7$ equal to 5. Solving $3x - 7 = 5$ gives $x = 4$. So, if we let $x = 4$ in our equation, we have $y = 2g(3 \cdot 4 - 7) + 4 = 6 + 4 = 10$. When $x = 4$, we have $y = 10$, so $\boxed{(4, 10)}$ is on our graph.

17.2.3 Let $f(x) = x^2$. Then, we have $f(x - 3) = (x - 3)^2$. The graph of $y = f(x - 3)$ is a rightward shift of the graph of $y = f(x)$ by 3 units. We also have $f(x - 3) + 4 = (x - 3)^2 + 4$. The graph of $y = f(x - 3) + 4$ is an upward shift of the graph of $y = f(x - 3)$ by 4 units. Putting these two shifts together, we see that the graph of $y = f(x - 3) + 4 = (x - 3)^2 + 4$ is the result of shifting the graph of $y = x^2$ right 3 units, then up 4 units. This shouldn't be much of a surprise; notice that the vertex of the graph of $y = x^2$ is $(0, 0)$ and the vertex of the graph of $y = (x - 3)^2 + 4$ is $(3, 4)$. Moreover, if we view these equations as quadratics in standard form, both have $a = 1$. Since a is a measure of the "wideness", or shape, of the parabola, we see that these two parabolas have the same shape.

17.2.4 The graphs of $y = f(x)$ and $y = f(x + 3)$ can intersect. Consider the example we showed in the text, which is reproduced at right. As we see, the graphs of $y = f(x)$ and $y = f(x + 3)$, a leftward shift of the graph of $y = f(x)$, do intersect.

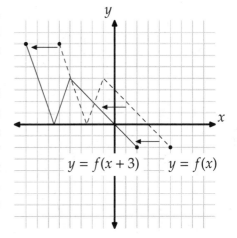

On the other hand, the graphs of $y = f(x)$ and $y = f(x) + 3$ cannot intersect. To see why, note that each vertical line intersects each of these graphs in no more than one point. Because the graph of $y = f(x) + 3$ is a 3-unit vertical shift of the graph of $y = f(x)$, any point at which a vertical line hits the graph of $y = f(x) + 3$ must be 3 units above the point where that line hits the graph $y = f(x)$. Therefore, no vertical line meets both graphs at the same point, so there is no point through which both graphs pass.

17.2.5 As we saw in the text, when $k > 1$, the graph of $y = kf(x)$ is a vertical stretch of the graph of $y = f(x)$, taking it farther from the x-axis wherever y is nonzero. When $0 < k < 1$, the graph of $y = kf(x)$ is a vertical compression of the graph of $y = f(x)$, bringing it closer to the x-axis wherever y is nonzero. When $k = 0$, we take this compression to the extreme, because then we have $y = kf(x) = 0 \cdot f(x) = 0$. So, whatever the graph of $y = f(x)$ is, the graph of $y = kf(x)$ when $k = 0$ is the result of compressing the entire graph down to the x-axis.

Exercises for Section 17.3

17.3.1

(a) There are many horizontal lines that pass through more than one point on the graph, so the function graphed in this part does not have an inverse.

(b) No horizontal line passes through more than one point on the graph, so the function graphed in this part does have an inverse.

(c) There are many horizontal lines that pass through more than one point on the graph, so the function graphed in this part does not have an inverse.

(d) No horizontal line passes through more than one point on the graph, so the function graphed in this part does have an inverse. Make sure you see why the line $y = -2$ passes through just one point on the graph. The open circle at $(-2, -2)$ means that this point is not on the graph.

17.3.2   As we learned in the text, the graph of $y = f^{-1}(x)$ is the result of reflecting the graph of $y = f(x)$ over the line $y = x$. The result in this problem is shown at left below.

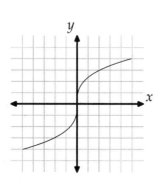

Figure 17.1: Diagram for Problem 17.3.2

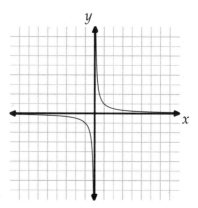

Figure 17.2: Diagram for Problem 17.3.3

17.3.3   The graph of f^{-1} is the result of reflecting the graph of f over the line $y = x$. If f^{-1} and f are the same, then their two graphs must be the same. In other words, the graph of f must be its own mirror image when reflected over the line $y = x$. Some example functions for which this is the case are $f(x) = x$, $f(x) = -x$, and $f(x) = 1/x$. The last of these is graphed at right above.

17.3.4   We learned in the previous chapter that a function has an inverse if, for every value y in the range, there is only one input x such that $y = f(x)$. In other words, for every possible output from f, there is only one input to f that produces that output. If the graph of $y = f(x)$ is such that no horizontal line passes through more than one point of the graph, then no two points on the graph have the same y-coordinate. In other words, there is no output of f that is produced by two different inputs. So, such a function f must have an inverse.

Review Problems

17.15   Part (a) is a horizontal line, and (b) is a linear equation. Part (c) is a quadratic. For parts (d) and (e), we pick different values of x to get a feel for the shape of the graph, then plot them as shown. For part (f), we can write $f(x) = 2g(x + 3) = 2[-(x + 3)^2] = -2(x + 3)^2$, then graph the resulting parabola. Or, we could start from the graph of $y = -x^2$, then shift it to the left 3 units, then stretch it vertically by a factor of 2. The graphs of the six functions are shown on the next page.

(a)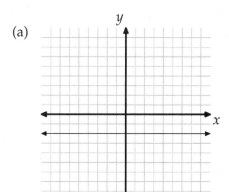

(d)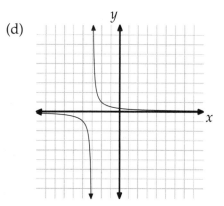

(b)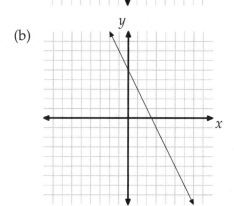

(e)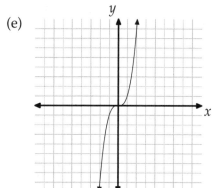

(c)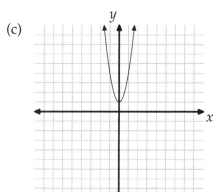

(f)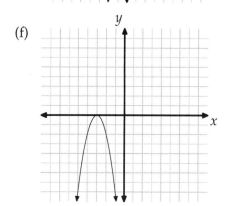

17.16

(a) No vertical line intersects the graph in more than one point, so the graph is the graph of a function. Similarly, no horizontal line intersects the graph in more than one point, so the function has an inverse.

(b) Several vertical lines intersect the graph in more than one point. (For example, consider the line $x = 0$.) So, this is not the graph of a function.

(c) No vertical line intersects the graph in more than one point, so the graph is the graph of a function. However, there are horizontal lines that intersect the graph in more than one point, so this function does not have an inverse.

(d) No vertical line intersects the graph in more than one point, so the graph is the graph of a function. The horizontal line $y = -1$ intersects the graph at two points, so this function does not have an inverse.

17.17

(a) The point $(2,3)$ is on the graph, so $f(2) = \boxed{3}$.

(b) The point $(5,4)$ is on the graph, so $f(5) = \boxed{4}$.

(c) The point $(-3,-5)$ is on the graph, so $f(-3) = -5$. Therefore, $f(f(-3)) = f(-5)$. The point $(-5,-1)$ is on the graph, so $f(-5) = -1$. So, we have $f(f(-3)) = f(-5) = \boxed{-1}$.

(d) The point $(5,4)$ is on the graph, so $f(5) = 4$. There are no other points on the graph with a y-coordinate of 4, so $\boxed{5}$ is the only value of a such that $f(a) = 4$.

(e) First, we find the values of c such that $(c,-1)$ is on the graph of f. In our graph, we see that $(-5,-1)$ and $(-1,-1)$ are the only points on the graph with y-coordinate equal to -1. Therefore, if $f(c) = -1$, then c is -5 or -1.

So, if $f(f(f(b))) = -1$, then $f(f(b))$ equals -1 or -5. We'll consider each possibility separately:

- *Case 1:* $f(f(b)) = -5$. The only point on the graph with y-coordinate -5 is $(-3,-5)$. Therefore, if $f(f(b)) = -5$, then $f(b) = -3$. There are two points on the graph with y-coordinate equal to -3, namely, $(-2,-3)$ and $(-4,-3)$. So, if $f(b) = -3$, then b equals -2 or -4.

- *Case 2:* $f(f(b)) = -1$. As we saw above, if $f(f(b)) = -1$, then $f(b)$ equals -1 or -5. As we saw in Case 1, if $f(b) = -5$, then $b = -3$. If $f(b) = -1$, then b is -1 or -5.

Putting Case 1 and Case 2 together, we see that $f(f(f(b))) = -1$ if b is $\boxed{-1, -2, -3, -4, \text{ or } -5}$. Just to make sure, we test each:

$$f(f(f(-1))) = f(f(-1)) = f(-1) = -1,$$
$$f(f(f(-2))) = f(f(-3)) = f(-5) = -1,$$
$$f(f(f(-3))) = f(f(-5)) = f(-1) = -1,$$
$$f(f(f(-4))) = f(f(-3)) = f(-5) = -1,$$
$$f(f(f(-5))) = f(f(-1)) = f(-1) = -1.$$

17.18

(a) The graph of $y = f(x)$ is the line below. The graph of $y = g(x)$ is the parabola below.

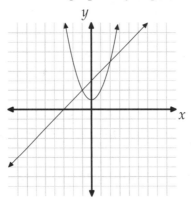

(b) Rearranging the equation $x + 3 = x^2 + 1$ gives $x^2 - x - 2 = 0$. Factoring gives $(x - 2)(x + 1) = 0$, so the solutions to this equation are $\boxed{x = 2 \text{ and } x = -1}$. This equation is equivalent to the equation $f(x) = g(x)$, where f and g are the functions graphed in part (a). We see that these two graphs intersect at $(2, 5)$, so $f(2) = g(2)$. Similarly, the two graphs meet at $(-1, 2)$, so $f(-1) = g(-1)$. In other words, the x-coordinates of the intersection points of the two graphs tell us the values of x for which $f(x) = g(x)$.

(c) We graph $f(x) = 1/x$ and $g(x) = x^2 - 7$ on the graph below. The graph of g is the parabola, and the other curves together are the graph of f.

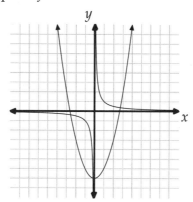

The graphs of f and g meet at 3 points, so there are $\boxed{3}$ values of x for which $\frac{1}{x} = x^2 - 7$.

17.19

(a) $\boxed{\text{False}}$. It is possible that the graph is the graph of a function, but there exist vertical lines that pass through 0 points on the graph.

(b) $\boxed{\text{True}}$, as explained in the text.

(c) $\boxed{\text{False}}$. Just because no horizontal line passes through more than one point on the graph does not mean the graph is the graph of a function. However, if we already know the graph is the graph of a function, then...

(d) $\boxed{\text{True}}$. As explained in the text, if the graph of a function is such that no horizontal line passes through more than one point on the graph, then the function has an inverse.

(e) $\boxed{\text{True}}$. As explained in the text, the graph of $y = f(x + 2)$ is the result of a leftward shift of the graph of $y = f(x)$.

(f) $\boxed{\text{False}}$. The graph of $y = f(3x)$ is the result of scaling the graph *horizontally* by a factor of $1/3$, not a result of stretching the graph vertically.

17.20 The graph of $y = g(2x)$ is a horizontal scaling of the graph of $y = g(x)$ by a factor of $1/2$. As discussed in the text, the graph of $y = g(-x)$ is the result of reflecting the graph of $y = g(x)$ over the y-axis. Note that this just gives us the same graph as $y = g(x)$! The graph of $y = -g(2x)$ is the result of reflecting the graph of $y = g(2x)$ over the x-axis. Finally, the graph of $y = g(x)/3$ is vertical scaling of the graph of $y = g(x)$ by a factor of $1/3$. The four graphs are shown below.

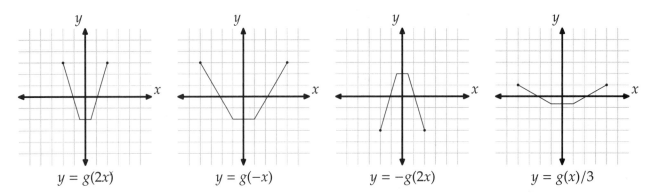

$$y = g(2x) \qquad y = g(-x) \qquad y = -g(2x) \qquad y = g(x)/3$$

17.21 The graph of $y = h(x) + 4$ is a 4-unit upward shift of the graph of $y = h(x)$. The graph of $y = h(x+4)$ is a 4-unit leftward shift of the graph of $y = h(x)$. The graph of $y = h(x - 2)$ is a 2-unit rightward shift of the graph of h, and the graph of $y = h(x - 2) + 3$ is a 3-unit upward shift of the graph of $y = h(x - 2)$. Therefore, the graph of $y = h(x - 2) + 3$ is the result of shifting the graph of $y = h(x)$ to the right 2 units and upward 3 units. Our three graphs are shown below.

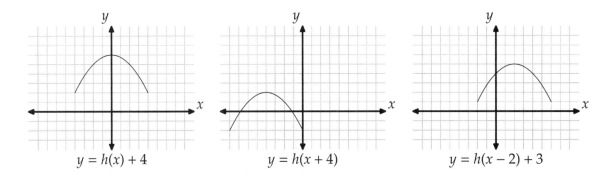

$$y = h(x) + 4 \qquad y = h(x + 4) \qquad y = h(x - 2) + 3$$

17.22 Because $(2, 3)$ is on the graph of $y = f(x)$, we have $f(2) = 3$. In each of the following parts, we find the value of x that makes the input to f equal to 2.

(a) When $x = 1$, we have $y = f(2 \cdot 1) = f(2) = 3$. So, $\boxed{(1, 3)}$ is on the graph of $y = f(2x)$.

(b) When $x = -1$, we have $y = f(-1 + 3) = f(2) = 3$, so $\boxed{(-1, 3)}$ is on the graph of $y = f(x + 3)$. (We also might note that this graph is a 3-unit leftward shift of the graph of $y = f(x)$, thus taking $(2, 3)$ to $(-1, 3)$.)

(c) When $x = 2$, we have $y = f(2) - 5 = -2$, so $\boxed{(2, -2)}$ is on the graph of $y = f(x) - 5$. We also might note that this graph is a 5-unit downward shift of the graph of $y = f(x)$, thus taking $(2, 3)$ to $(2, -2)$.

(d) We know the output when 2 is inputted to f. So, we first find the value of x such that $2x + 3 = 2$. Solving this equation gives $x = -1/2$. So, when $x = -1/2$, we have $y = f(2x+3) - 5 = f(2) - 5 = -2$. Therefore, $\boxed{(-1/2, -2)}$ is on the graph of $y = f(2x + 3) - 5$.

17.23

(a) $\boxed{\text{False}}$. The graph of $y = f(2x)$ is a horizontal scaling of the graph of $y = f(x)$ by a factor of $1/2$. So, each x-intercept of the graph of $y = f(x)$ is twice as far from the origin as the corresponding x-intercept of $y = f(2x)$.

(b) $\boxed{\text{False}}$. The graph of $y = f(x + 2)$ is a leftward shift of the graph of $y = f(x)$. Therefore, each x-intercept of $y = f(x + 2)$ is 2 units to the left of the corresponding x-intercept of $y = f(x)$.

(c) $\boxed{\text{True}}$. Any x-intercept of a graph has y-coordinate equal to 0. Therefore, the x-intercepts of $y = f(x)$ have x-coordinates that are solutions to the equation $f(x) = 0$. Similarly, the x-intercepts of $y = 3f(x)$ have x-coordinates that are solutions to the equation $3f(x) = 0$. Dividing this equation by 3 gives $f(x) = 0$, so the x-intercepts of $y = f(x)$ and $y = 3f(x)$ are the same.

   We also can note that the graph of $y = 3f(x)$ is a vertical scaling of $y = f(x)$ by a factor of 3. A vertical scaling of a point on the x-axis doesn't move the point at all. So, the x-intercepts of $y = 3f(x)$ are the same as the x-intercepts of $y = f(x)$.

(d) $\boxed{\text{False}}$. The graph of $y = f(x) - 3$ is a downward shift of the graph of $y = f(x)$. Suppose $(a, 0)$ is an x-intercept of $y = f(x)$, so that $f(a) = 0$. Then, when $x = a$, we have $y = f(x) - 3 = f(a) - 3 = -3$. Therefore, $(a, -3)$ is on the graph of $y = f(x) - 3$, so $(a, 0)$ is not on the graph of $y = f(x) - 3$. This means $(a, 0)$ is not an x-intercept of the graph of $y = f(x) - 3$.

(e) $\boxed{\text{True}}$. Suppose $f(x) = 0$. Then, the graphs of $y = f(x)$ and $y = f(x - 3)$ have the same x-intercepts. This isn't the only possible function that has this property. Suppose that f is a function such that $f(x) = 0$ for all values of x that are multiples of 3, but $f(x) \neq 0$ for all other values of x. Since the graph of $y = f(x - 3)$ is a rightward shift of the graph of $y = f(x)$, we have $f(x - 3) = 0$ for all values of x that are multiples of 3 (but $f(x - 3) \neq 0$ for all other values of x). So the graphs of $y = f(x)$ and $y = f(x - 3)$ have the same x-intercepts.

17.24   The graph of $\boxed{y = f(2x)}$ is a horizontal scaling of the graph of $y = f(x)$ by a factor of $1/2$. So, if the point $(0, a)$ is on the graph of $y = f(x)$, it is also on the graph of $y = f(2x)$. So, the graphs of $y = f(x)$ and $y = f(2x)$ must have the same y-intercept.

   The graph of $y = 2f(x)$ is a vertical scaling of $y = f(x)$ by a factor of 2. So, the y-intercept of $y = 2f(x)$ is twice as far from the x-axis as the y-intercept of $y = f(x)$. Therefore, these graphs only have the same y-intercept if the graph of $y = f(x)$ passes through the origin.

   The graph of $y = f(x + 2)$ is a 2-unit leftward shift of the graph of $y = f(x)$. So, the y-intercept of $y = f(x + 2)$ depends on the value of $f(2)$, which we know nothing about. Therefore, the graphs of $y = f(x)$ and $y = f(x + 2)$ do not necessarily have the same y-intercept.

   The graph of $y = f(x) + 2$ is a 2-unit vertical shift of the graph of $y = f(x)$. So, the y-intercept of $y = f(x) + 2$ is 2 units above the y-intercept of $y = f(x)$. Therefore, they cannot be the same point.

17.25

(a) If $f(2) = 5$, then $(2, 5)$ is on the graph of $y = f(x)$. Such a graph is at left below.

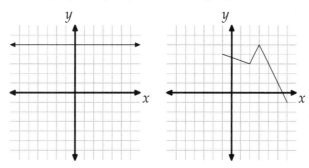

(b) Suppose $f(2) = a$. If $f(f(2)) = 5$, then we must have $f(a) = 5$. Therefore, the graph of $y = f(x)$ must pass through $(2, a)$ and $(a, 5)$. We can pick any value of a except $a = 2$ and $a = 5$. (If $a = 2$, then the graph would pass through $(2, 2)$ and $(2, 5)$, which the graph of a function cannot do. We are also told that $f(2) \neq 5$, so we cannot have $a = 5$.) A sample graph is shown above in which $a = 3$. The graph passes through $(2, 3)$ and $(3, 5)$, so $f(f(2)) = f(3) = 5$, as desired.

17.26

(a) The value of x such that $x = f^{-1}(3)$ is the solution to $f(x) = 3$. Since $(4, 3)$ is on the graph of $y = f(x)$, we have $f(4) = 3$. So, $f^{-1}(3) = \boxed{4}$.

(b) Since $(-3, -5)$ is on the graph of $y = f(x)$, we have $f(-3) = -5$. Therefore, $f^{-1}(-5) = \boxed{-3}$.

(c) If $f^{-1}(a) = 5$, we have $f(f^{-1}(a)) = f(5)$. Since f^{-1} is the inverse of f, we have $f(f^{-1}(a)) = a$. Therefore, we have $a = f(5) = \boxed{4}$.

17.27 The graph of $y = f(x)$ is a parabola that opens upward or downward. In either case, there exists a horizontal line that passes through more than one point of the graph. Therefore, the function $\boxed{\text{does not}}$ have an inverse.

Challenge Problems

17.28 $\boxed{\text{Yes}}$. Any function of the form $f(x) = ax$, where a is a constant, satisfies $3f(x) = f(3x)$.

17.29 The x-intercepts are the points on the graph where $y = 0$. Solving $(x - 3)^2(x + 2) = 0$ gives $x = 3$ and $x = -2$ as solutions. So, the x-intercepts of the graph are $(3, 0)$ and $(-2, 0)$. The y-intercept is the point where $x = 0$. When $x = 0$, we have $y = (0 - 3)^2(0 + 2) = 18$. Therefore, the y-intercept is $(0, 18)$. Because the y-intercept is 18 units above the x-axis and our two x-intercepts are both on the x-axis, the height of our triangle is 18 units. The two x-intercepts are 5 units apart, so the length of the base of the triangle is 5.

The area of a triangle is half the product of a side of the triangle and the height of the triangle drawn to that side. So, the area of our triangle in this problem is $(5)(18)/2 = \boxed{45}$ square units.

17.30 The negative before the x makes this problem considerably trickier than other function transformations we have studied. We know how the graph of $y = f(2 + x)$ is related to the graph of $y = f(x)$; it is a 2-unit leftward shift of the graph of $y = f(x)$. However, what about the graph of $y = f(2 - x)$? While it may not be clear how to relate this graph to the graph of $y = f(x)$, the only difference between $y = f(2 + x)$ and $y = f(2 - x)$ is that we have replaced x with $-x$. In other words, the graph of $y = f(2 - x)$ is the graph of $y = f(2 + x)$ reflected over the y-axis. Putting our two steps together, we see that we get the graph of $y = f(2 - x)$ by shifting the graph of $y = f(x)$ to the left 2 units, and then reflecting the result over the y-axis.

17.31

(a) We can show f has an inverse by considering the result of reflecting the graph of f over the line $y = x$. Because f is a function, every vertical line intersects $y = f(x)$ in at most one point. When we reflect a vertical line over the line $y = x$, we get a horizontal line. Since each vertical line meets at most one point on $y = f(x)$, we know that each horizontal line meets at most one point on the reflection of $y = f(x)$ over $y = x$. Since the reflection of the graph of $y = f(x)$ over $y = x$ is again the graph of $y = f(x)$, we therefore know that each horizontal line meets at most one point on the graph of $y = f(x)$. So, f has an inverse.

(b) $\boxed{\text{Yes}}$. We found in the text that if a function has an inverse, then the reflection of its graph over the line $y = x$ is the graph of the inverse of the function. Here, the reflection of the graph of $y = f(x)$ over $y = x$ is the graph of $y = f(x)$. Therefore, f is its own inverse! Because f is its own inverse, we must have $f(f(x)) = x$ for all x in the domain of f.

17.32 We saw in an earlier problem that if a function and its inverse are the same function, then the graph of the function is its own image when reflected over the line $x = y$. Suppose the point (r, s) is on the graph of $y = mx + b$. As described in the text, when we reflect (r, s) over the graph of $y = x$, the result is (s, r). If both of these points are on the graph of $y = mx + b$, then there are two possibilities.

First, the two points (r, s) and (s, r) could be the same point, in which case $r = s$, so the graph of $y = mx + b$ is the same as the graph of $y = x$, giving $m = 1$ as one possibility.

Second, we consider the possibility that (r, s) and (s, r) are different points on the graph of $y = mx + b$. Then, the slope between the two points (r, s) and (s, r) must be m, which gives us $m = (r - s)/(s - r) = -1$. A line with slope -1 is perpendicular to the graph of $y = x$, so it is its own image when we reflect it over the graph of $y = x$, as desired. Therefore, the possible values of m are $\boxed{-1 \text{ and } 1}$.

17.33 We don't have tools to find all the exact solutions to this equation, but we can graph $y = 2^x$ and $y = x^3 + 1$. For each value of x such that $2^x = x^3 + 1$ intersect, these two graphs will intersect at a point with that value of x as the x-coordinate. If we graph both of these for small values of x, we get graphs like those shown at right. It looks like the graphs intersect in exactly $\boxed{3}$ points, so Jerry writes down 3 as the answer to the question on his test. However, if $x = 9$, we have $2^x = 512$ and $x^3 + 1 = 730$, so $2^x < x^3 + 1$ for $x = 9$. But when $x = 10$, we have $2^x = 1024$ and $x^3 + 1 = 1001$, which means $2^x > x^3 + 1$ for $x = 10$. Since $2^x < x^3 + 1$ for $x = 9$ but $2^x > x^3 + 1$ for $x = 10$, we know that $2^x = x^3 + 1$ for some value of x between $x = 9$ and $x = 10$. Therefore, there is a fourth solution to the equation $2^x = x^3 + 1$. Be very careful when using a graph to answer a question!

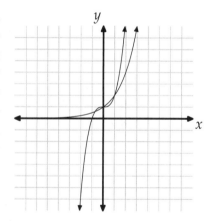

17.34 We are told that $f(x) = 8$ for $0 \le x < 3$, so that part of the graph of $y = f(x)$ is easy to produce. Because $f(x) = 2f(x - 3)$, when x is such that $3 \le x < 6$, the value of $f(x)$ is twice the value of $f(x)$ when x is between 0 and 3. For example, $f(5) = 2f(5 - 3) = 2f(2) = 2 \cdot 8 = 16$. Therefore, $f(x) = 16$ for $3 \le x < 6$.

Similarly, when we go the other direction, we see that for $-3 \le x < 0$, we have $f(x) = 4$. For example, we have $f(2) = 2f(2 - 3) = 2f(-1)$. Since $f(2) = 8$, we have $2f(-1) = 8$, or $f(-1) = 4$. Similarly, for $-6 \le x < -3$, the value of $f(x)$ is half that of $f(x)$ when $-3 \le x < 0$. So, we have $f(x) = 2$ for $-6 \le x < -3$. Similarly, for $-9 \le x < -6$, we have $f(x) = 1$ and for $-12 \le x < -9$, we have $f(x) = 1/2$. The resulting graph is shown at right.

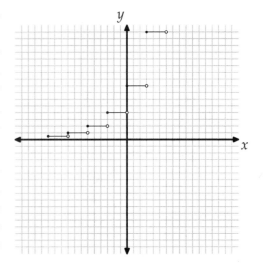

17.35 We let $g(x) = x^2 + bx + ac$. The roots of $f(x)$ are the x-coordinates of the x-intercepts of the graph of $y = f(x)$. Similarly, the roots of $g(x)$ are the x-coordinates of the x-intercepts of the graph of $y = g(x)$. Therefore, if we can find a way to relate the graphs of f and g, we may be able to find a way to relate the roots of $f(x)$ and $g(x)$.

Note that $g(ax) = a^2x^2 + bax + ac = a(ax^2 + bx + c)$. Therefore, we see that $g(ax) = af(x)$. So, $f(x) = g(ax)/a$. So, the graph of $y = f(x)$ is the graph of $y = g(ax)/a$. The graph of $y = g(ax)$ is a horizontal scaling of the graph of $y = g(x)$ by a factor of $1/a$. This moves the x-intercepts of the graph closer to the origin by a factor of $1/a$. In other words, the roots of $g(ax)$ are $1/a$ times the roots of $g(x)$. The graph of $y = g(ax)/a$ is a vertical scaling of the graph of $y = g(ax)$, and this leaves the x-intercepts unchanged. So, the x-coordinates of the x-intercepts of the graph of $y = f(x) = g(ax)/a$ are $1/a$ times the x-coordinates of the graph of $y = g(x)$. Therefore, the roots of $f(x)$ are $1/a$ times the roots of $g(x)$. This means that the roots of $g(x)$ are a times the roots of $f(x)$, as desired.

CHAPTER 18

Polynomials

Exercises for Section 18.1

18.1.1

(a) $f(1) = 1^4 - 3 \cdot 1^3 + 1 - 3 = 1 - 3 + 1 - 3 = \boxed{-4}$. $g(3) = 3^3 + 7 \cdot 3^2 - 2 = \boxed{88}$.

(b) $f(y) + g(y) = y^4 - 3y^3 + y - 3 + y^3 + 7y^2 - 2 = \boxed{y^4 - 2y^3 + 7y^2 + y - 5}$.

(c) Since $f(y) + h(y) = g(y)$, we can subtract $f(y)$ from both sides to have $h(y) = g(y) - f(y)$. So, we have
$$h(y) = y^3 + 7y^2 - 2 - (y^4 - 3y^3 + y - 3) = \boxed{-y^4 + 4y^3 + 7y^2 - y + 1}.$$

18.1.2 $\boxed{\text{No}}$. If two monic polynomials have the same degree, then their sum is not monic. For example, if $f(x) = x^2 + 1$ and $g(x) = x^2 - 3x$, then we have $f(x) + g(x) = 2x^2 - 3x + 1$, which is not monic. However, if two monic polynomials have different degrees, then their sum is monic, since there is only one term of the highest degree in the sum, and we know its coefficient is 1 because it is the leading coefficient of one of the monic polynomials.

18.1.3 If we get rid of all the terms of $f(x)$ that have an x in them, we are left with the constant term of $f(x)$. We can get rid of all these terms by letting $x = 0$. So, $f(0)$ equals the constant term of $f(x)$, because all the other terms equal 0 when $x = \boxed{0}$.

18.1.4 $\boxed{\text{Yes}}$. When we multiply a polynomial by a constant, we multiply each coefficient by that constant. The result is another polynomial. We can also see this by considering the general form of a polynomial:
$$f(x) = a_n x^n + a_{n-1} x^{n-1} + a_{n-2} x^{n-2} + \cdots + a_1 x + a_0.$$

Multiplying this by a constant, c, gives
$$cf(x) = c(a_n x^n + a_{n-1} x^{n-1} + a_{n-2} x^{n-2} + \cdots + a_1 x + a_0) = ca_n x^n + ca_{n-1} x^{n-1} + ca_{n-2} x^{n-2} + \cdots + ca_1 x + ca_0,$$

which is also a polynomial.

18.1.5 $\boxed{\text{Yes}}$. Suppose f has a greater degree than g, and that $\deg f = n$, so that the leading term of $f(x)$ is $a_n x^n$ for some constant a_n. Because the degree of f is greater than the degree of g, the term with the highest power of x in the sum $f(x) + g(x)$ is also $a_n x^n$. Therefore, $\deg(f + g) = \deg f$. Similarly, if the degree of g is greater than the degree of f, then the degree of the sum of f and g equals the degree of g.

If the degrees of f and g are the same, however, then it is possible for the degree of the sum $f + g$ to be less than the degrees of f and g. This occurs when the leading coefficients of f and g are opposites. For example, if $f(x) = x^2 + x - 3$ and $g(x) = -x^2 + 2x - 7$, then $f(x) + g(x) = 3x - 10$, so $\deg f = \deg g = 2$, but $\deg(f + g) = 1$. So, the only possible way for the degree of $f + g$ to be less than both $\deg f$ and $\deg g$ occurs when f and g have the same degree. Therefore, if $\deg(f + g)$ is less than both $\deg f$ and $\deg g$, we know that the degrees of f and g are the same.

Exercises for Section 18.2

18.2.1

$$
\begin{array}{rrrrr}
 & z^3 & & +4z & -2 \\
\times & & z^2 & -3z & +2 \\
\hline
 & +2z^3 & & +8z & -4 \\
-3z^4 & & -12z^2 & +6z & \\
+ \; z^5 & & +4z^3 & -2z^2 & \\
\hline
z^5 & -3z^4 & +6z^3 & -14z^2 & +14z & -4
\end{array}
$$

18.2.2

$$
\begin{array}{rrrrrrrr}
 & & 3x^5 & -3x^4 & & & & +1 \\
\times & & & & x^3 & -2x^2 & +x & +6 \\
\hline
 & & +18x^5 & -18x^4 & & & & +6 \\
 & +3x^6 & -3x^5 & & & & +x & \\
-6x^7 & +6x^6 & & & & -2x^2 & & \\
+ \; 3x^8 & -3x^7 & & & +x^3 & & & \\
\hline
3x^8 & -9x^7 & +9x^6 & +15x^5 & -18x^4 & +x^3 & -2x^2 & +x & +6
\end{array}
$$

18.2.3

$$
\begin{array}{rrrrr}
 & x^3 & +5x^2 & -7x & +9 \\
\times & & x^2 & +8x & +5 \\
\hline
 & +5x^3 & +25x^2 & -35x & +45 \\
+8x^4 & +40x^3 & -56x^2 & +72x & \\
+ \; x^5 & +5x^4 & -7x^3 & +9x^2 & \\
\hline
x^5 & +13x^4 & +38x^3 & -22x^2 & +37x & +45
\end{array}
$$

Yep, that matches the polynomial given in the problem, so our answer must be right.

18.2.4 We look at the coefficient of x in the expansion of the product on the left. We get an x term when we multiply $(+4)(+ax)$ and when we multiply $(-3x)(+7)$ in the expansion. So, on the left the x term is $4ax - 21x$. Since this term must equal $-41x$, we have $4ax - 21x = -41x$, so $a = \boxed{-5}$.

We can check our answer (and check that it is indeed possible to find a solution to this problem) by multiplying out the left when $a = -5$:

$$(x^2 - 3x + 4)(2x^2 - 5x + 7) = x^2(2x^2 - 5x + 7) - 3x(2x^2 - 5x + 7) + 4(2x^2 - 5x + 7)$$
$$= 2x^4 - 11x^3 + 30x^2 - 41x + 28.$$

This matches the polynomial given in the problem, so our answer is correct.

18.2.5

(a) Since the product of f and a polynomial with degree 1 equals a polynomial with degree 4, we know that f is a polynomial with degree $4 - 1 = \boxed{3}$.

(b) The product of the leading term of $x - 1$ and the leading term of $f(x)$ equals the leading term of the polynomial on the right, namely, $3x^4$. Since $(x)(3x^3) = 3x^4$, we see that the leading term of $f(x)$ is $\boxed{3x^3}$.

(c) The product of the constant term of $(x - 1)$ and the constant term of $f(x)$ is the constant term of the right side, -17. Since $(-1)(17) = -17$, we see that the constant term of $f(x)$ is $\boxed{17}$.

(d) From the first three parts, we see that $f(x) = 3x^3 + ax^2 + bx + 17$, for some constants a and b. So, we have

$$(x - 1)(3x^3 + ax^2 + bx + 17) = 3x^4 + x^3 - 25x^2 + 38x - 17.$$

When we expand the left side, the x term is $(-1)(bx) + (x)(17) = -bx + 17x$. This must equal $38x$, so we have $-bx + 17x = 38x$, so $-b + 17 = 38$. Therefore, $b = -21$ and we have

$$(x - 1)(3x^3 + ax^2 - 21x + 17) = 3x^4 + x^3 - 25x^2 + 38x - 17.$$

When we expand the left side, the x^3 term is $(-1)(3x^3) + (x)(ax^2) = -3x^3 + ax^3$. This must equal the x^3 term on the right, so we have $-3 + a = 1$, or $a = 4$. So, we have $f(x) = \boxed{3x^3 + 4x^2 - 21x + 17}$. We can check this answer by multiplying out $(x - 1)f(x)$:

$$(x - 1)(3x^3 + 4x^2 - 21x + 17) = x(3x^3 + 4x^2 - 21x + 17) - 1(3x^3 + 4x^2 - 21x + 17)$$
$$= 3x^4 + 4x^3 - 21x^2 + 17x - 3x^3 - 4x^2 + 21x - 17$$
$$= 3x^4 + x^3 - 25x^2 + 38x - 17.$$

This matches the polynomial on the right in the problem, so our answer is correct.

IMPORTANT NOTE: We can't always find a polynomial as we've found $f(x)$ in this problem. For example, try finding a polynomial $g(x)$ such that $(x + 1)g(x) = 3x^4 + x^3 - 25x^2 + 38x - 17$. It's impossible, just like finding an integer such that $3n = 14$ is impossible. We'll explore this in much more detail in *Intermediate Algebra*.

Review Problems

18.8

(a) We have

$$p(x) + 3q(x) = 2x - 5 + 3(3x^2 + 7x - 4) = 2x - 5 + 9x^2 + 21x - 12 = \boxed{9x^2 + 23x - 17}.$$

(b) $\boxed{\text{No}}$. Because b is nonzero, the polynomial $b \cdot q(x)$ must have an x^2 term. Meanwhile, because a is a constant and p has degree 1, the polynomial $a \cdot p(x)$ has degree 1, so its highest power of x is 1. Therefore, the sum $a \cdot p(x) + b \cdot q(x)$ must have an x^2 term.

(c) We have

$$(2x - 5)(3x^2 + 7x - 4) = 2x(3x^2 + 7x - 4) - 5(3x^2 + 7x - 4)$$
$$= 6x^3 + 14x^2 - 8x - 15x^2 - 35x + 20 = \boxed{6x^3 - x^2 - 43x + 20}.$$

18.9 We have $f(3) = 3^4 - 2 \cdot 3^3 + a \cdot 3^2 + 3 + 3 = 9a + 33$, so we must have $9a + 33 = 2$. Therefore, $a = \boxed{-31/9}$.

18.10

(a) $f(x) + g(x) = x^3 + 3x^2 - 2x + 7 + 4x^3 + 17x^2 + 2x - 9 = \boxed{5x^3 + 20x^2 - 2}$.

(b) In order for $f(x) + c \cdot g(x)$ to have degree 2, the x^3 terms must cancel. The x^3 term of $f(x)$ is just x^3. The x^3 term of $c \cdot g(x)$ is $4cx^3$. Therefore, we must have $x^3 + 4cx^3 = 0$, from which we find $c = \boxed{-1/4}$.

(c)

				x^3	$+3x^2$	$-2x$	$+7$
$\times$				$4x^3$	$+17x^2$	$+2x$	-9
				$-9x^3$	$-27x^2$	$+18x$	-63
			$2x^4$	$+6x^3$	$-4x^2$	$+14x$	
		$17x^5$	$+51x^4$	$-34x^3$	$+119x^2$		
$+$	$4x^6$	$+12x^5$	$-8x^4$	$+28x^3$			
	$4x^6$	$+29x^5$	$+45x^4$	$-9x^3$	$+88x^2$	$+32x$	-63

18.11 Because $h(x) = f(x) \cdot g(x)$, the constant term of h equals the product of the constant terms of $f(x)$ and $g(x)$. We are told that the constant terms of $h(x)$ and $f(x)$ are 3 and -4, respectively. Let the constant term of $g(x)$ be c. When we evaluate $g(0)$, all terms with x in them equal 0, so we are left with the constant term, c. Therefore, $g(0) = c$. So, we must have $3 = (-4) \cdot c$, from which we find $c = \boxed{-3/4}$.

18.12 If the degrees of f and g are the same, then the sum $f(x) + g(x)$ will combine the leading terms of the two polynomials, giving the leading term a coefficient of 2, not 1. Therefore, if the degrees of f and g are the same, then the sum $f(x) + g(x)$ is not monic. We found in an earlier exercise (18.1.2) that if the degrees of monic polynomials $f(x)$ and $g(x)$ are different, then the sum $f(x) + g(x)$ is a monic polynomial. Therefore, if $f(x)$, $g(x)$, and $f(x) + g(x)$ are all monic, then we know that the degrees of f and g are different.

18.13

(a) $g(2) = 2^4 - 3 \cdot 2^2 + 9 = \boxed{13}$ and $g(-2) = (-2)^4 - 3 \cdot (-2)^2 + 9 = \boxed{13}$. Similarly, $g(5) = 5^4 - 3 \cdot 5^2 + 9 = \boxed{559}$ and $g(-5) = (-5)^4 - 3 \cdot (-5)^2 + 9 = \boxed{559}$.

(b) $\boxed{\text{Yes}}$. Because $g(-x) = (-x)^4 - 3(-x)^2 + 9 = x^4 - 3x^2 + 9 = g(x)$, we always have $g(x) = g(-x)$.

(c) If $f(x)$ and $f(-x)$ are the same polynomial, then for each term in $f(x)$, the same term also exists in $f(-x)$. For each term of $f(x)$ with an even power of x, that term is the same in $f(x)$ as it is in $f(-x)$. For example $4x^4 = 4(-x)^4$. However, if $f(x)$ has a term with an odd power, then the negative of that term appears in $f(-x)$. For example if $f(x)$ has the term $6x^3$, then $f(-x)$ will have the term $6(-x)^3 = -6x^3$. Therefore, if $f(x)$ and $f(-x)$ are the same, then $f(x)$ has no terms with an odd power of x. Unsurprisingly, we call such a function an even function.

(d) As we saw in the previous part, if $f(x)$ has a term with an even power of x, then that term will also exist in $f(-x)$. Moreover, if $f(x)$ has a term with an odd power of x, then the opposite of that term will exist in $f(-x)$. Therefore, if $f(x) = -f(-x)$, we know that $f(x)$ has no terms with an even power of x. We call such a function an odd function.

18.14 $\boxed{\text{Yes}}$. Because f is a polynomial, each term of $f(x)$ is of the form ax^n for some constant a and nonnegative integer n. In evaluating $f(x^2)$, we substitute x^2 for x in $f(x)$. Therefore, each of these terms of the form ax^n becomes a term of the form $a(x^2)^n = ax^{2n}$. In other words, $f(x^2)$ consists of a sum of terms that are each a constant times an integer power of x. So, $f(x^2)$ is indeed a polynomial.

18.15 $\boxed{\text{Yes}}$. Because $h = f \cdot g$, we must have $h(1) = f(1) \cdot g(1)$. Since $f(1) = 0$, we have $h(1) = 0 \cdot g(1) = 0$, no matter what $g(1)$ is.

18.16

(a) Because the constant terms of both polynomials in the product are positive, are the same, and multiply to 16, they must each equal $\sqrt{16} = \boxed{4}$.

(b) The coefficients of x in the two polynomials are the same and the polynomials are monic, so the product is of the form
$$(x^3 + ax^2 + cx + 4)(x^3 + bx^2 + cx + 4).$$

The x term of this product is $(cx)(4) + (4)(cx) = 8cx$. This must equal the x term of the polynomial given in the problem, so $8cx = 40x$, or $c = \boxed{5}$.

(c) Combining the first two parts, we know that the product is of the form
$$(x^3 + ax^2 + 5x + 4)(x^3 + bx^2 + 5x + 4).$$

The x^5 term in the expansion is $(x^3)(bx^2) + (ax^2)(x^3) = (a + b)x^5$. Comparing this to the given polynomial, we have $a + b = 2$. Similarly, the x^4 term in the expansion is
$$(x^3)(5x) + (ax^2)(bx^2) + (5x)(x^3) = (10 + ab)x^4,$$

so we must have $10 + ab = 2$, which gives $ab = -8$. Solving the system $a + b = 2$, $ab = -8$ tells us that $(a, b) = (4, -2)$ or $(a, b) = (-2, 4)$. In either case, the two polynomials are
$$\boxed{x^3 + 4x^2 + 5x + 4 \text{ and } x^3 - 2x^2 + 5x + 4}.$$

Notice that we can't tell who has which polynomial.

Challenge Problems

18.17

(a) We have $(x - p)(x - q) = x^2 - px - qx + pq$, so

$$(x - p)(x - q)(x - r) = (x^2 - px - qx + pq)(x - r)$$
$$= (x^2 - px - qx + pq)(x) + (x^2 - px - qx + pq)(-r)$$
$$= x^3 - px^2 - qx^2 + pqx - x^2r + prx + qrx - pqr$$
$$= x^3 + (-p - q - r)x^2 + (pq + qr + rp)x - pqr.$$

(b) The product $(x - p)(x - q)(x - r)$ equals 0 when any one of the factors equals zero. Therefore, the product equals 0 when $\boxed{x = p, x = q, \text{ or } x = r}$.

(c) We compare $a_3x^3 + a_2x^2 + a_1x + a_0$ to the expansion found in part (a):

$$x^3 + (-p - q - r)x + (pq + qr + rp)x - pqr.$$

We have

$$\boxed{a_3 = 1, \; a_2 = -p - q - r, \; a_1 = pq + qr + rp, \; a_0 = -pqr}.$$

(d) The coefficient of x^3 is 1, so this polynomial matches the form of $h(x)$ from the previous three parts. Therefore, from part (c), we see that the coefficient of x^2 is the opposite of the sum of the roots. So, the sum of the roots of our polynomial is $\boxed{-6}$. Similarly, the constant term is the opposite of the product of the roots, so the product of the roots is $\boxed{10}$.

Challenge: What if the coefficient of x^3 is not 1? What if the polynomial has a higher degree?

18.18 Since f is a polynomial, we have

$$f(x) = a_nx^n + a_{n-1}x^{n-1} + \cdots + a_1x + a_0.$$

We would like to find the value of k such that

$$f(k) = a_n + a_{n-1} + \cdots + a_1 + a_0.$$

In other words, we want the x's in $f(x)$ to disappear from each term. These x's are multiplied by the a's in each term, so to make the x's disappear and leave only the coefficients, we set each equal to $\boxed{1}$:

$$f(1) = a_n \cdot 1^n + a_{n-1} \cdot 1^{n-1} + \cdots + a_1 \cdot 1 + a_0 = a_n + a_{n-1} + \cdots + a_1 + a_0.$$

So, if f is a polynomial, then $f(1)$ equals the sum of its coefficients.

18.19 From our previous problem, we know that the sum of the coefficients of $G(y)$ equals $G(1)$. We can use our expression for $F(G(y))$ to write $F(G(1)) = 12 - 62 + 81 = 31$. If we let $G(1) = k$, then we have $F(k) = 31$. We also know that $F(k) = 3k^2 - k + 1$. So, we must have $3k^2 - k + 1 = 31$, so $3k^2 - k - 30 = 0$. Factoring gives $(3k - 10)(k + 3) = 0$, so our possible values of k are $\boxed{10/3 \text{ and } -3}$.

We can also find the two polynomials $G(y)$ that fit the problem statement. Because $F(y) = 3y^2 - y + 1$, the leading term of $F(G(y))$ is 3 times the square of the leading term of $G(y)$. Therefore, the leading term of $G(y)$ is either $2y^2$ or $-2y^2$. We know that $G(y)$ does not have a linear term, because if it did, then $F(G(y))$ would have a y^3 term. So, our two options for $G(y)$ are $G(y) = 2y^2 + c$ or $G(y) = -2y^2 + d$. We can find $c = -5$ and $d = 16/3$ by substituting our two expressions for $G(y)$ into our equation for $F(G(y))$. So, our two possible polynomials $G(y)$ are $G(y) = 2y^2 - 5$ and $G(y) = -2y^2 + 16/3$. The sum of the coefficients of these are $\boxed{-3 \text{ and } 10/3}$, as expected.

CHAPTER 19

Exponents and Logarithms

Exercises for Section 19.1

19.1.1 We can input any real number, so the domain is all real numbers. As output, 5^x can be any positive number, so $3 \cdot 5^x$ can be any positive number. Because $3 \cdot 5^x$ can be any value greater than 0, the expression $3 \cdot 5^x - 4$ can equal any number greater than -4. So, the range of f is all real numbers greater than -4.

19.1.2 We write both sides with the same base, 5. This gives us $5^{2r-3} = 5^2$. Since the bases of both sides are the same, the exponents must be equal. Therefore, we have $2r - 3 = 2$, so $r = \boxed{5/2}$.

19.1.3 We write both sides with the same base, 6. This gives us $6^{3t-1} = (6^2)^{t-3}$, so $6^{3t-1} = 6^{2t-6}$. Since the bases are the same, the exponents must be the same. Therefore, we have $3t - 1 = 2t - 6$, or $t = \boxed{-5}$.

19.1.4 Because we know what 5^x is, we write 5^{2x+3} in terms of 5^x:

$$5^{2x+3} = 5^{2x} \cdot 5^3 = (5^{2x})(125) = (5^x)^2(125).$$

Because $5^x = 3$, we have $(5^x)^2(125) = (3)^2(125) = \boxed{1125}$.

19.1.5 We write everything with the same base, 5, and simplify the right side. This gives us $25^{-2} = (5^2)^{-2} = 5^{-4}$ on the left, so our equation is

$$5^{-4} = \frac{5^{48/x}}{5^{26/x} \cdot 25^{17/x}} = \frac{5^{48/x}}{5^{26/x} \cdot (5^2)^{17/x}} = \frac{5^{48/x}}{5^{26/x} \cdot 5^{34/x}} = 5^{\frac{48}{x} - \frac{26}{x} - \frac{34}{x}}.$$

So, we have

$$5^{-4} = 5^{-\frac{12}{x}}.$$

The bases are the same, so the exponents must be the same. This gives $-4 = -12/x$, so $x = \boxed{3}$.

19.1.6 We start by writing the terms that have variables in their exponents with the same base, 3. Because $9^t = (3^2)^t = 3^{2t}$, this makes our equation

$$3 \cdot 3^{2t} - 82 \cdot 3^t + 27 = 0.$$

We get a better handle on this equation by making the substitution $x = 3^t$, so that $3^{2t} = (3^t)^2 = x^2$ and our equation is

$$3x^2 - 82x + 27 = 0.$$

Factoring gives us $(3x - 1)(x - 27) = 0$, so we must have $x = 1/3$ or $x = 27$. Because $x = 3^t$, we have $3^t = 1/3$ or $3^t = 27$. Therefore, our solutions are $\boxed{t = -1 \text{ and } t = 3}$.

19.1.7 We have $3^{2x+3} = (3^{2x})(3^3) = 27 \cdot 3^{2x}$. Therefore, we just have to find 3^{2x}. However, $3^{2x} = (3^2)^x = 9^x$, and we can find 9^x from our given equation. Because $9^{x-1} = 9^x/9$, we have $9^x/9 = 7$, so $9^x = 63$. Therefore, we have $3^{2x+3} = 27 \cdot 3^{2x} = 27 \cdot 9^x = 27 \cdot 63 = \boxed{1701}$.

Exercises for Section 19.2

19.2.1

(a) If the interest is simple interest, then she earns $0.1(\$10,0000) = \$1,000$ each year. Therefore, at the end of 5 years, she has earned $5(\$1,000) = \$5,000$. So, her investment is now worth $\$10,000 + \$5,000 = \boxed{\$15,000}$.

(b) At the end of the first year, Paula's investment is worth $\$10,000 + 0.10(\$10,000) = (1.1)(\$10,000)$. The interest in the second year is calculated based on this amount, so at the end of the second year, the investment is worth

$$(1.10)(\$10,000) + 0.10(1.10)(\$10,000) = (1.10)(1.10)(\$10,000) = (1.10)^2(\$10,000).$$

Continuing in this way, after 5 years, the investment is worth $(1.10)^5(\$10,000) = \boxed{\$16,105.10}$.

(c) In the first quarter, Paula earns $\frac{0.10}{4}(\$10,000)$ in interest, so her investment is worth $\$10,000 + \frac{0.10}{4}(\$10,000) = (1 + \frac{0.10}{4})(\$10,000)$. Similarly, the value of her investment is multiplied by $1 + \frac{0.10}{4}$ each quarter, so after 5 years, which is $5 \cdot 4 = 20$ quarters, her investment is worth

$$\left(1 + \frac{0.10}{4}\right)^{5 \cdot 4}(\$10,000) \approx \boxed{\$16,386.16}.$$

19.2.2 If the interest compounds quarterly, she owes $(1 + \frac{0.12}{4})^{4 \cdot 4}(\$6,000) \approx \$9,628.24$. If it compounds annually, she owes $(1 + 0.12)^4(\$6,000) \approx \$9,441.12$. Therefore, if the interest compounds quarterly, she owes $\$9,628.24 - \$9,441.12 = \boxed{\$187.12}$ more.

19.2.3 For each year that an amount of money is invested at an annually compounded interest rate of $r\%$, its value is multiplied by $1 + \frac{r}{100}$. So, after the first two years, Bill's investment is worth $(1.05)^2(\$2,500)$. For each of the next three years, the value of this investment is multiplied by 1.08, so at the end of the five years, it is worth $(1.08)^3(1.05)^2(\$2,500)$.

After the first three years of being invested at 8%, Debbie's investment is worth $(1.08)^3(\$2,500)$. During the next two years, the value of Debbie's investment is multiplied by 1.05 each year, to give her a total of $(1.05)^2(1.08)^3(\$2,500)$.

The expressions we have for Bill's investment after five years and Debbie's investment after five years consist of the product of the same factors, so the two investments are equally valuable after 5 years. This happens because both investments involve the initial investment being multiplied by 1.05 twice and by 1.08 three times.

19.2.4 After n years, Beth's investment is worth $(1.06)^n(\$4{,}200)$ and Josey's is worth $(1 + \frac{rn}{100})(\$4{,}200)$. Setting these equal gives us $1.06^n = 1 + \frac{rn}{100}$.

(a) We have $n = 1$, so $1.06 = 1 + \frac{r}{100}$, so $r = \boxed{6}$.

(b) We have $n = 2$, so $1.06^2 = 1 + \frac{2r}{100}$. Solving this linear equation gives $r = \boxed{6.18}$

(c) We have $n = 10$, so $1.06^{10} = 1 + \frac{10r}{100}$. Solving this linear equation gives $r \approx \boxed{7.91}$.

19.2.5 In the first year, the interest is $\frac{r}{100}(\$k)$, so at the end of the first year, the investment is worth $\$k + \frac{r}{100}(\$k) = \left(1 + \frac{r}{100}\right)(\$k)$. In the next year, the interest is earned on this amount, so the interest is

$$\frac{r}{100}\left(1 + \frac{r}{100}\right)(\$k).$$

Adding this to the amount we have at the beginning of the year, at the end of the second year the investment is worth

$$\left(1 + \frac{r}{100}\right)(\$k) + \frac{r}{100}\left(1 + \frac{r}{100}\right)(\$k) = \left(1 + \frac{r}{100}\right)\left(1 + \frac{r}{100}\right)(\$k) = \left(1 + \frac{r}{100}\right)^2(\$k).$$

Similarly, in the next year, the interest is earned on the amount from the end of the second year. So, the investment earns

$$\frac{r}{100}\left(1 + \frac{r}{100}\right)^2(\$k)$$

in interest. Adding this to the value of the investment at the beginning of the year gives us a total value at the end of year 3 of

$$\left(1 + \frac{r}{100}\right)^2(\$k) + \frac{r}{100}\left(1 + \frac{r}{100}\right)^2(\$k) = \left(1 + \frac{r}{100}\right)\left(1 + \frac{r}{100}\right)^2(\$k) = \left(1 + \frac{r}{100}\right)^3(\$k).$$

Continuing in this way, the value of the investment is multiplied by $1 + \frac{r}{100}$ each year, so after n years, it is worth $\left(1 + \frac{r}{100}\right)^n(\$k)$.

19.2.6 There are m compounding periods each year. The interest earned during the first period is $\frac{1}{m}$ of the interest that would have been earned the whole year. In the first year, the investment would have earned $\frac{r}{100}(\$k)$ in interest. So, in the first compounding period, the investment earns $\frac{1}{m}$ of this interest, or $\frac{r}{100m}(\$k)$. Adding this to the principal, at the end of the first compounding period the investment is worth

$$\$k + \frac{r}{100m}(\$k) = \left(1 + \frac{r}{100m}\right)(\$k).$$

In the next time period ($1/m$ of a year), the interest is earned on this amount, so the interest is

$$\frac{r}{100m}\left(1 + \frac{r}{100m}\right)(\$k),$$

where the m in the denominator comes from the fact that during this $1/m$ of a year, the investment earns $1/m$ of the interest it would earn in a full year. Adding this to the amount of the investment at the beginning of the time period, the investment is worth

$$\left(1 + \frac{r}{100m}\right)(\$k) + \frac{r}{100m}\left(1 + \frac{r}{100m}\right)(\$k) = \left(1 + \frac{r}{100m}\right)^2(\$k)$$

at the end of the time period. Similarly, in each time period, the value of the investment is multiplied by $1 + \frac{r}{100m}$. There are m time periods in each of the n years, so there are a total of mn time periods. Therefore, the value of the investment at the end of n years is

$$\left(1 + \frac{r}{100m}\right)^{mn} (\$k).$$

Exercises for Section 19.3

19.3.1 Suppose the interest rate is r. Because the interest compounds twice a year, at the end of 4 years, he owes $(1 + \frac{r}{2 \cdot 100})^{2 \cdot 4}(\$3,000)$. We are told that this equals \$3,950.43, so we have the equation

$$\left(1 + \frac{r}{2 \cdot 100}\right)^{2 \cdot 4} (\$3,000) = \$3,950.43.$$

Dividing both sides by \$3,000 gives

$$\left(1 + \frac{r}{2 \cdot 100}\right)^{8} = 1.31681.$$

Raising both sides to the 1/8 power to get rid of the exponent on the left side gives $1 + \frac{r}{200} = 1.035$. Therefore, $r/200 = 0.035$, and $r = 7$. So, the interest rate of the loan is $\boxed{7\%}$.

19.3.2 This question is equivalent to asking, "What is the present value of \$500,000 paid 10 years from now if the annually compounded interest rate is 5%?" This present value is

$$\frac{\$500,000}{(1 + 0.05)^{10}} \approx \boxed{\$306,956.63}.$$

19.3.3 To compare the four options, we find the present value of each one. The annually compounded interest rate is 8% in each case.

(a) The present value of \$100,000 paid 20 years from now is $\dfrac{\$100,000}{(1 + .08)^{20}} \approx \$21,454.82$.

(b) The present value of \$50,000 paid 10 years from now is $\dfrac{\$50,000}{(1 + .08)^{10}} \approx \$23,159.67$.

(c) The present value of \$30,000 paid 10 years from now is $\dfrac{\$30,000}{(1 + .08)^{10}} \approx \$13,895.80$. The present value of \$50,000 paid 20 years from now is $\dfrac{\$50,000}{(1 + .08)^{20}} \approx \$10,727.41$. Combining these, the present value of this option is \$13,895.80 + \$10,727.41 = \$24,623.21.

(d) Finally, an easy one. The present value of \$25,000 paid today is \$25,000.

The most valuable option is the one with the highest present value, which is option (d). We use these present values to order the options from most valuable to least: $\boxed{\text{(d), (c), (b), (a)}}$.

19.3.4 The value of Greta's investment at the end of the first three years is $(1.03)^3(\$10,000)$. In each of the next three years, the value of the investment is multiplied by 1.09, so the value of her investment at the end of all 6 years is $(1.09)^3(1.03)^3(\$10{,}000)$. Rui's investment earns $r\%$, compounded annually, in each of the 6 years, so his investment is worth $\left(1 + \frac{r}{100}\right)^6 (\$10{,}000)$ at the end of the 6 years. Since these investments must be equal in value at the end of the 6 years, we have

$$\left(1 + \frac{r}{100}\right)^6 (\$10{,}000) = (1.09)^3(1.03)^3(\$10{,}000).$$

Dividing both sides by $\$10{,}000$ gives

$$\left(1 + \frac{r}{100}\right)^6 = (1.09)^3(1.03)^3.$$

Raising both sides to the 1/6 power to get rid of the exponent on the left gives

$$1 + \frac{r}{100} = [(1.09)^3(1.03)^3]^{1/6} = (1.09)^{1/2}(1.03)^{1/2} \approx 1.0596.$$

Therefore, $r \approx 5.96$, so it is $\boxed{\text{less than}}$ 6.

19.3.5 Suppose Stefan pays $\$x$ at the end of each of the four years. Because the interest rate is 8.5% compounded annually, the present value of his first payment is $\frac{\$x}{1.085}$, the present value of his second payment is $\frac{\$x}{1.085^2}$, the present value of his third payment is $\frac{\$x}{1.085^3}$, and the present value of his fourth payment is $\frac{\$x}{1.085^4}$. The sum of the present values of the four payments must equal the present value of the amount Stefan borrowed today, which is $\$12{,}000$, so we have

$$\frac{\$x}{1.085} + \frac{\$x}{1.085^2} + \frac{\$x}{1.085^3} + \frac{\$x}{1.085^4} = \$12{,}000.$$

This is just a linear equation! Solving this equation gives $x \approx 3663.45$, so Stefan must pay $\boxed{\$3{,}663.45}$ at the end of each year.

Exercises for Section 19.4

19.4.1

(a) We have $4^3 = 64$, so $\log_4 64 = \boxed{3}$.

(b) We have $6^4 = 1296$, so $\log_6 1296 = \boxed{4}$.

19.4.2

(a) Let $x = \log_2 \frac{1}{16}$. Then, we must have $2^x = \frac{1}{16} = 2^{-4}$, so $x = \boxed{-4}$.

(b) Let $x = \log_{1/3} 27$. Then, we must have $(1/3)^x = 27$. Writing both 1/3 and 27 with 3 as the base, we have $(3^{-1})^x = 3^3$, so $3^{-x} = 3^3$. Therefore, we must have $x = \boxed{-3}$.

19.4.3

(a) Let $x = \log_2 8\sqrt{2}$. Then, we must have $2^x = 8\sqrt{2}$. Since $8 = 2^3$ and $\sqrt{2} = 2^{1/2}$, we have $2^x = 2^3 \cdot 2^{1/2} = 2^{7/2}$. Therefore, $x = \boxed{7/2}$.

(b) Let $x = \log_{\sqrt{3}} 9$. Then, we have $(\sqrt{3})^x = 9$. We write both $\sqrt{3}$ and 9 with 3 as the base to find

$$\left(3^{1/2}\right)^x = 3^2 \quad \Rightarrow \quad 3^{x/2} = 3^2.$$

Therefore, $x/2 = 2$, so $x = \boxed{4}$.

19.4.4

(a) Let $x = \log_4 32$. Then, we must have $4^x = 32$. Writing both 4 and 32 with 2 as the base gives $(2^2)^x = 2^5$, so $2^{2x} = 2^5$. Therefore, we must have $2x = 5$, so $x = \boxed{5/2}$.

(b) Let $x = \log_{27} 3\sqrt{3}$. Then, we must have $27^x = 3\sqrt{3}$. We have $27 = 3^3$ and $3\sqrt{3} = 3^1 \cdot 3^{1/2} = 3^{3/2}$, so our equation becomes $(3^3)^x = 3^{3/2}$. Therefore, we must have $3^{3x} = 3^{3/2}$, so $x = \boxed{1/2}$. (We might also simply have noticed that $3\sqrt{3}$ is the square root of 27.)

19.4.5 Writing the equation $\log_{81}(2r-1) = -1/2$ in exponential notation gives $2r-1 = 81^{-1/2} = (9^2)^{-1/2} = 9^{-1} = 1/9$. Solving $2r - 1 = 1/9$ gives $r = \boxed{5/9}$.

19.4.6

(a) Writing the equation in exponential notation gives $2^y = x - 4$, so $\boxed{x = 2^y + 4}$.

(b) We first divide by 3 to isolate the logarithm. This gives $\frac{y}{3} = \log_4(2x)$. Writing this equation in exponential notation gives $4^{y/3} = 2x$, and dividing by 2 gives $\boxed{x = (4^{y/3})/2}$. We can also write this as $x = 2^{2y/3-1}$.

(c) We first isolate the logarithm and find that $\log_7(3 - x) = \frac{4-y}{2}$. Putting this in exponential notation gives $3 - x = 7^{(4-y)/2}$. Solving for x gives $\boxed{x = 3 - 7^{(4-y)/2}}$.

19.4.7

(a) We cannot take the logarithm of a nonpositive number, so we must have $5 - x > 0$. Solving this gives $x < 5$, so the domain is all real numbers less than 5. The output of a logarithm can be any real number, so the range is all real numbers. We can also see this by letting $f(x) = y$, so we have $y = 2\log_3(5 - x)$. Therefore, we have $\frac{y}{2} = \log_3(5 - x)$. Writing this in exponential notation gives $3^{y/2} = 5 - x$, so $x = 5 - 3^{y/2}$. We can place any value of y in this equation to find the corresponding value of x (which is less than 5) such that $y = 2\log_3(5 - x)$.

(b) We cannot take the logarithm of a nonpositive number, so we must have $2x - 5 > 0$, or $x > 5/2$. Therefore, the domain is all real numbers greater than 5/2. The output of a logarithm can be anything, so g can be equal to anything. (We could also solve $y = 7 - 3\log_8(2x - 5)$ for x in terms of y to get $8^{(7-y)/3} = 2x - 5$, or $x = (8^{(7-y)/3} + 5)/2$. We can place any value of y in this equation to find the corresponding x such that $y = 7 - 3\log_8(2x - 5)$.)

19.4.8 Converting the given equation to exponential form gives $b^{5/2} = 5\sqrt{5} = 5^{3/2}$. Raising both sides to the 2/5 power to isolate b gives $b = (5^{3/2})^{2/5} = 5^{(3/2)(2/5)} = \boxed{5^{3/5}}$, or $\sqrt[5]{125}$.

19.4.9 For any positive value of a (besides $a = 1$, which cannot be the base of a logarithm), we have $\log_a 1 = 0$, because $a^0 = 1$ for all positive values of a. Therefore, the graph of $y = \log_a x$ passes through $(1, 0)$ for all positive values of a.

19.4.10

(a) We have $\log_2 4 = \boxed{2}$, $\log_2 4^2 = \log_2(2^2)^2 = \log_2 2^4 = \boxed{4}$, $\log_2 4^3 = \log_2(2^2)^3 = \log_2 2^6 = \boxed{6}$, and $\log_2 4^4 = \log_2(2^2)^4 = \log_2 2^8 = \boxed{8}$. Hmmm... it looks like there's a pattern.

(b) $a^x = b$.

(c) $a^y = b^c$.

(d) If we raise the equation in part (b) to the c power, we have $a^{cx} = b^c$. Because we also have $a^y = b^c$, we have $a^{cx} = a^y$. Because the bases in this equation are the same, the exponents must be the same as well. So, we have $y = cx$. Going back to our definitions of x and y, we have $\log_a b^c = c \log_a b$, as desired.

Review Problems

19.19 We can raise 4 to any real number power, so the domain is all real numbers. The expression 4^x must be positive, and can take on any positive value, so the range is all real numbers less than 7. We can also see this by letting $y = 7 - 4^x$, so $4^x = 7 - y$. Writing this in logarithmic notation, we have $x = \log_4(7 - y)$. We can take the logarithm of any positive number, but not of any nonpositive number, so we have $7 - y > 0$. Therefore, for any $y < 7$ we can use $x = \log_4(7 - y)$ to find the value of x such that $y = g(x) = 7 - 4^x$.

19.20 We write the right side with 6 as the base and we have $6^{3c-1} = 6^{-2}$. Therefore, we must have $3c - 1 = -2$, so $c = \boxed{-1/3}$.

19.21 We have $4^{x+3} = 4^x \cdot 4^3 = (2^2)^x \cdot 4^3 = 2^{2x} \cdot 64 = (2^x)^2 \cdot 64 = 9^2 \cdot 64 = \boxed{5184}$.

19.22 Writing all terms with base 3 and combining the terms on the left side, we have $3^{x^2-2x} = 3^3$. Because the bases on both sides are the same, we must have $x^2 - 2x = 3$, so $x^2 - 2x - 3 = 0$. Therefore, we have $(x - 3)(x + 1) = 0$, so our solutions are $\boxed{x = 3 \text{ and } x = -1}$.

19.23 We let $y = 2^x$, because then we can write $2^{2x} = (2^x)^2 = y^2$, and our equation is $y^2 - 8y + 12 = 0$. We factor the left side to find $(y - 6)(y - 2) = 0$. Therefore, we have $y = 6$ or $y = 2$. So, we have $2^x = 6$ or $2^x = 2$. Solving these two equations gives us the solutions $\boxed{x = \log_2 6 \text{ and } x = 1}$.

19.24 Adisa owes $\left(1 + \frac{0.14}{2}\right)^{2 \cdot 8} (\$5{,}000) = \boxed{\$14{,}760.82}$.

19.25 Let Jake's life savings be x. Then we must have $(1.06)^{10}(x) = \$531{,}402$, so $x = \$531{,}402/(1.06)^{10} = \boxed{\$296{,}732.10}$. (Notice that Jake's life savings is the present value of \$531,402.)

19.26 Suppose we invest \$$x$ in each investment. We compute how much each will be worth, in terms of x, at the end of ten years:

(a) This investment earns $0.1(\$x)$ each year, so it is worth $\$x + (10)(0.1)(\$x) = [1 + 0.1(10)](\$x) = 2(\$x)$ at the end of 10 years.

(b) This investment is worth $(1 + 0.075/4)^{4 \cdot 10} \approx 2.10(\$x)$ after 10 years.

(c) This investment is worth $(1 + 0.08/2)^{2 \cdot 10} \approx 2.19(\$x)$ after 10 years.

(d) This investment earns 0.11(x) in interest each year for the first 5 years, so it is worth 1.55(x) after 5 years. In each of the next 5 years, its value is multiplied by 1.07, so at the end of the 10 years, it is worth $(1.07)^5(1.55)(\$x) \approx 2.17(\$x)$.

We use these results to rate the investments from best to worst: (c), (d), (b), (a).

19.27 Suppose Gert invests $x at the end of 2000.

(a) After n years of investing her money at 5% simple interest, she'll have $(1 + 0.05n)(\$x)$. We seek the smallest value of n for which $1 + 0.05n \geq 2$. Solving this inequality gives us $n \geq 20$, so the earliest year she'll find she's doubled her money is 2020.

(b) After n years of investing her money at 5% compounded annually, she'll have $(1 + 0.05)^n(\$x)$. We seek the smallest value of n for which $(1 + 0.05)^n \geq 2$. Using a calculator and some trial-and-error, we find that the smallest integer n that satisfies this inequality is $n = 15$. So, the earliest year she'll find she's doubled her money is 2015.

(c) After n years of investing her money at 5% compounded twice a year, she'll have $(1 + 0.05/2)^{2n}(\$x)$. We seek the smallest value of n for which $(1 + 0.05/2)^{2n} \geq 2$. Using a calculator and some trial-and-error, we find that the smallest integer n that satisfies this inequality is $n = 15$. So, the earliest year she'll find she's doubled her money is 2015.

(d) After n years of investing her money at 5% compounded monthly, she'll have $(1 + 0.05/12)^{12n}(\$x)$. We seek the smallest value of n for which $(1 + 0.05/12)^{12n} \geq 2$. Using a calculator and some trial-and-error, we find that the smallest integer n that satisfies this inequality is $n = 14$. So, the earliest year she'll find she's doubled her money is 2014.

19.28

(a) The present value of the lottery ticket is $(\$1{,}000{,}000)/(1.09)^{10} = \$422{,}410.81$. Therefore, the ticket is worth much more today than I am offered. So, I shouldn't accept the offer.

(b) Let the desired interest rate be $r\%$. If the present value of the lottery ticket is $300,000, then we must have $(\$1{,}000{,}000)/(1 + r/100)^{10} = \$300{,}000$. Rearranging this equation gives

$$\left(1 + \frac{r}{100}\right)^{10} = \frac{\$1{,}000{,}000}{\$300{,}000} \approx 3.33.$$

Taking the tenth root of both sides gives $1 + \frac{r}{100} \approx 1.128$. Solving this equation gives $r = 12.8$, so the desired interest rate is 12.8%.

19.29

(a) $2,500 paid today is worth $2,500 today.

(b) The present value of $5,000 paid 9 years from now is $(\$5{,}000)/(1.08)^9 = \$2{,}501.24$.

(c) The present value of $4,500 paid 8 years from now is $(\$4{,}500)/(1.08)^8 = \$2{,}431.21$.

(d) The present value of $4,000 paid 5 years from now is $(\$4{,}000)/(1.08)^5 = \$2{,}722.33$.

Putting these in order from most valuable to least gives us (d), (b), (a), (c).

19.30 Let Claire's first payment be x, so her second one is $2x$. The present value of the first payment is $(x)/(1.085)^3$ and the present value of the second is $(2x)/(1.085)^6$. The sum of these present values must equal \$5,000, the amount Claire borrows today:

$$\frac{x}{1.085^3} + \frac{2x}{1.085^6} = \$5,000 \quad \Rightarrow \quad x = \boxed{\$2,489.05}.$$

19.31

(a) Because $3^5 = 243$, we have $\log_3 243 = \boxed{5}$.

(b) Because $(1/2)^3 = 1/8$, we have $\log_{1/2} \frac{1}{8} = \boxed{3}$.

(c) We have $4\sqrt[3]{2} = 2^2 \cdot 2^{\frac{1}{3}} = 2^{\frac{7}{3}}$, so $\log_2 4\sqrt[3]{2} = \log_2 2^{\frac{7}{3}} = \boxed{\dfrac{7}{3}}$.

(d) Let $x = \log_9 27$, so $9^x = 27$. Writing both sides with 3 as the base gives us $3^{2x} = 3^3$, so we must have $2x = 3$, or $x = \boxed{3/2}$.

(e) Because $7^3 = 343$, we have $7^{-3} = \frac{1}{343}$, so $\log_7 \frac{1}{343} = \log_7 7^{-3} = \boxed{-3}$.

(f) Let $x = \log_{\sqrt{5}} 125\sqrt{5}$. Putting this in exponential notation gives $(\sqrt{5})^x = 125\sqrt{5}$. Writing both sides with 5 as the base gives us $5^{\frac{x}{2}} = 5^3 \cdot 5^{\frac{1}{2}} = 5^{\frac{7}{2}}$, so $x/2 = 7/2$. Therefore, $x = \boxed{7}$.

19.32 Putting the equation in exponential notation gives $n^{10} = 4\sqrt{2}$. We write the right side with 2 as the base to find $n^{10} = 2^2 \cdot 2^{\frac{1}{2}} = 2^{\frac{5}{2}}$. Raising both sides to the $1/10$ power gives $n = (2^{5/2})^{1/10} = 2^{1/4} = \boxed{\sqrt[4]{2}}$.

19.33 Suppose $a = 1$, so that $f(x) = \log_1 x$. Then, what is $f(1)$? Since 1 raised to any power equals 1, there is not a single unique value that we can say $\log_1 1$ equals. Therefore, if $a = 1$, then $f(x) = \log_a x$ is not a function. So, we cannot have $a = 1$.

19.34 Writing the equation in exponential form gives us $2x - 7 = 9^{\frac{3}{2}} = (9^{\frac{1}{2}})^3 = 3^3 = 27$. Solving $2x - 7 = 27$ gives us $x = \boxed{17}$.

19.35 Let $\log_b 729 = n$, so we must have $b^n = 729 = 3^6$. We must have both b and n be integers. We can solve this equation for b by raising both sides to the $1/n$ power, which gives $b = 3^{6/n}$. Since b is a positive integer, we must have $6/n$ be a positive integer. We also know that n must be an integer, so our possibilities for n are 1, 2, 3, and 6. These give us $\boxed{4}$ possible values for b.

Challenge Problems

19.36 If we let $y = \log_2(\log_2(x))$, then our given equation is just $\log_2 y = 2$. Putting this in exponential notation gives us $y = 2^2 = 4$, so our equation is now $\log_2(\log_2(x)) = 4$. We let $z = \log_2(x)$, making our equation $\log_2(z) = 4$. Putting this in exponential notation gives us $z = 2^4 = 16$. Therefore, we have $\log_2(x) = 16$, so $x = 2^{16}$. To find the number of digits in 2^{16}, we note that $2^{16} = 2^{10} \cdot 2^6 = (1024)(64)$, which is a little more than 64,000. So, x has $\boxed{5}$ digits.

19.37

(a) Suppose the interest rate is $r\%$. After the first ten years, the value of the investment has been multiplied by $(1+r/100)^{10}$. We also know that the value of the investment has doubled over these 10 years, so we have $(1+r/100)^{10} = 2$. We could now solve for r, but we don't have to. We want to find out how long it takes for the investment to double twice. Because the annually compounded interest rate stays the same throughout the investment, we multiply by $(1+r/100)^{10}$ for the first ten years, and also for the second ten years. In other words, my investment doubles once in the first ten years, then again in the next ten years. So, after $\boxed{20}$ years, its value is multiplied by $2 \cdot 2 = 4$.

(b) My investment doubles in the first ten years, then doubles again in the next ten years, then doubles yet again in the third ten years. So, my investment has doubled 3 times, which means it is multiplied by $2^3 = 8$. Therefore, after 30 years, my $5,000 investment has grown to $8(\$5,000) = \boxed{\$40,000}$.

19.38 We cannot take the logarithm of a nonpositive number, so we must have $x^2 - 4x - 5 > 0$. Factoring gives us $(x-5)(x+1) > 0$, so $x > 5$ or $x < -1$. Therefore, our domain consists of all real numbers that are either greater than 5 or less than -1. Since $x^2 - 4x - 5$ can equal any positive number (which we can see by considering the graph of $y = x^2 - 4x - 5$), we see that the range is all real numbers.

19.39 We can raise 3 to any real power, but we must make sure we don't make our denominator equal to zero. Because 3^x is always positive, there is no value of x that makes our denominator equal to 0. So, the domain is all real numbers.

Because 3^x is always positive, the denominator of $f(x)$ is always greater than 2. Therefore, $f(x)$ is always less than 5/2. Because 3^x is always positive, we must have $f(x) > 0$ for all x. To see that $f(x)$ can equal any number between 0 and 5/2, we let $y = f(x) = 5/(2 + 3^x)$ and solve for x in terms of y. We start by multiplying both sides by $2 + 3^x$, which gives $2y + 3^x y = 5$. Subtracting $2y$ from both sides, then dividing by y, gives

$$3^x = \frac{5 - 2y}{y} = \frac{5}{y} - 2.$$

Putting this in logarithmic notation gives $x = \log_3\left(\frac{5}{y} - 2\right)$. Since we can only take the logarithm of positive numbers, we must have $\frac{5}{y} - 2 > 0$, or $(5 - 2y)/y > 0$. This inequality is only true if $0 < y < 5/2$, so only for values of y in this interval can we find a value of x such that $y = 5/(2 + 3^x)$. So, the range of f is all numbers between 0 and 5/2.

19.40 The graph passes through $(2,0)$ and $(6,1)$. Letting $x = 2$ and $y = 0$ in the equation $y = \log_a(xb)$ gives us $0 = \log_a(2b)$ Putting this in exponential notation gives $a^0 = 2b$, so $2b = 1$. Therefore, $\boxed{b = 1/2}$, and our graph is the graph of $y = \log_a(x/2)$. Letting $x = 6$ and $y = 1$ in this equation gives us $1 = \log_a(6/2) = \log_a 3$. Therefore, we have $a^1 = 3$, so $\boxed{a = 3}$.

19.41 Because $8^x = 27$, we have $(2^3)^x = 27$, so $2^{3x} = 27$. Raising both sides to the 1/3 power gives us $2^x = 3$. Therefore, we have $4^{2x-3} = (4^{2x})(4^{-3}) = (2^2)^{2x}/64 = 2^{4x}/64 = (2^x)^4/64 = 3^4/64 = \boxed{81/64}$.

19.42 We write all the terms in the first equation with 2 as the base, and we have

$$\frac{2^{2x}}{2^{x+y}} = 2^3 \quad \Rightarrow \quad 2^{2x-(x+y)} = 2^3 \quad \Rightarrow \quad 2^{x-y} = 2^3.$$

Therefore, we must have $x - y = 3$. We write all the terms in the second equation with 3 as the base, and we have

$$\frac{(3^2)^{x+y}}{3^{5y}} = 3^5 \quad \Rightarrow \quad 3^{2x+2y-5y} = 3^5 \quad \Rightarrow \quad 3^{2x-3y} = 3^5.$$

The bases of both sides are the same, so the exponents must be the same. Therefore, we have $2x - 3y = 5$. Solving the system of equations $x - y = 3$, $2x - 3y = 5$ gives us $(x, y) = \boxed{(4, 1)}$.

19.43

(a) We have $\log_2 8 = \boxed{3}$, $\log_2 16 = \boxed{4}$ and $\log_2(8 \cdot 16) = \log_2(2^3 \cdot 2^4) = \log_2 2^7 = \boxed{7}$.

(b) We have $\log_3 \frac{1}{9} = \boxed{-2}$, $\log_3 \sqrt{3} = \boxed{\frac{1}{2}}$, and $\log_3\left(\frac{1}{9} \cdot \sqrt{3}\right) = \log_3\left(3^{-2} \cdot 3^{1/2}\right) = \log_3 3^{-3/2} = \boxed{-\frac{3}{2}}$.

(c) It looks like $\log_a b + \log_a c = \log_a(bc)$. We don't know much about logs, but we know a lot about exponents, so we let $x = \log_a b$, $y = \log_a c$, and $z = \log_a(bc)$, and write these in exponential notation. This gives us $a^x = b$, $a^y = c$, and $a^z = bc$. We can also get an expression for bc by multiplying our first two equations, which gives $a^{x+y} = bc$. Therefore, we have $a^{x+y} = bc = a^z$. The ends of this equation have the same base, so their exponents must be equal. So, we have $x + y = z$, which means $\log_a b + \log_a c = \log_a(bc)$.

19.44 Suppose the amount of money they each invest is m dollars. Let $A(x)$ be the amount of money Alice has after x years and $B(x)$ be the amount of money Bob has after x years. Then, we have

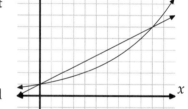

$$A(x) = \left(1 + \frac{r}{100}\right)^x (m) \qquad \text{and} \qquad B(x) = \left(1 + x \cdot \frac{s}{100}\right)(m).$$

Since r, s, and m are all constants, this means that $A(x)$ is an exponential function and $B(x)$ is a linear function. Moreover, we know that $A(0) = B(0) = m$, and we are told that $A(10) = B(10)$, so the graphs of $y = A(x)$ and $y = B(x)$ meet at two points, one where $x = 0$ and one where $x = 10$. The graph of a line and the graph of an exponential function $y = a^x$, where a is positive and not equal to 1, can only meet in at most 2 points. (An example is shown above; here, we have $y = \frac{x}{2} + 1$, and $y = (\sqrt[10]{6})^x$, which meet where $x = 0$ and where $x = 10$.) In between these two points, the linear function is higher than the exponential function, but in the 'outside' regions beyond these two points, the exponential function is higher. Therefore, Bob has more money than Alice in years 1-9, but $\boxed{\text{Alice}}$ has more after year 10.

Note: this is exactly like the Meena-Arial comparison at the start of the chapter. Arial's grain is essentially simple interest, while Meena's is compounded.

19.45 Those huge powers of x suggest we try factoring x^{98} out of the first two terms, which gives us

$$x^{98}(x^2 - 4^x) - x^2 + 4^x = 0.$$

Seeing $x^2 - 4^x$ and $-x^2 + 4^x$, we discover that we can factor again:

$$x^{98}(x^2 - 4^x) - 1(x^2 - 4^x) = (x^{98} - 1)(x^2 - 4^x) = 0.$$

Therefore, we must have $x^{98} - 1 = 0$ or $x^2 - 4^x = 0$. The first equation is only true for real values of x when $x = 1$ or $x = -1$.

To analyze the equation $x^2 - 4^x = 0$, we consider the graphs of $y = x^2$ and $y = 4^x$. Our solutions to $x^2 - 4^x = 0$ are the points where these graphs meet. The graphs meet in one point (we know they won't meet again to the right because $y = 4^x$ rises much faster than $y = x^2$ when x is positive). While we don't know for sure what that one point is, we don't need to. We only need to know how many solutions there are to the original equation. Since there are two solutions to $x^{98} - 1 = 0$ and one different solution to $x^2 - 4^x = 0$, there are $\boxed{3}$ solutions to the product of these two equations.

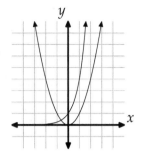

_____Special Functions

Exercises for Section 20.1

20.1.1

(a) The input to the radical must be nonnegative, so we must have $2 + x \geq 0$, or $x \geq -2$. Therefore, our domain is all values of x such that $x \geq -2$. The output from the radical must be nonnegative, so $\sqrt{2 + x} - 5 > 0 - 5 = -5$. Since the output from the radical can be any nonnegative number, the range is all real numbers greater than or equal to -5.

(b) The input to the radical must be nonnegative, so we must have $3 - x \geq 0$, or $x \leq 3$. Therefore, our domain is all values of x such that $x \leq 3$. The output from the radical must be nonnegative, so $-2\sqrt{3 - x}$ must be nonpositive. Therefore, $-2\sqrt{3 - x} + 7 \leq 0 + 7 = 7$. Since $-2\sqrt{3 - x}$ can take on any negative value, the range is all real numbers less than or equal to 7.

20.1.2 Squaring both sides of the equation gives $-r^2 + 8r + 1 = 16$, so $r^2 - 8r + 15 = 0$. Factoring the left side gives $(r - 5)(r - 3) = 0$, so $r = 5$ or $r = 3$. Because we squared an equation as a step in our solution, we must check for extraneous solutions. When $r = 5$, we have $\sqrt{1 + 8r - r^2} = \sqrt{1 + 40 - 25} = \sqrt{16} = 4$. When $r = 3$, we have $\sqrt{1 + 8r - r^2} = \sqrt{1 + 24 - 9} = 4$. So, both $\boxed{r = 3 \text{ and } r = 5}$ are solutions.

20.1.3 We have $\sqrt{\dfrac{x}{1 - \frac{x-1}{x}}} = \sqrt{\dfrac{x}{\frac{x}{x} - \frac{x-1}{x}}} = \sqrt{\dfrac{x}{\frac{1}{x}}} = \sqrt{x^2} = \boxed{x}$. Why did the question state that $x > 0$?

20.1.4 Starting from the graph of $y = \sqrt{x}$, we can use transformations to produce the graph of the equation $y = -\sqrt{x - 2}$. First, we shift the graph of $y = \sqrt{x}$ exactly 2 units to the right to produce the graph of $y = \sqrt{x - 2}$. Then, we flip this graph over the x-axis to produce the graph of $y = -\sqrt{x - 2}$. These transformations are shown below.

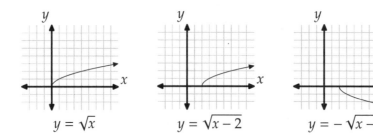

20.1.5 Squaring both sides gives $2 + \sqrt{x} = 9$. So, we have $\sqrt{x} = 7$. Squaring both sides of this equation gives $x = \boxed{49}$. We check to see if this solution is extraneous by substituting it into the original equation. We find $\sqrt{2 + \sqrt{49}} = \sqrt{2 + 7} = 3$, so the solution is correct.

20.1.6 Squaring both sides gives

$$\left(\sqrt{x} + \sqrt{x+4}\right)^2 = 4(4x - 5)$$
$$\Rightarrow \quad x + 2\sqrt{x^2 + 4x} + x + 4 = 16x - 20$$
$$\Rightarrow \quad \sqrt{x^2 + 4x} = 7x - 12.$$

Squaring both sides of this equation gives $x^2 + 4x = 49x^2 - 168x + 144$. Rearranging this gives us

$$48x^2 - 172x + 144 = 0 \quad \Rightarrow \quad 12x^2 - 43x + 36 = 0 \quad \Rightarrow \quad (3x - 4)(4x - 9) = 0.$$

Therefore, we have $x = 4/3$ or $x = 9/4$. When $x = 4/3$, we have

$$\sqrt{x} + \sqrt{x+4} = \sqrt{\frac{4}{3}} + \sqrt{\frac{16}{3}} = \frac{2}{\sqrt{3}} + \frac{4}{\sqrt{3}} = \frac{6}{\sqrt{3}} = 2\sqrt{3}$$

and

$$2\sqrt{4x - 5} = 2\sqrt{\frac{16}{3} - 5} = 2\sqrt{\frac{1}{3}} = \frac{2\sqrt{3}}{3}.$$

So, $x = 4/3$ is an extraneous solution. Testing $x = 9/4$ gives

$$\sqrt{x} + \sqrt{x+4} = \sqrt{\frac{9}{4}} + \sqrt{\frac{25}{4}} = \frac{3}{2} + \frac{5}{2} = 4$$

and

$$2\sqrt{4x - 5} = 2\sqrt{9 - 5} = 4.$$

So, $\boxed{x = 9/4}$ is the only solution to our equation.

Exercises for Section 20.2

20.2.1

(a) Adding 7 to both sides gives $|r + 3| = 16$. Here are a couple solutions from this point:

Solution 1: Casework. If $r \geq -3$, we have $|r + 3| = r + 3$, so our equation is $r + 3 = 16$. This gives us $r = 13$, which is indeed greater than -3. If $r < -3$, we have $|r + 3| = -(r + 3) = -r - 3$, so our equation is $-r - 3 = 16$. Solving this gives us $r = -19$, which is less than -3. So, our solutions are $\boxed{r = -19 \text{ and } r = 13}$.

Solution 2: The number line. Since $|r + 3| = 16$, we know that r is 16 units from -3 on the number line. The two numbers that are 16 units from -3 on the number line are $-3 - 16 = -19$ and $-3 + 16 = 13$. Therefore, our solutions are $\boxed{r = -19 \text{ and } r = 13}$.

(b) Subtracting 7 from both sides gives $|r + 8| = -3$. The absolute value of an expression cannot be negative, so there are $\boxed{\text{no solutions}}$ to this equation.

20.2.2 If $x < 2$, then $|x - 2| = -(x - 2) = -x + 2$. Therefore, we must have $-x + 2 = p$, so $x = 2 - p$. So, we have $x - p = 2 - p - p = 2 - 2p$, which is choice $\boxed{(C)}$.

20.2.3

(a) Subtracting 3 from each side gives $|6x - 7| = 9$. We can consider cases in order to get rid of the absolute value signs. When $6x - 7 \geq 0$, or $x \geq 7/6$, we have $|6x - 7| = 6x - 7$. Therefore, we have $6x - 7 = 9$, so $x = 8/3$. Since $8/3 > 7/6$, this is a valid solution. When $6x - 7 < 0$, or $x < 7/6$, we have $|6x - 7| = -(6x - 7) = -6x + 7$. Therefore, we have $-6x + 7 = 9$, so $x = -1/3$. Since $-1/3 < 7/6$, this is also a valid solution. So, we have two solutions, $\boxed{x = -1/3 \text{ and } x = 8/3}$.

(b) Dividing both sides by 2 gives $|3 - 5x| = 7/2$. Again, we turn to cases. If $3 - 5x \geq 0$, so that $x \leq 3/5$, we have $|3 - 5x| = 3 - 5x$. This makes our equation $3 - 5x = 7/2$, from which we have $x = -1/10$. Since $-1/10 \leq 3/5$, this is a valid solution. If $3 - 5x < 0$, so that $x > 3/5$, we have $|3 - 5x| = -(3 - 5x) = -3 + 5x$. Therefore, we have the equation $-3 + 5x = 7/2$, from which we find $x = 13/10$. Since $13/10 > 3/5$, this solution is also valid. So, our solutions are $\boxed{x = -1/10 \text{ and } 13/10}$.

20.2.4 We consider two cases, $y \geq 6$ and $y < 6$.

- *Case 1: $y \geq 6$:* If $y \geq 6$, then $|y - 6| = y - 6$ and our equation is $y - 6 + 2y = 9$. So, we have $3y = 15$, or $y = 5$. However, $y = 5$ does not satisfy $y \geq 6$. Testing $y = 5$, we have $|5 - 6| + 2 \cdot 5 = 11$, not 9, and we see that $y = 5$ is not a solution.

- *Case 2: $y < 6$:* If $y < 6$, then $|y - 6| = -(y - 6) = -y + 6$, so our equation is $-y + 6 + 2y = 9$, from which we have $y = 3$. This is a valid solution, since $y = 3$ satisfies the restriction $y < 6$.

So, our only solution is $\boxed{y = 3}$.

20.2.5 In order to get rid of the absolute value signs, we consider three cases: both $2z - 9$ and $z - 3$ are nonnegative, one is nonnegative and the other negative, and both are negative. Since $2z - 9$ is nonnegative for $z \geq 9/2$ (and negative for $z < 9/2$) and $z - 3$ is nonnegative for $z \geq 3$ (and negative for $z < 3$), our three cases are:

- *Case 1: $z \geq 9/2$.* Here, we have $|2z - 9| = 2z - 9$ and $|z - 3| = z - 3$, so our equation is $2z - 9 + z - 3 = 15$. So, we have $3z = 27$, or $z = 9$. Since $z = 9$ satisfies the restriction for this case, it is a valid solution.

- *Case 2: $3 \leq z < 9/2$.* Here, we have $|2z - 9| = -(2z - 9) = -2z + 9$ and $|z - 3| = z - 3$, so our equation is $-2z + 9 + z - 3 = 15$. Therefore, we have $-z + 6 = 15$, so $z = -9$. This value of z does not satisfy the restrictions of this case, so we discard this solution.

- *Case 3: $z < 3$.* Here, we have $|2z - 9| = -(2z - 9) = -2z + 9$ and $|z - 3| = -(z - 3) = -z + 3$, so our equation is $-2z + 9 - z + 3 = 15$, or $-3z = 3$. Therefore, we have $z = -1$, which does satisfy the restriction of this case.

Combining the valid solutions from our three cases, the solutions to the original equation are
$\boxed{z = 9 \text{ and } z = -1}$.

20.2.6

(a) We get rid of the absolute value signs by considering cases. When $x \geq -4$, our equation is $y = x + 4$, so the graph of $y = |x + 4|$ matches the graph of $y = x + 4$ when $x \geq -4$. Similarly, when $x < 4$, we have $|x + 4| = -(x + 4) = -x - 4$, so the graph of $y = |x + 4|$ matches the graph of $y = -x - 4$ when $x < 4$. Putting these two cases together gives us the graph at left below.

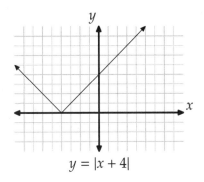

$y = |x + 4|$

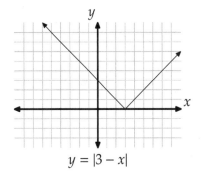

$y = |3 - x|$

(b) Again, we consider cases. Solving $3 - x \geq 0$, we find that $3 - x$ is nonnegative when $x \leq 3$. So, when $x \leq 3$, the graph of $y = |3 - x|$ matches the graph of $y = 3 - x$. Similarly, when $x > 3$, the graph of $y = |3 - x|$ matches the graph of $y = -3 + x$. Combining these cases gives us the graph at right above.

(c) Our first two parts tell us the three cases we have to consider:

- *Case 1:* $x \geq 3$. When $x \geq 3$, we have $|x + 4| = x + 4$ and $|3 - x| = -(3 - x) = -3 + x$. So, when $x \geq 3$, the graph of $y = |x + 4| + |3 - x|$ matches the graph of $y = x + 4 + (-3 + x) = 2x + 1$.

- *Case 2:* $-4 \leq x < 3$. When $-4 \leq x < 3$, we have $|x + 4| = x + 4$ and $|3 - x| = 3 - x$, so the graph of $y = |x + 4| + |3 - x|$ matches the graph of $y = x + 4 + 3 - x = 7$.

- *Case 3:* $x < -4$. When $x < -4$, we have $|x + 4| = -(x + 4) = -x - 4$ and $|3 - x| = 3 - x$, so the graph of $y = |x + 4| + |3 - x|$ matches the graph of $y = -x - 4 + 3 - x = -1 - 2x$.

Combining these three cases gives us the graph at left below.

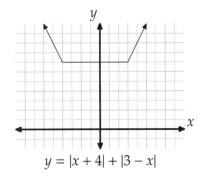

$y = |x + 4| + |3 - x|$

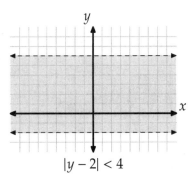

$|y - 2| < 4$

(d) The equation $|y - 2| < 4$ tells us that the distance between y and 2 is less than 4. In other words, y is between -2 and 6. On the Cartesian plane, the points for which y is between -2 and 6 is the region of points between the horizontal lines $y = -2$ and $y = 6$. This region is shaded at right above.

20.2.7 We get rid of the absolute value signs by considering two cases:

- *Case 1:* $r^2 - 5r \geq 0$. This gives us $r^2 - 5r = 6$, so $r^2 - 5r - 6 = 0$. Factoring the left side gives $(r - 6)(r + 1) = 0$, so $r = 6$ and $r = -1$. Both of these satisfy the restriction of this case, so both are solutions.

- *Case 2:* $r^2 - 5r < 0$. This gives us $-r^2 + 5r = 6$, so $r^2 - 5r + 6 = 0$. Factoring the left side gives us $(r - 2)(r - 3) = 0$, so $r = 2$ or $r = 3$. Both satisfy the restriction of this case, so both are solutions.

Combining the two cases gives us the solutions $r = \boxed{-1, 2, 3, 6}$. (We could also square both sides of the original equation, but this would produce an equation that we haven't learned how to solve yet!)

Exercises for Section 20.3

20.3.1

(a) $\lfloor 3.2 \rfloor = \boxed{3}$.

(b) The greatest integer less than -21.8 is -22, so $\lfloor -21.8 \rfloor = \boxed{-22}$.

(c) Since $\frac{230}{7} = 32\frac{6}{7}$, the smallest integer greater than $\frac{230}{7}$ is 33. So, $\left\lceil \frac{230}{7} \right\rceil = \boxed{33}$.

(d) Since $\sqrt{23}$ is between $\sqrt{16} = 4$ and $\sqrt{25} = 5$, we know that $-5 < -\sqrt{23} < -4$. Therefore, the smallest integer greater than $-\sqrt{23}$ is -4. So, we have $\left\lceil -\sqrt{23} \right\rceil = \boxed{-4}$.

(e) Since $\sqrt{26}$ is a little more than $\sqrt{25} = 5$ and $\sqrt{8}$ is a little less than $\sqrt{9} = 3$, the difference $\sqrt{26} - \sqrt{8}$ is a little more than 2. Therefore, $\left\lfloor \sqrt{26} - \sqrt{8} \right\rfloor = \boxed{2}$.

To be more precise, we have $5.5 > \sqrt{26} > 5$ because $5.5^2 > 26 > 5^2$. We also have $3 > \sqrt{8} > 2.5$. From $5.5 > \sqrt{26}$, we have $5.5 - \sqrt{26} > 0$. From $\sqrt{8} > 2.5$, we have $0 > 2.5 - \sqrt{8}$. Combining $5.5 - \sqrt{26} > 0$ and $0 > 2.5 - \sqrt{8}$ gives us $5.5 - \sqrt{26} > 2.5 - \sqrt{8}$. Rearranging this inequality gives $5.5 - 2.5 > \sqrt{26} - \sqrt{8}$. Similarly, we can show that $\sqrt{26} - \sqrt{8} > 5 - 3$. Combining these two inequalities gives us $3 > \sqrt{26} - \sqrt{8} > 2$. Since $\sqrt{26} - \sqrt{8}$ is between 2 and 3, we have $\left\lfloor \sqrt{26} - \sqrt{8} \right\rfloor = 2$.

(f) Since $104/5 = 20.4$ and $\sqrt{20.4}$ is between $\sqrt{16} = 4$ and $\sqrt{25} = 5$, we have $\left\lfloor \sqrt{\frac{104}{5}} \right\rfloor = \boxed{4}$.

20.3.2 We have
$$1 = \sqrt{1} < \sqrt{2} < \sqrt{3} < \sqrt{4} = 2,$$
so $\lfloor \sqrt{1} \rfloor = \lfloor \sqrt{2} \rfloor = \lfloor \sqrt{3} \rfloor = 1$. So, these three expressions add to $3(1) = 3$. We have
$$2 = \sqrt{4} < \sqrt{5} < \cdots < \sqrt{8} < \sqrt{9} = 3,$$
so $\lfloor \sqrt{4} \rfloor = \lfloor \sqrt{5} \rfloor = \cdots = \lfloor \sqrt{8} \rfloor = 2$. There are 5 of these expressions, so they add to $5(2) = 10$. We have
$$3 = \sqrt{9} < \sqrt{10} < \cdots < \sqrt{15} < \sqrt{16} = 4,$$

so $\lfloor \sqrt{9} \rfloor = \lfloor \sqrt{10} \rfloor = \cdots = \lfloor \sqrt{15} \rfloor = 3$. There are 7 of these expressions, so they add to $7(3) = 21$.

Putting these together with $\sqrt{16} = 4$, we have

$$\lfloor \sqrt{1} \rfloor + \lfloor \sqrt{2} \rfloor + \lfloor \sqrt{3} \rfloor + \cdots + \lfloor \sqrt{16} \rfloor = 3 + 10 + 21 + 4 = \boxed{38}.$$

20.3.3 The graph looks a lot like a transformation of the graph of $y = g(x)$, where $g(x) = \lfloor x \rfloor$. (The fact that the points on the left ends of each segment on the graph are included, but the right ones aren't, tells us that we are rounding down instead of up. This means our graph is like that of $y = \lfloor x \rfloor$, not $y = \lceil x \rceil$.)

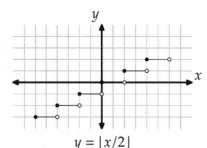

$y = \lfloor x/2 \rfloor$

Each of the intervals of our given graph is twice as long as the intervals in the graph of $y = \lfloor x \rfloor$, so we start by scaling the graph of $y = g(x)$ horizontally by a factor of 2. This gives us the graph of $y = g(x/2) = \lfloor x/2 \rfloor$, part of which is shown at right. This graph is just 1 unit lower than the graph given in the problem. The graph that is a 1-unit vertical shift of the graph of $y = g(x/2)$ is the graph of $y = g(x/2) + 1$. So, our desired function is $f(x) = g(x/2) + 1$, or $\boxed{f(x) = \lfloor x/2 \rfloor + 1}$.

See if you can find another solution by noting that the graph in the problem is a 2-unit leftward shift of our graph of $y = g(x/2) = \lfloor x/2 \rfloor$ shown above.

20.3.4 We get rid of the floor function by considering the cases $\lfloor x \rfloor = 0$, $\lfloor x \rfloor = 1$, $\lfloor x \rfloor = 2$, etc.:

- *Case 1:* $\lfloor x \rfloor = 0$; $0 < x < 1$. This gives us division by 0 on the right side of our equation, so there are no solutions in this case.

- *Case 2:* $\lfloor x \rfloor = 1$; $1 \le x < 2$. This makes our equation is $x - 1 = 1$, which gives $x = 2$. But $x = 2$ does not satisfy $1 \le x < 2$, so this is not a valid solution.

- *Case 3:* $\lfloor x \rfloor = 2$; $2 \le x < 3$. Our equation is $x - 2 = \frac{1}{2}$, so $x = 2\frac{1}{2}$.

- *Case 4:* $\lfloor x \rfloor = 3$; $3 \le x < 4$. Our equation is $x - 3 = \frac{1}{3}$, so $x = 3\frac{1}{3}$.

- *Case 5:* $\lfloor x \rfloor = 4$; $4 \le x < 5$. Our equation is $x - 4 = \frac{1}{4}$, so $x = 4\frac{1}{4}$.

All other positive solutions are larger than 5, so the sum of the three smallest positive solutions to our equation is $2\frac{1}{2} + 3\frac{1}{3} + 4\frac{1}{4} = \boxed{10\frac{1}{12}}$.

Exercises for Section 20.4

20.4.1 We start by getting rid of the fractions. We do so by multiplying both sides by $(x - 5)(x - 3)$. This gives us

$$\frac{2x}{x - 5} \cdot (x - 5)(x - 3) = \left(3 + \frac{1 - x}{x - 3}\right)(x - 5)(x - 3) \quad \Rightarrow \quad 2x(x - 3) = 3(x - 5)(x - 3) + (1 - x)(x - 5).$$

Expanding the products of binomials gives $2x^2 - 6x = 3(x^2 - 8x + 15) - x^2 + 6x - 5$. Simplifying this equation gives $12x = 40$, so $x = \boxed{\dfrac{10}{3}}$.

20.4.2

(a) We can input any value of x except $x = 7$, since $x = 7$ causes division by 0. Therefore, the domain is $\boxed{\text{all real numbers except } 7}$.

(b) To find the range, we let $y = g(x)$ and solve the equation

$$y = \frac{3 - 2x}{x - 7}$$

for x in terms of y. Multiplying both sides by $x - 7$ gives $xy - 7y = 3 - 2x$. Isolating the terms with x gives $xy + 2x = 3 + 7y$, so $x(y + 2) = 3 + 7y$. Therefore, we have

$$x = \frac{3 + 7y}{y + 2}.$$

So, we cannot have $y = -2$, because this will cause division by 0. Substituting any other value of y into this equation produces the corresponding x such that $y = g(x)$. So, this equation for x in terms of y tells us that the range of g is $\boxed{\text{all real numbers except } -2}$.

(c) As x gets very close to 7, the value of y in

$$y = \frac{3 - 2x}{x - 7}$$

becomes either a very large positive number (when x is a little less than 7) or a very small negative number (when x is a little more than 7). Therefore, the line $\boxed{x = 7}$ is a vertical asymptote. Similarly, as y gets very close to -2 in

$$x = \frac{3 + 7y}{y + 2},$$

the value of x becomes either a very large positive number (when y is a little less than -2) or a very small negative number (when y is a little more than -2). Therefore, the line $\boxed{y = -2}$ is a horizontal asymptote.

(d) Using our asymptotes as guides, we produce the graph shown below. The asymptotes are the dashed lines in the diagram.

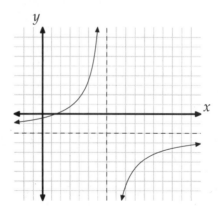

20.4.3 We factor the denominator on the left to make our equation

$$\frac{x+7}{(x-7)(x+5)} = \frac{A}{x-7} + \frac{B}{x+5}.$$

Multiplying both sides by $(x-7)(x+5)$ gives

$$x + 7 = A(x+5) + B(x-7).$$

This equation must hold for all values of x. We let $x = -5$ to eliminate A. This gives us $-5 + 7 = A(-5+5) + B(-5-7)$, so $2 = -12B$, and $B = -1/6$. Similarly, we let $x = 7$ to eliminate B. This gives us $7+7 = A(7+5)+B(7-7)$, so $14 = 12A$, which means $A = 7/6$. Therefore, our solution is $\boxed{A = 7/6, B = -1/6}$.

20.4.4

(a) Factoring the numerator and denominator gives $g(x) = \dfrac{(3x-4)(x-1)}{(x+5)(x-1)}$.

(b) Because we cannot have 0 in the denominator, the domain is $\boxed{\text{all real numbers except } -5 \text{ and } 1}$.

(c) If we cancel the common factors in the numerator and denominator, we have $g(x) = (3x-4)/(x+5)$, which is the same expression as $f(x) = (3x-4)/(x+5)$ that we graphed in the text. So, for all values of x *except* $x = 1$, the two functions are the same. For $x = 1$, the function $g(x)$ in this problem is not defined, whereas $f(x)$ is defined for $x = 1$.

(d) As we saw in the previous problem, g is the same as f everywhere except for where $x = 1$. For $x = 1$, $g(x)$ is not defined but $f(1) = -1/6$. Therefore, the graph of g is the same as the graph of f, except the point $\boxed{(1, -1/6)}$ must be omitted from the graph of g.

(e) We found in the text that the range of f is all real numbers except 3. We also know that $(1, -1/6)$ is on the graph of f but not on the graph of g. When we omit $(1, -1/6)$ from the graph of f, there are no remaining points with y-coordinate equal to $-1/6$. Since the graphs of f and g are otherwise the same, we see that the range of g is $\boxed{\text{all real numbers except 3 and } -1/6}$.

Exercises for Section 20.5

20.5.1

(a) Because $3 \geq -2$, we use the second case to determine that $f(3) = 5 - 2(3) = \boxed{-1}$.

(b) Because $-7 < -2$, we use the first case to determine that $f(-7) = 2(-7) + 9 = \boxed{-5}$.

(c) We check both cases. If $k < -2$ and $f(k) = 3$, then $2k + 9 = 3$, so $2k = -6$, or $k = -3$, which is less than -2, as required for this case. If $k \geq -2$ and $f(k) = 3$, then $5 - 2k = 3$, so $k = 1$, which is greater than -2, as required for this case. Therefore, both $\boxed{k = -3 \text{ and } k = 1}$ give us $f(k) = 3$.

(d) $\boxed{\text{No}}$. Because two different values of k give us the same value of $f(k)$ (in part (c), we saw that $f(-3) = f(1) = 3$), the function f does not have an inverse.

20.5.2

(a) Let f be our function. For $x \geq 1$, the graph is a line that passes through $(2, 1)$ and $(3, 2)$. The slope of this line is $(2 - 1)/(3 - 2) = 1$, so its equation is $y - 1 = 1(x - 2)$, or $y = x - 1$. Therefore, when $x \geq 1$, we have $f(x) = x - 1$. When $x < 1$, the graph is a line that passes through $(0, 1)$ and $(-1, 2)$. This line has slope -1, so its equation is $y - 1 = -1(x - 0)$, or $y = -x + 1 = -(x - 1)$. Therefore, if $x < 1$, we have $f(x) = -x + 1$. Putting these together, we have

$$f(x) = \begin{cases} -x + 1 & \text{if } x < 1, \\ x - 1 & \text{if } x \geq 1. \end{cases}$$

That V-shaped graph makes us think of absolute value. Our cases do, as well. We can also define f as $\boxed{f(x) = |x - 1|}$, which gives us the same cases as shown in the piecewise defined notation above.

(b) Let g be our function. It's not clear at first what the point at the bottom of the V is, but we can find two points on each side of the V. The right side passes through $(3, 2)$ and $(4, 5)$, so the right side is part of a line with slope $(5 - 2)/(4 - 3) = 3$ that passes through $(3, 2)$. A point-slope equation of this line is $y - 2 = 3(x - 3)$, which gives us $y = 3x - 7$. The bottom point of the V is the x-intercept of this line. Solving $0 = 3x - 7$ gives us the x-coordinate of this x-intercept, which is $x = 7/3$. Therefore, when $x \geq 7/3$, we have $g(x) = 3x - 7$.

Turning to the left branch, we see that it passes through $(2, 1)$ and $(1, 4)$, so it has slope -3 and is part of the line with equation $y - 4 = -3(x - 1)$. Simplifying this equation gives us $y = -3x + 7$, so when $x < 7/3$, we have $g(x) = -3x + 7$. Putting our two cases together gives us

$$g(x) = \begin{cases} -3x + 7 & \text{if } x < 7/3, \\ 3x - 7 & \text{if } x \geq 7/3. \end{cases}$$

Again, both the V shape and the fact that $-3x + 7$ is the opposite of $3x - 7$ make us think of absolute value. We can also define our function g as $\boxed{g(x) = |3x - 7|}$, which gives us the same cases as shown above.

20.5.3 The price per book is the same for up to 24 books, but it suddenly steps down from 12 dollars to 11 if we buy 25. Therefore, it's possible for the total cost of 24 books to be more than the total cost of 25 books. The cost of 25 books is $11(25) = 275$ dollars, but the cost of 24 books is $12(24) = 288$. Since 288 is 13 more than 275, we see that buying 23 books, at a cost of $288 - 12 = 276$ dollars, is also more expensive than buying 25 books. So, if $n = 23$ or $n = 24$, then it is possible to buy more than n books more cheaply than buying exactly n books.

We also must check the next point at which the price per book drops. It costs $10(49) = 490$ dollars to buy 49 books. If we buy n books, and $25 \leq n \leq 48$, then it costs $11n$ dollars. So, we seek the integer values of n for which $11n > 490$ and $n \leq 48$. Dividing both sides of $11n > 490$ by 11 gives $n > 44\frac{6}{11}$. So, we have four more values of n (45, 46, 47, and 48) for which it is cheaper to buy more than n books than to buy exactly n books.

Combining these two cases, we have $\boxed{6}$ values of n for which it is cheaper to buy more than n books than to buy exactly n books.

20.5.4 If the graph of f is continuous, then the graphs of the two cases must meet when $x = 2$, which (loosely speaking) is the dividing point between the two cases. Therefore, we must have $2(2^2)-3 = 2a+4$. Solving this equation gives $a = \boxed{1/2}$.

20.5.5 Clearly, f is not so simple as $f(x) = |x + a|$ for some value of a, because the constants in the two cases are not opposites. So, we guess that $f(x) = |x + a| + b$, for some values of a and b. We guess that the coefficient of x is 1 because the coefficients of x in our cases are -1 and 1. Because the "dividing point" between the cases is $x = -3$, we guess that a is 3, so that $f(x) = |x + 3| + b$. When we write such an $f(x)$ in piecewise defined notation, we get rid of the absolute value by considering the two cases $x \ge -3$ and $x < -3$, which matches the cases we are given in the problem.

If $f(x) = |x + 3| + b$, then when $x \ge -3$, we have $f(x) = x + 3 + b$. Since this must equal the expression $x + 5$ given in the problem for $x \ge -3$, we have $b = 2$. We must check that the other case matches, too. If $f(x) = |x + 3| + 2$ and $x < -3$, then we have $f(x) = -(x + 3) + 2 = -x - 1$, which matches the expression for $f(x)$ when $x < -3$ in the problem. So, we have $\boxed{f(x) = |x + 3| + 2}$.

Review Problems

20.23 We square both sides to get rid of the square root sign. This gives us $2 - 3z = 81$. Solving for z gives $z = \boxed{-79/3}$. We squared an equation, so we have to test our solution to make sure it isn't extraneous. We have

$$\sqrt{2 - 3\left(-\frac{79}{3}\right)} = \sqrt{2 + 79} = 9,$$

so our solution is valid.

20.24 *Solution 1: Casework.* If $|3 - 2r| = 7$, then $3 - 2r = -7$ or $3 - 2r = 7$. In the former case, we have $2r = 10$, so $r = 5$. In the latter case, we have $2r = -4$, so $r = -2$. Therefore, our solutions are $\boxed{r = -2 \text{ and } r = 5}$.

Solution 2: Think of the number line. Because $|3 - 2r| = 7$, the distance between 3 and $2r$ on the number line is 7. The two numbers that are 7 away from 3 on the number line are $3 + 7 = 10$ and $3 - 7 = -4$. So, $2r$ must be 10 or -4. If $2r = 10$, then $r = 5$. If $2r = -4$, then $r = -2$. Therefore, our solutions are $\boxed{r = -2 \text{ and } r = 5}$.

20.25

(a) The number inside the square root symbol must be nonnegative, so we must have $2x - 3 \ge 0$. Therefore, $x \ge 3/2$, so our domain is all real numbers greater than or equal to $3/2$. The range of f is all nonnegative real numbers because $\sqrt{2x - 3}$ can take on any nonnegative real value. (To see why, let $y = \sqrt{2x - 3}$, so $x = (y^2 + 3)/2$. So, for any nonnegative value of y we can find a corresponding x such that $y = f(x)$.)

(b) Although the square root of a negative number is not real, the cube root of a negative number is. Therefore, we have no restriction on the input to f, so the domain is all real numbers. The range of f is all real numbers, as well. To see why, let $y = -3\sqrt[3]{3x - 1}$. Cubing both sides gives $y^3 = -27(3x - 1)$. Solving for x in terms of y gives $x = 1/3 - y^3/81$. So, for any value of y we can find a corresponding x such that $y = f(x)$.

(c) The input to the square root must be nonnegative, so $x - 3 \geq 0$. Therefore, $x \geq 3$, so the domain is all real numbers greater than or equal to 3. The output of the square root can equal any nonnegative number. So, for any input, $f(x)$ equals 7 minus 2 times some nonnegative number. This means $f(x)$ cannot exceed 7. Since $\sqrt{x-3}$ can equal any nonnegative number, the range of f is all real numbers less than or equal to 7.

(d) We can input any number to f, so the domain is all real numbers. Since the floor function can only output integers, and it can output any integer, the range is all integers.

(e) We can input any number to f, so the domain is all real numbers. Because the absolute value of any expression is nonnegative, f equals some nonnegative number plus 6. Therefore, the range of f is all real numbers greater than or equal to 6.

(f) The input to the square root must be nonnegative, so $x + 3 \geq 0$. Therefore, the domain is all real numbers greater than or equal to -3. The output of $2\sqrt{x+3}$ can be any nonnegative number, so the expression inside the absolute value, $2\sqrt{x+3}+4$, can be any real number greater than or equal to 4. Therefore, $3|2\sqrt{x+3}+4|$ can be any real number greater than or equal to 12. Finally, this means that $f(x) = 3|2\sqrt{x+3}+4| - 7$ can be any number greater than or equal to $12 - 7 = 5$. So, the range of f is all real numbers greater than or equal to 5.

20.26 We get rid of the cube root sign by cubing both sides. This gives us $2 - \frac{x}{2} = -27$. Solving this equation gives $x = \boxed{58}$.

20.27

(a) Since $-3 \leq 0$, we have $|-3| = -(-3) = \boxed{3}$.

(b) Because $10 = \sqrt{100} < \sqrt{109} < \sqrt{121} = 11$, the greatest integer less than or equal to $\sqrt{109}$ is $\boxed{10}$.

(c) The smallest integer greater than -54.3 is -54, so $\lceil -54.3 \rceil = \boxed{-54}$.

(d) We have $|-34.1| = 34.1$, so $\lfloor |-34.1| \rfloor = \lfloor 34.1 \rfloor = \boxed{34}$.

(e) Because $-\frac{14}{3} = -4\frac{2}{3}$, we have $\lceil -\frac{14}{3} \rceil = \lceil -4\frac{2}{3} \rceil = -4$. Therefore, $\left| \lceil -\frac{14}{3} \rceil \right| = |-4| = \boxed{4}$.

(f) Because $4 < 7.3 < 9$, we have $2 < \sqrt{7.3} < 3$. Therefore, $\lfloor \sqrt{7.3} \rfloor = 2$. Whether we round 2 up or down, the result is still 2, so $\lfloor \lceil \lfloor \sqrt{7.3} \rfloor \rceil \rfloor = \boxed{2}$.

20.28 $\boxed{\text{No}}$. Suppose $x = 1.3$, for example. Then, $\lfloor \lceil x \rceil \rfloor = \lfloor \lceil 1.3 \rceil \rfloor = 2$ but $\lfloor 1.3 \rfloor = 1$. Extra challenge: Is it true that $\lfloor \lceil x \rceil \rfloor = \lceil x \rceil$?

20.29 $\boxed{\text{Yes}}$. The output of $\sqrt{x}$ must be nonnegative, so $|\sqrt{x}| = \sqrt{x}$.

20.30

(a) Because 2 is between -2 and 4, we have $h(2) = 2 + 4 = \boxed{6}$. Because -1 is between -2 and 4, we have $h(-1) = -1 + 4 = \boxed{3}$. Because -4 is between -2 and -7, we have $h(-4) = -2(-4) + 3 = \boxed{11}$.

(b) Because h is defined for all values of x such that $-7 \leq x \leq 4$, these values are the domain of h.

(c) For $-7 \le x \le -2$, we graph $y = -2x + 3$, and for $-2 < x \le 4$, we graph $y = x + 4$. Note that we have an open circle at $(-2, 2)$, to show that this point is not on the graph of h. The graph is shown below.

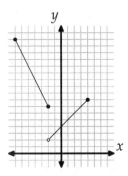

(d) We can read the range of h off our graph. The possible y-coordinates of points on the graph of h are those values of y such that $2 < y \le 17$, so the range of h is all values of y such that $2 < y \le 17$.

(e) Let $k = h(x)$, so that $h(h(x)) = h(k)$. Therefore, $h(h(x))$ exists only if $h(k)$ exists. In other words, k must be in the domain of h. Since $h(x) = k$, k must also be in the range of h. Therefore, $h(h(x))$ exists only when $h(x)$ is in both the domain and the range of h. The domain of h are those values of x such that $-7 \le x \le 4$ and the range of h are those values of y such that $2 < y \le 17$. Since k must be in both the domain and range of h, we must have $2 < k \le 4$.

So, because $k = h(x)$, $h(h(x))$ only exists if $2 < h(x) \le 4$. We consider our two cases separately to find those values of x for which $2 < h(x) \le 4$. If $-7 \le x \le -2$, we must have $2 \le -2x + 3 \le 4$. Solving this inequality chain, we have $1/2 \ge x \ge -1/2$. However, because we must have $-7 \le x \le -2$, we cannot also have $1/2 \ge x \ge -1/2$, so there are no solutions for this case. If $-2 < x \le 4$, we have $h(x) = x + 4$, so $2 < h(x) \le 4$ becomes $2 \le x + 4 \le 4$. Therefore, we have $-2 \le x \le 0$. We cannot have $x = -2$ for this case, but all other values of x such $-2 < x \le 0$ are acceptable for this case. So, the values of x for which $h(h(x))$ exists are $\boxed{-2 < x \le 0}$.

See if you can also solve this part by considering the graph of h.

20.31 We first isolate the square root, so we can then square both sides to get rid of it. Subtracting 4 from both sides gives $x - 4 = \sqrt{11 - 2x}$. Squaring both sides gives $x^2 - 8x + 16 = 11 - 2x$, or $x^2 - 6x + 5 = 0$. Factoring gives $(x - 5)(x - 1) = 0$, so $x = 5$ or $x = 1$. Because we squared the equation, we must check if our solutions are extraneous. For $x = 5$, the equation reads $5 = \sqrt{11 - 10} + 4$, which is true. If $x = 1$, we have $1 = \sqrt{11 - 2} + 4$, which is not true, so $x = 1$ is extraneous. Therefore, our only solution is $\boxed{x = 5}$.

20.32 We get rid of the absolute values by considering cases. Since $x + 3$ is nonnegative for $x \ge -3$ and negative for $x < -3$, and $x + 9$ is nonnegative for $x \ge -9$ and negative for $x < -9$, we have three cases:

- *Case 1:* $x \ge -3$. When $x \ge -3$, we have $|x + 3| = x + 3$ and $|x + 9| = x + 9$, so our equation is $y = |x + 3| - |x + 9| = x + 3 - (x + 9) = -6$.

- *Case 2:* $-9 \le x < -3$. When $-9 \le x < -3$, we have $|x + 3| = -(x + 3) = -x - 3$ and $|x + 9| = x + 9$, so our equation is $y = |x + 3| - |x + 9| = -x - 3 - (x + 9) = -2x - 12$.

- *Case 3:* $x < -9$. When $x < -9$, we have $|x + 3| = -(x + 3) = -x - 3$ and $|x + 9| = -(x + 9) = -x - 9$, so our equation is $y = |x + 3| - |x + 9| = -x - 3 - (-x - 9) = 6$.

Reading through our cases, we produce the graph below.

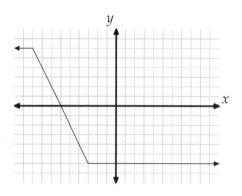

20.33

(a) We can input any value to g except $x = 3$, since $x = 3$ will make the denominator of $(2 + 5x)/(3 - x)$ equal to 0. Therefore, the domain of g is all real numbers except 3.

(b) To find the range of g, we let $y = g(x)$, then solve the equation $y = (2 + 5x)/(3 - x)$ for x in terms of y. Multiplying both sides by $3 - x$, we have $3y - xy = 2 + 5x$. Isolating the terms with x gives $5x + xy = 3y - 2$, so $x(5 + y) = 3y - 2$. Therefore,

$$x = \frac{3y - 2}{5 + y}.$$

So, for any value of y except $y = -5$, we can use this equation to find the value of x such that $y = g(x)$. So, the range of g is all real numbers except -5.

(c) From our first two parts, we see that $x = 3$ is a vertical asymptote of the graph while $y = -5$ is a horizontal asymptote.

(d) Using our asymptotes as a guide (the dashed lines below), we graph $y = g(x)$ below.

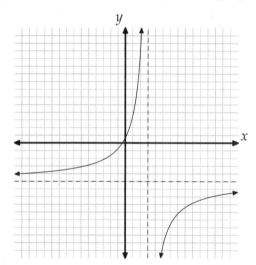

20.34 *Solution 1: Casework.*

- *Case 1: $x \geq 2$.* If $x \geq 2$, then $|x - 2| = x - 2$ and $|x + 2| = x + 2$, so we have $x + 2 = 2(x - 2)$. Solving this equation gives $x = 6$, which satisfies the constraints of this case.

- *Case 2: $-2 \leq x < 2$.* If $-2 \leq x < 2$, then $|x - 2| = -(x - 2) = -x + 2$ and $|x + 2| = x + 2$, so we have $x + 2 = 2(-x + 2)$. Solving this equation gives $x = 2/3$, which satisfies the constraints of this case.

- *Case 3: $x < -2$.* Here, we have $|x - 2| = -(x - 2) = -x + 2$ and $|x + 2| = -(x + 2) = -x - 2$, so we have $-x - 2 = 2(-x + 2)$. Solving this equation gives $x = 6$, which doesn't satisfy the constraints of this case. (But we already found $x = 6$ in the first case, anyway.)

Our three cases give us the solutions $x = 6$ and $x = 2/3$, which have sum $6 + 2/3 = \boxed{20/3}$.

Solution 2: Square both sides. Because $|x|^2 = x^2$, we can square both sides to give

$$(x + 2)^2 = 4(x - 2)^2 \quad \Rightarrow \quad x^2 + 4x + 4 = 4x^2 - 16x + 16 \quad \Rightarrow \quad 3x^2 - 20x + 12 = 0 \quad \Rightarrow \quad (3x - 2)(x - 6) = 0.$$

This gives us the solutions $x = 2/3$ and $x = 6$. Because we squared the original equation, we have to check our answers. Both satisfy the original equation, so neither is extraneous. The sum of these solutions is $20/3$, as before. Notice that we have to find the solutions and test them rather than just using the fact that the sum of the solutions of $3x^2 - 20x + 12 = 0$ is $-(-20/3) = 20/3$ to find our final answer. This is because one or both of the solutions might be extraneous.

See if you can also solve the problem using the number line. (Hint: Because $|x + 2| = 2|x - 2|$, the distance between x and -2 is double the distance between x and 2.)

20.35 Because $\lfloor 3x \rfloor = 6$, the value of $3x$ is greater than or equal to 6 and less than 7. So, we have $6 \leq 3x < 7$. Dividing by 3 gives $\boxed{2 \leq x < 2\frac{1}{3}}$. In interval notation, this is $x \in [2, 2\frac{1}{3})$.

20.36 First, we note that r must be positive, since otherwise $\lfloor r \rfloor + r$ is nonpositive. Next, because $\lfloor r \rfloor$ is an integer and $\lfloor r \rfloor + r = 12.2$, the decimal part of r must be 0.2. Therefore, $r = n + 0.2$ for some integer n, so that $\lfloor r \rfloor = n$ and $\lfloor r \rfloor + r = 2n + 0.2 = 12.2$. Therefore, $n = 6$, and the only value of r that satisfies the equation is $\boxed{r = 6.2}$.

20.37 Factoring the quadratic in the denominator on the right side, we have

$$\frac{4}{x - 3} + 1 = \frac{6}{(x - 3)(x + 2)}.$$

Multiplying both sides by $(x - 3)(x + 2)$ gives

$$(x - 3)(x + 2)\left(\frac{4}{x - 3} + 1\right) = (x - 3)(x + 2) \cdot \frac{6}{(x - 3)(x + 2)} \quad \Rightarrow \quad 4(x + 2) + (x - 3)(x + 2) = 6.$$

Therefore, we have $4x + 8 + x^2 - x - 6 = 6$, or $x^2 + 3x - 4 = 0$. Factoring gives $(x + 4)(x - 1) = 0$, which has solutions $\boxed{x = -4 \text{ and } x = 1}$.

20.38

(a) The right branch of the V goes through $(-3, 1)$ and $(-2, 2)$, so it is part of the line with slope $(2-1)/[-2-(-3)] = 1$ through $(-3, 1)$. A point-slope equation of this line is $y-1 = 1[x-(-3)] = x+3$, so $y = x+4$. Similarly, the left branch passes through $(-5, 1)$ and $(-6, 2)$, which is part of the line with slope $(2-1)/[-6-(-5)] = -1$ through $(-6, 2)$. The equation of this line is $y-2 = -1[x-(-6)] = -x-6$, or $y = -x - 4$. The dividing point between the two branches is $x = -4$. When $x \geq -4$, we have $y = x + 4$; when $x < -4$, we have $y = -x - 4$. So, if we let f be the function that is graphed in this part, we have

$$f(x) = \begin{cases} x + 4 & \text{if } x \geq -4, \\ -x - 4 & \text{if } x < -4. \end{cases}$$

We can also express this in terms of absolute value, $\boxed{f(x) = |x + 4|}$.

(b) Let g be the function that is graphed in this part. The right branch passes through $(2, 1)$ and $(3, 3)$, so it is part of a line with slope $(3 - 1)/(3 - 2) = 2$ through $(2, 1)$. An equation of this line is $y - 1 = 2(x - 2) = 2x - 4$, so $y = 2x - 3$. The x-coordinate of the x-intercept is the solution to the equation $0 = 2x - 3$, which is $x = 3/2$. So, the dividing point between the two branches is $(3/2, 0)$. Therefore, when $x \geq 3/2$, we have $g(x) = 2x - 3$.

The left branch passes through $(1, 1)$ and $(-1, 5)$, so it is part of a line with slope $(5-1)/(-1-1) = -2$ through $(1, 1)$. An equation of this line is $y - 1 = -2(x - 1) = -2x + 2$, so $y = -2x + 3$. Therefore, when $x < 3/2$, we have $g(x) = -2x + 3$.

Putting these two parts together, we have

$$g(x) = \begin{cases} -2x + 3 & \text{if } x < 3/2, \\ 2x - 3 & \text{if } x \geq 3/2. \end{cases}$$

We can express g with absolute value as $\boxed{g(x) = |2x - 3|}$, since $|2x - 3| = 2x - 3$ when $x \geq 3/2$ and $|2x - 3| = -(2x - 3) = -2x + 3$ when $x < 3/2$.

20.39 We get rid of the absolute value sign by considering two cases:

- *Case 1: $x \geq 0$.* If $x \geq 0$, then $|x| = x$, so our expression is $(1 - x)(1 + x) = 1 - x^2$. This expression is positive when $x^2 < 1$. Since x is nonnegative in this case, we must have $0 \leq x < 1$.

- *Case 2: $x < 0$.* If $x < 0$, we have $|x| = -x$, so our expression is $(1 + x)(1 + x) = (1 + x)^2$, which is positive for all x except $x = -1$.

Combining these cases, the values of x for which $(1 - |x|)(1 + x)$ is positive are all real numbers less than 1 except -1. In interval notation, we can write this as $\boxed{x \in (-\infty, -1) \cup (-1, 1)}$.

Challenge Problems

20.40

(a) We have $f(3) = 2(3^2) = \boxed{18}$, so $f(f(3)) = f(18) = 2 - \sqrt{18} = \boxed{2 - 3\sqrt{2}}$ and $f(f(f(3))) = f(f(18)) = f(2 - 3\sqrt{2})$. Since $2 - 3\sqrt{2}$ is negative, and $2 - 3\sqrt{2} > 2 - 3 \cdot 2 = -4$, we have $-6 \leq 2 - 3\sqrt{2} \leq 5$, so we must use the middle rule to find $f(2 - 3\sqrt{2}) = 2(2 - 3\sqrt{2})^2 = 8 - 24\sqrt{2} + 36 = \boxed{44 - 24\sqrt{2}}$.

(b) We check each case:

- *Case 1: $x < -6$.* If $x < -6$, then $f(x) = |2x + 3|$. We also have $2x + 3 < 0$ if $x < -6$, so $|2x + 3| = -(2x + 3) = -2x - 3$. Solving the equation $-2x - 3 = x$ gives $x = -1$, which does not satisfy the constraint $x < -6$ of this case.

- *Case 2: $-6 \leq x \leq 5$.* In this case, $f(x) = 2x^2$, so we must have $2x^2 = x$, or $2x^2 - x = 0$. Factoring gives $x(2x - 1) = 0$, so $x = 0$ or $x = 1/2$. Both satisfy $-6 \leq x \leq 5$, so both are solutions.

- *Case 3: $x > 5$.* In this case, we have $f(x) = 2 - \sqrt{x}$, so we must have $2 - \sqrt{x} = x$. Isolating $\sqrt{x}$ gives $\sqrt{x} = 2 - x$. We could square both sides and find x, but we don't even have to. In this case, we must have $x > 5$, so $2 - x$ is negative. However, $\sqrt{x}$ must be positive, so there are no solutions to $\sqrt{x} = 2 - x$ for $x > 5$.

  Therefore, the values of x for which $f(x) = x$ are $\boxed{x = 0 \text{ and } x = 1/2}$.

(c) Again, we check each case:

- *Case 1: $x < -6$.* If $x < -6$, then $f(x) = |2x + 3| = -2x - 3$, as before. Therefore, we must have $-2x - 3 = 9$, which gives us $x = -6$. Since $x = -6$ does not satisfy $x < -6$, we have no solutions for this case.

- *Case 2: $-6 \leq x \leq 5$.* In this case, $f(x) = 2x^2$, so we must have $2x^2 = 9$, so $x = \pm\sqrt{9/2} = \pm 3\sqrt{2}/2$. Both of these satisfy $-6 \leq x \leq 5$.

- *Case 3: $x > 5$.* In this case, $f(x) = 2 - \sqrt{x}$, so we must have $2 - \sqrt{x} = 9$. Isolating $\sqrt{x}$ gives $\sqrt{x} = -7$, which has no solutions.

  So, the values of x such that $f(x) = 9$ are $x = \boxed{\pm 3\sqrt{2}/2}$.

20.41 We factor both denominators to find $\dfrac{t - 7}{(t - 3)(t + 1)} = \dfrac{t + 2}{(t + 5)(t - 3)}$. Multiplying both sides by $(t + 1)(t - 3)(t + 5)$ gives

$$(t - 7)(t + 5) = (t + 2)(t + 1) \quad \Rightarrow \quad t^2 - 2t - 35 = t^2 + 3t + 2 \quad \Rightarrow \quad 5t = -37 \quad \Rightarrow \quad \boxed{t = -\frac{37}{5}}.$$

20.42 We get rid of the denominators by multiplying both sides by $6\sqrt{x + 1}$, which gives

$$6\sqrt{x + 1}\left(\sqrt{x + 1} + \frac{1}{\sqrt{x + 1}}\right) = 13\sqrt{x + 1} \quad \Rightarrow \quad 6(x + 1) + 6 = 13\sqrt{x + 1} \quad \Rightarrow \quad 6x + 12 = 13\sqrt{x + 1}.$$

Squaring both sides gets rid of the square root sign and gives us

$$36x^2 + 144x + 144 = 169x + 169 \quad \Rightarrow \quad 36x^2 - 25x - 25 = 0 \quad \Rightarrow \quad (9x + 5)(4x - 5) = 0.$$

Therefore, we have $x = -5/9$ or $x = 5/4$. Because we squared the equation, we must test the solutions. For $x = -5/9$, we have $\sqrt{x+1} = \sqrt{4/9} = 2/3$, and $2/3 + 3/2 = 4/6 + 9/6 = 13/6$, so $x = -5/9$ is a valid solution. For $x = 5/4$, we have $\sqrt{x+1} = \sqrt{9/4} = 3/2$ and $3/2 + 2/3 = 9/6 + 4/6 = 13/6$, so $x = 5/4$ is also a valid solution. So, the solutions to the equation are $\boxed{x = -5/9 \text{ and } x = 5/4}$.

20.43 We start by squaring both sides to get rid of some of the square root signs. This gives

$$4x + 1 - 2\sqrt{(4x+1)(x-11)} + x - 11 = 2x - 4 \quad \Rightarrow \quad 3x - 6 = 2\sqrt{4x^2 - 43x - 11}.$$

Squaring both sides again gives

$$(3x - 6)^2 = \left(2\sqrt{4x^2 - 43x - 11}\right)^2 \quad \Rightarrow \quad 9x^2 - 36x + 36 = 4(4x^2 - 43x - 11) = 16x^2 - 172x - 44.$$

Therefore, we have $7x^2 - 136x - 80 = 0$. Factoring gives $(x - 20)(7x + 4) = 0$, which has solutions $x = 20$ and $x = -4/7$. Because we squared the equation, we must make sure the solutions are not extraneous. Since $x = 20$ satisfies the equation, it is not extraneous. On the other hand, $x = -4/7$ is extraneous (because $x = -4/7$ makes $\sqrt{x-11}$ in the original equation undefined), so $\boxed{x = 20}$ is the only solution.

20.44 Dividing all parts of the inequality chain by 4 gives $\frac{1}{4} \le \{x\} < \frac{3}{4}$. The values of x that satisfy this inequality are all values of x that have a fractional part greater than or equal to $1/4$ and less than $3/4$. These values of x are highlighted on the number line below; the pattern continues in both directions.

20.45 The expression $|6x^2 - 2x|$ is nonnegative because absolute value is always nonnegative. Therefore, to minimize $|6x^2 - 2x|$, we must find the values of x that make $6x^2 - 2x = 0$. Factoring gives $2x(3x - 1) = 0$. Therefore, the values of x that minimize $|6x^2 - 2x|$ are $\boxed{x = 0 \text{ and } x = 1/3}$.

20.46 We have $\sqrt{1} = 1$, so $\lceil \sqrt{1} \rceil = 1$.

We have $1 < \sqrt{2} < \sqrt{3} < \sqrt{4} = 2$, so $\lceil \sqrt{2} \rceil = \lceil \sqrt{3} \rceil = \lceil \sqrt{4} \rceil = 2$. These three combine in our sum for a total of $3 \cdot 2 = 6$.

We have $2 < \sqrt{5} < \sqrt{6} < \cdots < \sqrt{9} = 3$, so $\lceil \sqrt{5} \rceil = \lceil \sqrt{6} \rceil = \cdots = \lceil \sqrt{9} \rceil = 3$. These 5 combine in our sum for a total of $5 \cdot 3 = 15$.

We have $3 < \sqrt{10} < \sqrt{11} < \cdots < \sqrt{16} = 4$, so $\lceil \sqrt{10} \rceil = \lceil \sqrt{11} \rceil = \cdots = \lceil \sqrt{16} \rceil = 4$. These 7 combine in our sum for a total of $7 \cdot 4 = 28$.

Similarly, $\lceil \sqrt{17} \rceil = \lceil \sqrt{18} \rceil = \cdots = \lceil \sqrt{25} \rceil = 5$, so these 9 combine in our sum for a total of $9 \cdot 5 = 45$.

Similarly, $\lceil \sqrt{26} \rceil = \lceil \sqrt{27} \rceil = \cdots = \lceil \sqrt{36} \rceil = 6$, so these 11 combine in our sum for a total of $11 \cdot 6 = 66$.

Similarly, $\lceil \sqrt{37} \rceil = \lceil \sqrt{38} \rceil = \cdots = \lceil \sqrt{49} \rceil = 7$, so these 13 combine in our sum for a total of $13 \cdot 7 = 91$.

We make sure we don't forget $\lceil \sqrt{1} \rceil = 1$ at the beginning and $\lceil \sqrt{50} \rceil = 8$ at the end, and our sum is

$$1 + 3 \cdot 2 + 5 \cdot 3 + 7 \cdot 4 + 9 \cdot 5 + 11 \cdot 6 + 13 \cdot 7 + 8 = \boxed{260}.$$

20.47 We get rid of the cube root by cubing both sides. Since $(a+b)^3 = a^3 + 3a^2b + 3ab^2 + b^3$, we have

$$x^3 - x^2 - 10 = (x-1)^3 = x^3 + 3(x^2)(-1) + 3(x)(-1)^2 + (-1)^3 = x^3 - 3x^2 + 3x - 1.$$

Bringing all the terms to the left side gives $2x^2 - 3x - 9 = 0$. Factoring gives $(2x+3)(x-3) = 0$, so our solutions are $\boxed{x = -3/2 \text{ and } x = 3}$.

20.48 Factoring the denominator on the left side gives us

$$\frac{3x-2}{x(x+1)^2} = \frac{A}{x} + \frac{B}{x+1} + \frac{C}{(x+1)^2}.$$

Multiplying both sides by $x(x+1)^2$ to get rid of the fractions gives us

$$3x - 2 = A(x+1)^2 + Bx(x+1) + Cx.$$

This must hold for all values of x, so we can choose values of x that allow us to compute A, B, and C easily. Letting $x = 0$ gets rid of B and C on the right, leaving $-2 = A(0+1)^2 = A$. Letting $x = -1$ gets rid of A and B on the right, leaving $-3 - 2 = -C$, or $C = 5$. Using $A = -2$ and $C = 5$ and letting $x = 1$, we have $3 - 2 = -2(1+1)^2 + B(1)(1+1) + 5 \cdot 1$. Solving this equation for B gives us $B = 2$. Therefore, $\boxed{(A, B, C) = (-2, 2, 5)}$.

Note that we also could have found A, B, and C by expanding our equation above and matching the coefficients on the left side with the coefficients on the right side.

20.49 Since $f(f(x)) = |2|2x - 1| - 1|$, our equation is $|2|2x - 1| - 1| = x$. We get rid of the innermost absolute value signs by considering when $2x - 1 \geq 0$ and when $2x - 1 < 0$. Since $2x - 1 \geq 0$ when $x \geq 1/2$ and $2x - 1 < 0$ when $x < 1/2$, we have

- *Case 1:* $x \geq 1/2$. Here, we have $|2x - 1| = 2x - 1$, so our equation is $|2(2x - 1) - 1| = x$, so $|4x - 3| = x$. Once again, we consider two cases to get rid of the absolute value signs.

  - *Case 1A:* $x \geq 3/4$. If $x \geq 3/4$, then $|4x - 3| = 4x - 3$, so our equation is $4x - 3 = x$. Solving this equation gives us $x = 1$. This satisfies the constraints of this case.

  - *Case 1B:* $1/2 \leq x < 3/4$. If $x < 3/4$, then $|4x - 3| = -(4x - 3) = -4x + 3$, so our equation is $-4x + 3 = x$. Therefore, $x = 3/5$, which satisfies the constraints of this case.

- *Case 2:* $x < 1/2$. If $x < 1/2$, then $|2x - 1| = -(2x - 1) = -2x + 1$, so our equation is $|2(-2x + 1) - 1| = x$, so $|-4x + 1| = x$. We break this case into two subcases to get rid of the absolute value signs.

  - *Case 2A:* $1/4 < x < 1/2$. If $x > 1/4$, then $|-4x + 1| = -(-4x + 1) = 4x - 1$, so our equation is $4x - 1 = x$, from which we have $x = 1/3$, which satisfies the constraints of this case.

  - *Case 2B:* $x \leq 1/4$. If $x \leq 1/4$, then $|-4x + 1| = -4x + 1$, so our equation is $-4x + 1 = x$, which gives $x = 1/5$, which satisfies the constraints of this case.

Putting all four subcases together, our solutions are $\boxed{1/5, 1/3, 3/5, \text{ and } 1}$.

20.50 Thinking about graphs helped us learn how to minimize quadratic expressions, so we try that here, by graphing $y = |x + 3| + |x - 7|$. We first break it into three cases:

- *Case 1:* $x \geq 7$. This gives us $y = x + 3 + x - 7 = 2x - 4$.

- *Case 2:* $-3 \leq x < 7$. This gives us $y = x + 3 + [-(x - 7)] = 10$.

- *Case 3:* $x < -3$. This gives us $y = -(x + 3) + [-(x - 7)] = -2x + 4$.

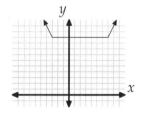

We might be able to determine our minimum just by thinking about the results of these three cases, but drawing the graph makes it very clear. At right is our graph. We see that the smallest value y takes on in the graph is $\boxed{10}$, so the minimum possible value of $|x + 3| + |x - 7|$ is 10.

20.51 In each part, let f be the function that is graphed.

(a) This graph has the V-shape of an absolute value graph, but the graph is upside-down! We know that the graph of $y = -f(x)$ is the result of flipping the graph of $y = f(x)$ over the x-axis, so we guess that this graph is the graph of $y = -|x + a|$ for some value of a. (We guess that the coefficient of x in this expression is 1 because the branches have slopes 1 and -1.) Finally, the graph goes through $(-4, 0)$, so $-|x + a| = 0$ when $x = -4$. Therefore, $a = 4$. Graphing $y = -|x + 4|$ produces the graph in the problem, so our function is $\boxed{f(x) = -|x + 4|}$.

(b) The upward opening V-shape makes us first guess that it is the graph of $y = |x + a|$ for some a, because the slopes of our two branches are 1 and -1. However, the graph of $y = |x + a|$ meets the x-axis at $(-a, 0)$, but the graph in the problem never gets lower than $y = 4$. So, we guess that it is the graph of $y = |x + a| + 4$, because the lowest point on this graph is $(-a, 4)$. Because absolute value is always nonnegative, all other points on the graph of $y = |x + a| + 4$ will have $y > 4$. To find a, we note that $y = 4$ at $(-3, 4)$ on our graph, so we must have $4 = |-3 + a| + 4$, from which we find $a = 3$. So, our graph is the graph of the function $\boxed{f(x) = |x + 3| + 4}$.

(c) This graph resembles the graph of $y = |x + 3| + |x - 2|$ that we graphed in the book, so we guess that it is the graph of $y = |x + a| + |x + b|$ for some a and b. In graphing $y = |x + 3| + |x - 2|$, we found that the graph included a horizontal segment from $(-3, 5)$ to $(2, 5)$, where these x-coordinates are the values of x that make $x + 3 = 0$ and $x - 2 = 0$, respectively. In other words, these values of x were the boundary points of our cases. In the graph in this problem, the boundary points of the cases are $(1, 3)$ and $(-2, 3)$, so we guess that a and b are -1 and 2. When we graph $y = |x - 1| + |x + 2|$, we get the graph shown in the problem. So, our graph is the graph of the function $\boxed{f(x) = |x - 1| + |x + 2|}$.

(d) Again, this graph resembles the graph of $y = |x + 3| + |x - 2|$ that we graphed in the book, so we guess that it is the graph of $y = |x + a| + |x + b|$ for some a and b. The boundary points of the "cases" in this graph are $(-1, 2)$ and $(3, 2)$, so we guess that our graph is the graph of $y = |x + 1| + |x - 3|$. However, when we graph this equation, it is everywhere exactly 2 higher than the graph in the problem. We correct for this by adding -2 to the right side: $y = |x + 1| + |x - 3| - 2$. The graph of this equation matches the graph in the problem, so our function is $\boxed{f(x) = |x + 1| + |x - 3| - 2}$.

20.52 We start by graphing the inequality $|x - 4| + |y - 3| \leq 4$. We graph this by considering 4 cases. The expression $x - 4$ is nonnegative for $x \geq 4$ and is negative for $x < 4$, so $|x - 4| = x - 4$ for $x \geq 4$ and $|x - 4| = -(x - 4) = -x + 4$ for $x < 4$. Similarly, $|y - 3| = y - 3$ for $y \geq 3$ and $|y - 3| = -(y - 3) = -y + 3$ for $y < 3$.

So, our cases and our graph are:

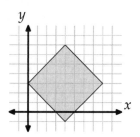

- $x \geq 4$, $y \geq 3$. We then have $x + y \leq 11$.

- $x \geq 4$, $y < 3$. We then have $x - y \leq 5$.

- $x < 4$, $y < 3$. We then have $-x - y \leq -3$.

- $x < 4$, $y \geq 3$. We then have $-x + y \leq 3$.

Combining these cases gives us the graph on the right above. The shaded region satisfies all four inequalities. As we discussed in Chapter 9, the maximum value of $4x + 5y$ must occur at one of the corners of this region. The corners of this region are $(8,3)$; $(4,7)$; $(4,-1)$; $(0,3)$. Testing the first two (which will clearly beat the last two), we find that our maximum is $4x + 5y = 4 \cdot 4 + 5 \cdot 7 = \boxed{51}$.

20.53 We consider two cases to get rid of the absolute value:

- *Case 1: $x - \log_{10} y \geq 0$.* If $x - \log_{10} y \geq 0$, then $|x - \log_{10} y| = x - \log_{10} y$, so our equation is $x - \log_{10} y = x + \log_{10} y$, or $2 \log_{10} y = 0$. Therefore, $\log_{10} y = 0$, so $y = 1$. So, the solutions (x, y) in this case are all ordered pairs $(x, 1)$ such that $x - \log_{10} 1 \geq 0$. Since $\log_{10} 1 = 0$, we have $x \geq 0$. So, all the solutions in this case are the ordered pairs $(x, 1)$ where x is nonnegative.

- *Case 2: $x - \log_{10} y < 0$.* If $x - \log_{10} y < 0$, then $|x - \log_{10} y| = -(x - \log_{10} y) = -x + \log_{10} y$. Therefore, we must have $-x + \log_{10} y = x + \log_{10} y$, which gives us $2x = 0$, or $x = 0$. So, in this case, our solutions are the ordered pairs $(0, y)$ such that $0 - \log_{10} y < 0$, or $\log_{10} y > 0$. We have $\log_{10} y > 0$ for all $y > 1$, so our solutions in this case are the ordered pairs $(0, y)$ with $y > 1$.

Combining these two cases, our solutions are ordered pairs $(x, 1)$ where x is nonnegative, and $(0, y)$ where y is greater than 1.

20.54 *Solution 1: Algebra.* Since the two graphs meet at $(2, 5)$ and $(8, 3)$, these two ordered pairs must satisfy both equations. This gives us a system of 4 equations:

$$5 = -|2 - a| + b,$$
$$3 = -|8 - a| + b,$$
$$5 = |2 - c| + d,$$
$$3 = |8 - c| + d.$$

Subtracting the second from the first eliminates b and leaves us an equation with just a:

$$2 = -|2 - a| + |8 - a|.$$

This is an absolute value equation – looks like a job for casework.

- *Case 1: $a > 8$.* If $a > 8$, then $|2 - a| = -2 + a$ and $|8 - a| = -8 + a$, so our equation is $2 = 2 - a - 8 + a$, which has no solution.

- *Case 2:* $2 < a \le 8$. If $2 < a \le 8$, then $|2-a| = -2+a$ and $|8-a| = 8-a$, so our equation is $2 = 2-a+8-a$, which gives $a = 4$. This value of a satisfies the constraints of this case.

- *Case 3:* $a \le 2$. If $a \le 2$, then $|2 - a| = 2 - a$ and $|8 - a| = 8 - a$, so our equation is $2 = -2 + a + 8 - a$, which has no solution.

So, $a = 4$.

Subtracting the fourth of our system of equations from the third gives

$$2 = |2 - c| - |8 - c|.$$

Working through cases as before, we find that $c = 6$. Therefore, $a + c = \boxed{10}$.

Solution 2: Graphing. The graph of $y = -|x - a| + b$ is an inverted V, in which the two branches have slopes of 1 and -1. Since we know that $(2, 5)$ and $(8, 3)$ are on different branches (because the slope between the two is neither 1 nor -1), we can start from those two points to create the graph. Since $(2, 5)$ is to the left of $(8, 3)$, it is on the left branch. This branch slopes upward to the right (the V is inverted, remember). The branch through $(8, 3)$ has slope -1, so it slopes upward to the left. As shown in the graph at right, these branches intersect at $(4, 7)$. The x-coordinate of this point gives us a, since $y = -|x - a| + b$ reaches its maximal value of y when $x = a$.

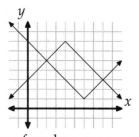

Similarly, the graph of $y = |x - c| + d$ is a V shape with branches of slopes 1 and -1. Only, this time, the left branch has slope -1 and the right has slope 1. As shown these branches meet at $(6, 1)$, so $c = 6$.

Therefore, $a + c = 4 + 6 = \boxed{10}$.

Notice that $a + c = 2 + 8$, where the 2 and the 8 are the x-coordinates of the intersection points. Is this a coincidence?

20.55 As usual, we get rid of the absolute value signs by considering cases. We ignore the case $x = 0$ because we want nonzero solutions.

- *Case 1:* $x > 0$. When $x > 0$, we have $|x| = x$, so our first equation becomes $x + y = 3$, or $y = 3 - x$. Our second equation becomes $xy + x^3 = 0$. Substitution then gives us

$$x(3 - x) + x^3 = 0 \quad \Rightarrow \quad x^3 - x^2 + 3x = 0 \quad \Rightarrow \quad x(x^2 - x + 3) = 0.$$

Since x must be positive in this case, we disregard the solution $x = 0$. The discriminant of the quadratic is negative, so there are no real solutions in this case.

- *Case 2:* $x < 0$. When $x < 0$, we have $|x| = -x$, so our first equation becomes $-x + y = 3$, or $y = 3 + x$. The second equation becomes $-xy + x^3 = 0$. Substitution then gives us

$$-x(3 + x) + x^3 = 0 \quad \Rightarrow \quad x^3 - x^2 - 3x = 0 \quad \Rightarrow \quad x(x^2 - x - 3) = 0.$$

Again, we disregard the $x = 0$ solution since we are told that x and y are nonzero. The discriminant of the quadratic is positive, and one of the roots of the quadratic is negative, because $-b - \sqrt{b^2 - 4ac}$ is negative. We don't have to find x and y to evaluate $x - y$, however. We can multiply the equation $-x + y = 3$ by -1 to find $x - y = \boxed{-3}$.

20.56 It looks nearly impossible to find the solutions to this equation, but fortunately, we don't have to find them. We only have to count them. Therefore, we try graphing both $y = \left| |x^2 - 1| - 1 \right|$ and $y = 2^x$ and counting the number of points at which the two graphs intersect. We know how to graph $y = 2^x$ already, but the other equation is more troublesome. We graph $y = \left| |x^2 - 1| - 1 \right|$ in steps, working from the inside out. First, we graph $y = x^2 - 1$, which is just a parabola with vertex $(0, -1)$, as shown in the first graph below. We use this graph to produce the graph of $y = |x^2 - 1|$ by reflecting over the x-axis the portion of the graph of $y = x^2 - 1$ for which y is negative. This gives us the second graph.

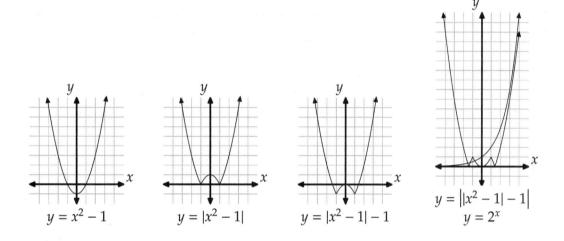

$$y = x^2 - 1 \qquad y = |x^2 - 1| \qquad y = |x^2 - 1| - 1$$

$$y = \left| |x^2 - 1| - 1 \right|$$
$$y = 2^x$$

The graph of $y = |x^2 - 1| - 1$ is a downward shift by 1 of the graph of $y = |x^2 - 1|$, and this is our third graph above. Finally, we produce $y = \left| |x^2 - 1| - 1 \right|$ from $y = |x^2 - 1| - 1$ by reflecting over the x-axis the portion of the graph of $y = |x^2 - 1| - 1$ for which y is negative. This gives us the fourth graph above. When we also graph $y = 2^x$ as shown in the fourth graph above, we see that our graphs intersect in $\boxed{3}$ points. (The graphs get very close on the right, but we can see that $(4, 16)$ is on $y = 2^x$ and $(4, 14)$ is on $y = \left| |x^2 - 1| - 1 \right|$. As x gets larger, the latter graph is the same as $y = x^2 - 2$, which goes upward much more slowly as x increases beyond 4 than the graph of $y = 2^x$.)

CHAPTER 21

Sequences & Series

Exercises for Section 21.1

21.1.1

(a) The first term is 1, and the common difference is 3. Therefore, to get to the 15th term, we must add 3 to the first term 14 times, to get $1 + 3(14) = \boxed{43}$.

(b) To get to the nth term, we must add 3 to the first term $n - 1$ times, so the nth term is $1 + 3(n - 1) = \boxed{3n - 2}$.

21.1.2 *Solution 1: Find the first term and the common difference.* Let the first term of the sequence be a and the common difference be d. Then, the third term is $a + 2d$ and the sixth term is $a + 5d$, so we have the system $a + 2d = 5$, $a + 5d = -1$. Subtracting the first equation from the second gives $3d = -6$, so $d = -2$. Substituting this into either of our original equations gives $a = 9$, so the twelfth term of the sequence is $a + 11d = 9 + 11(-2) = \boxed{-13}$.

Solution 2: Use our understanding of arithmetic sequences. The sixth term is 6 less than the third term. The twelfth term is twice as far from the sixth term (6 steps) as the sixth term is from the third term (3 steps). Therefore, the twelfth term is $2(6) = 12$ less than the sixth term, so it is $-1 - 12 = \boxed{-13}$.

21.1.3 Let there be n terms. The common difference of the sequence is $11 - 5 = 6$ and the first term is 5, so the nth term is $5 + 6(n - 1)$. Since this must equal our given last term, we have $5 + 6(n - 1) = 89$. Solving this equation gives $n = \boxed{15}$.

21.1.4 The even positive integers form the arithmetic sequence

$$2, 4, 6, 8, \ldots, 2n, \ldots$$

The odd positive integers form the arithmetic sequence

$$1, 3, 5, 7, \ldots, 2n - 1, \ldots$$

Therefore, the nth even number is $2n$ and the nth odd number is $2n - 1$. So, the 171st even number is $2(171) = 342$ and the 219th odd number is $2(219) - 1 = 437$. The difference of these is $437 - 342 = \boxed{95}$.

21.1.5 The term a_{17} is 9 steps after a_8, so we have $a_{17} = a_8 + 9d = 2001 + 9d$. We must have $a_{17} > 10000$, so we have $2001 + 9d > 10000$. Solving this inequality gives $d > 7999/9$, or $d > 888\frac{7}{9}$. The smallest integer that satisfies this inequality is $\boxed{889}$.

Exercises for Section 21.2

21.2.1

(a) We know the first and last term, so we only have to find the number of terms. The common difference between terms is 7. Let n be the number of terms. Since 21 is the first term and 105 is the last term (and is therefore $n-1$ steps from the first term), we have $21 + 7(n-1) = 105$, from which we find $n = 13$. Therefore, the sum equals $13(21 + 105)/2 = \boxed{819}$.

(b) The last term is 13 steps after the first term, so it equals $7 - 3(13) = -32$. Therefore, the sum equals $14(7 - 32)/2 = \boxed{-175}$.

(c) Let n be the number of terms. The common difference is $5/6 - 1/2 = 1/3$, the first term is $1/2$, and the last term is $19/2$. So, we must have $1/2 + (n-1)(1/3) = 19/2$. Solving this equation gives $n = 28$. Therefore, our sum is $28(1/2 + 19/2)/2 = \boxed{140}$.

21.2.2 Let d be the common difference. Then, the last term is $7 + (15-1)d = 7 + 14d$. In terms of d, the sum equals $15(7 + 7 + 14d)/2 = 15(7 + 7d) = 105 + 105d$. We are told that this sum equals -210, so we have $105 + 105d = -210$, from which we find $d = \boxed{-3}$.

21.2.3 Let the first term be a and the common difference be d. Because the sum of the first 5 terms is 70, we have $5(2a + 4d)/2 = 70$, so $a + 2d = 14$. Because the sum of the first 10 terms is 210, we have $10(2a + 9d)/2 = 210$, so $2a + 9d = 42$. Subtracting twice the equation $a + 2d = 14$ from the equation $2a + 9d = 42$ gives $5d = 14$, so $d = 14/5$. Therefore, $a = 14 - 2d = \boxed{42/5}$.

21.2.4 We showed that the sum of an arithmetic series equals the average of the first and last term times the number of terms. The average of the first and last term of an arithmetic series with an odd number of terms equals the middle term. To see why, let the first term be a, the common difference be d and the number of terms be $2n + 1$ (to make sure the number of terms is odd). Then, the first term is a and the last term is $a + (2n + 1 - 1)d = a + 2nd$. So, the average of the first and last terms is $(a + a + 2nd)/2 = a + nd$, which equals the $(n+1)^{\text{st}}$ term. The $(n+1)^{\text{st}}$ term of a series with $2n + 1$ terms is the middle term of the series. Therefore, the sum of an arithmetic series with an odd number of terms equals the number of terms times the middle term of the series.

21.2.5 As explained in the text, the sum of the first 8 consecutive odd counting numbers is $8^2 = 64$. Therefore, the sum of our 5 consecutive even integers must equal 60. As explained in the previous problem, the sum of a 5-term arithmetic series equals 5 times the middle term of the series. Therefore, the middle term of our series equals $60/5 = 12$. Since the middle term is 12, the smallest term, which is two steps away from the middle term, must be $12 - 2 - 2 = \boxed{8}$.

21.2.6 The sum of the first k positive integers is $k(k+1)/2$. Therefore, the sum of the first $3n$ positive integers is $3n(3n+1)/2$ and the sum of the first n positive integers is $n(n+1)/2$. So, the information in the problem gives us

$$\frac{3n(3n+1)}{2} - \frac{n(n+1)}{2} = 150 \quad \Rightarrow \quad \frac{9n^2 + 3n - n^2 - n}{2} = 150 \quad \Rightarrow \quad 8n^2 + 2n = 300.$$

So, we have $4n^2 + n - 150 = 0$. Factoring gives $(4n + 25)(n - 6) = 0$. Since n must be a positive integer, we have $n = 6$.

The problem then asks for the sum of the first $4n = 24$ positive integers, which is $24(24+1)/2 = \boxed{300}$.

21.2.7 Let d be the common difference, which equals the expression we seek, $a_2 - a_1$. We then have $a_{100} = a_1 + 99d$, so our first equation gives us $100(a_1 + a_1 + 99d)/2 = 100$, so $2a_1 + 99d = 2$. We also have $a_{101} = a_1 + 100d$ and $a_{200} = a_1 + 199d$, so the second equation gives us $100(a_1 + 100d + a_1 + 199d)/2 = 200$, which gives $2a_1 + 299d = 4$. Subtracting $2a_1 + 99d = 2$ from $2a_1 + 299d = 4$ gives $200d = 2$, so $d = \boxed{1/100}$.

We also could note that $a_{101} - a_1 = 100d$, $a_{102} - a_2 = 100d$, and so on, so that when we subtract the first equation from the second, we have $100(100d) = 100$, so $d = 1/100$.

21.2.8

(a) Because $a_1 \leq a_2 \leq \cdots \leq a_n$, we know that $a_1 + a_2 + \cdots + a_n \geq a_1 + a_1 + \cdots + a_1$, where there are n terms on each side. Therefore, $a_1 + a_2 + \cdots + a_n \geq na_1$, so dividing by n gives

$$\frac{a_1 + a_2 + \cdots + a_n}{n} \geq a_1.$$

This shows that the arithmetic mean of the n numbers is greater than or equal to the smallest of the numbers. Similarly, since no term is larger than a_n, we have $a_1 + a_2 + \cdots + a_n \leq na_n$. Dividing this by n tells us that

$$\frac{a_1 + a_2 + \cdots + a_n}{n} \leq a_n,$$

so the arithmetic mean of the n numbers is less than or equal to the largest of the numbers.

(b) The arithmetic series $a_1 + a_2 + \cdots + a_n$ equals the average of the first and last terms times the number of terms, or $n(a_1 + a_n)/2$. Therefore, the arithmetic mean of the numbers in the arithmetic sequence $a_1, a_2, \ldots, a_n$ is

$$\frac{a_1 + a_2 + \cdots + a_n}{n} = \frac{n(a_1 + a_n)/2}{n} = \frac{a_1 + a_n}{2},$$

which is the arithmetic mean of a_1 and a_n.

Exercises for Section 21.3

21.3.1

(a) The first term is 3, and the ratio between terms is $(9/2)/3 = 3/2$. Therefore, the eighth term of the sequence is $3 \cdot (3/2)^{8-1} = 3^8/2^7 = \boxed{6561/128}$.

(b) The first term is 3 and the ratio between terms is $3/2$, so the nth term is $3(3/2)^{n-1} = \boxed{3^n/2^{n-1}}$.

21.3.2

(a) The common ratio is $(-24)/16 = \boxed{-3/2}$.

(b) The first term is 16 and the common ratio is $(-3/2)$, so the nth term is $16(-3/2)^{n-1}$. Therefore, we must have $16(-3/2)^{n-1} = -273.375$. We write -273.375 as a fraction to help solve the equation:

$$16\left(\frac{-3}{2}\right)^{n-1} = -273\frac{3}{8} \quad \Rightarrow \quad 16 \cdot \frac{(-3)^{n-1}}{2^{n-1}} = -\frac{2187}{8} \quad \Rightarrow \quad \frac{(-3)^{n-1}}{2^{n-1}} = -\frac{2187}{128}.$$

Because $2^7 = 128$ and $(-3)^7 = -2187$, we see that $n = \boxed{8}$ is the solution to this equation.

21.3.3

(a) At the end of the first day, there are 2 amoebas. At the end of the second, there are $2 \cdot 2 = 2^2$ amoebas. At the end of the third day, there are $2 \cdot 2^2 = 2^3$ amoebas, and so on. So, after the seventh day, there are $2^7 = \boxed{128}$ amoebas.

(b) On the 23rd day, the number of amoebas doubles, just like it does every day. This doubling makes the puddle exactly full of amoebas. So, before this doubling, the puddle is half-full of amoebas. Therefore, it was the doubling the previous day that made the puddle half-full of amoebas. So, the puddle is first half-full of amoebas on the $\boxed{22^{\text{nd}}}$ day.

21.3.4 Let the common ratio between terms be r. We must multiply the fifth term by r twice to get to the seventh term, so we have $3r^2 = 9$. Therefore, we have $r = \pm\sqrt{3}$. We multiply the fifth term by r to get the sixth term, so the two possible sixth terms are $\boxed{3\sqrt{3} \text{ and } -3\sqrt{3}}$.

21.3.5

(a) We multiply a_1 by r exactly $n-1$ times to get a_n, so we have $a_n = a_1 r^{n-1}$. Similarly, we have $a_{n-k} = a_1 r^{n-k-1}$ and $a_{n+k} = a_1 r^{n+k-1}$.

(b) Using our expressions from part (a), we have

$$\sqrt{a_{n-k}a_{n+k}} = \sqrt{a_1 r^{n-k-1} \cdot a_1 r^{n+k-1}} = \sqrt{a_1^2 r^{n-k-1+n+k-1}} = \sqrt{a_1^2}\sqrt{r^{2n-2}} = a_1 r^{n-1}.$$

From part (a), we have $a_n = a_1 r^{n-1}$, so we have $\sqrt{a_{n-k}a_{n+k}} = a_n$. Therefore, if a term of a geometric sequence is exactly between two other terms in the sequence, then that term is the geometric mean of the other two terms. (Notice that our relationship only works when the sequence consists of nonnegative numbers, because the expression $\sqrt{a_{n-k}a_{n+k}}$ cannot be negative.)

21.3.6 If x, y, and z are allowed to be nonreal, then we cannot disregard the case $r^2 = -3$, as we did in the solution for which we assumed x, y, and z are real. If $r^2 = -3$, then $r = \pm i\sqrt{3}$. If $r = i\sqrt{3}$, then we have the sequence

$$3, 3i\sqrt{3}, -9, -9i\sqrt{3}, 27.$$

If $r = -i\sqrt{3}$, then we have the sequence

$$3, -3i\sqrt{3}, -9, 9i\sqrt{3}, 27.$$

Exercises for Section 21.4

21.4.1

(a) The first term is -1, the common ratio is 3, and there are 7 terms, so the sum equals

$$\frac{(-1)(3^7 - 1)}{3 - 1} = \frac{-2186}{2} = \boxed{-1093}.$$

(b) The first term is 3 and the common ratio is $(-6)/3 = -2$, but we'll have to do a little work to find the number of terms. Suppose there are n terms in the series. We must multiply the first term by

-2 exactly $n-1$ times to get to the last term, so we must have $3(-2)^{n-1} = 768$. Therefore, we have $(-2)^{n-1} = 256$, so $n-1 = 8$. This gives us $n = 9$, and our sum is

$$\frac{3[(-2)^9 - 1]}{-2 - 1} = \frac{3(-512 - 1)}{-3} = \boxed{513}.$$

(c) The common ratio of this infinite series is $10/100 = 1/10$, so its sum is $100/(1 - 1/10) = \boxed{1000/9}$.

(d) The common ratio of this infinite series is $(-6)/8 = -3/4$, so its sum is $8/[1 - (-3/4)] = \boxed{32/7}$.

21.4.2 This geometric series has first term 1, common ratio 2, and $n+1$ terms (make sure you see why there are $n+1$ terms, not n terms). Therefore, its sum is

$$\frac{1(2^{n+1} - 1)}{2 - 1} = \boxed{2^{n+1} - 1}.$$

21.4.3

(a) We have

$$0.\overline{4} = \frac{4}{10} + \frac{4}{100} + \frac{4}{1000} + \cdots.$$

This infinite geometric series has first term $4/10 = 2/5$ and common ratio $1/10$, so we have

$$0.\overline{4} = \frac{2/5}{1 - 1/10} = \boxed{\frac{4}{9}}.$$

(b) We have

$$0.\overline{273} = \frac{2}{10} + \frac{7}{10^2} + \frac{3}{10^3} + \frac{2}{10^4} + \frac{7}{10^5} + \frac{3}{10^6} + \cdots.$$

We could view this as three separate infinite series, one for the 2's, one for the 7's, and one for the 3's. Instead, we can combine the terms in groups of three, so that we have only one geometric series:

$$0.\overline{273} = \frac{273}{10^3} + \frac{273}{10^6} + \frac{273}{10^9} + \cdots.$$

This series has first term $273/1000$ and common ratio $1/10^3 = 1/1000$, so its sum is

$$\frac{\frac{273}{1000}}{1 - \frac{1}{1000}} = \frac{273}{999} = \boxed{\frac{91}{333}}.$$

(c) We have

$$0.63\overline{5} = \frac{6}{10} + \frac{3}{10^2} + \frac{5}{10^3} + \frac{5}{10^4} + \frac{5}{10^5} + \cdots.$$

After the first two terms, the series on the right is an infinite geometric series with first term $5/10^3$ and common ratio $1/10$. So, we have

$$0.63\overline{5} = \frac{6}{10} + \frac{3}{10^2} + \frac{\frac{5}{10^3}}{1 - \frac{1}{10}} = \frac{63}{100} + \frac{5}{900} = \frac{572}{900} = \boxed{\frac{143}{225}}.$$

(d) We have

$$0.8\overline{81} = \frac{8}{10} + \frac{8}{10^2} + \frac{1}{10^3} + \frac{8}{10^4} + \frac{1}{10^5} + \cdots .$$

As in the previous problem, we group terms corresponding to the repeating block. However, we must remember that the first term is not part of the repeating block:

$$0.8\overline{81} = \frac{8}{10} + \frac{81}{10^3} + \frac{81}{10^5} + \cdots .$$

After the first term, the rest of the series on the right is an infinite geometric series with first term $81/10^3$ and common ratio $1/100$. So, we have

$$0.8\overline{81} = \frac{8}{10} + \frac{81/1000}{1 - 1/100} = \frac{8}{10} + \frac{81}{990}$$

$$= \frac{8}{10} + \frac{9}{110} = \frac{88}{110} + \frac{9}{110} = \boxed{\frac{97}{110}}.$$

We could also have solved this problem by letting $x = 0.8\overline{81}$. Multiplying this by 100 gives $100x = 88.1\overline{81}$. Subtracting $x = 0.8\overline{81}$ from $100x = 88.1\overline{81}$ gives $99x = 87.3$. So, we have $x = 87.3/99 = 873/990 = 97/110$.

21.4.4 Shifting the decimal one place to the left is the same as dividing by 10. Therefore, we create each new term in the sequence by multiplying the previous term by 2 and dividing the result by 10. Combining these, this is equivalent to dividing the previous term by 5. So, the sum of all these terms forms an infinite geometric series with first term 1 and common ratio $1/5$. This sum equals $1/(1 - 1/5) = \boxed{5/4}$.

21.4.5 The series on the right is a geometric series with common ratio $-x$. The series is an infinite geometric series with first term 1 and common ratio $-x$, so if $-1 < x < 1$, we have

$$1 - x + x^2 - x^3 + x^4 - x^5 + \cdots = \frac{1}{1 - (-x)} = \frac{1}{1 + x}.$$

So, our equation is $x = \frac{1}{1+x}$. Multiplying both sides by $1 + x$ gives $x^2 + x = 1$, so $x^2 + x - 1 = 0$. Using the quadratic formula, we have $x = \frac{-1 \pm \sqrt{5}}{2}$. We must have $-1 < x < 1$, but $(-1 - \sqrt{5})/2 < -1$. So, the only solution is $x = \boxed{\dfrac{-1 + \sqrt{5}}{2}}$.

Exercises for Section 21.5

21.5.1 We could view the sum as the sum of a bunch of 6's, but a closer look simplifies the sum considerably. Both -745 and $+745$ appear in the expression, as do both -742 and $+742$, and both -739 and $+739$, and so on down to -496 and $+496$. These all cancel out, leaving $751 + 748 - 493 - 490 = \boxed{516}$.

21.5.2 The sum of the numbers from -25 to 25 is 0, because each number besides 0 cancels with its negative. Therefore, when we add -25 through 26, we have a total of 26. So, the desired smallest integer is $\boxed{26}$.

21.5.3 We have a bunch of differences of squares, so we factor them all. This gives

$$\frac{(100-99)(100+99)(100-98)(100+98)(100-97)(100+97)\cdots(100-1)(100+1)}{(99-98)(99+98)(99-97)(99+97)(99-96)(99+96)\cdots(99-1)(99+1)}$$

$$=\frac{(199)(198)(197)\cdots(101)(99)(98)(97)\cdots(3)(2)(1)}{(197)(196)(195)\cdots(101)(100)(98)(97)\cdots(3)(2)(1)}.$$

After a bunch of canceling, we're left with 199, 198, and 99 on top, and just 100 on the bottom, so our expression equals $199(198)(99)/100 = \boxed{39007.98}$.

21.5.4 The denominators have squares and the numerators are odd numbers. This makes us think of how odd numbers are related to squares. Specifically, we have $2^2 - 1^2 = 3$, $3^2 - 2^2 = 5$, $4^2 - 3^2 = 7$, and so on, so we can write

$$\frac{3}{1^2 \cdot 2^2} + \frac{5}{2^2 \cdot 3^2} + \frac{7}{3^2 \cdot 4^2} + \cdots = \frac{2^2 - 1^2}{1^2 \cdot 2^2} + \frac{3^2 - 2^2}{2^2 \cdot 3^2} + \frac{4^2 - 3^2}{3^2 \cdot 4^2} + \cdots$$

$$= \frac{1}{1^2} - \frac{1}{2^2} + \frac{1}{2^2} - \frac{1}{3^2} + \frac{1}{3^2} - \frac{1}{4^2} + \cdots.$$

All the terms cancel out except the first term, so the sum equals $\boxed{1}$.

Review Problems

21.25 The sixth term is exactly between the fourth and the eighth in the arithmetic sequence, so it is the average of the two terms. Therefore, the sixth term is $(200 + 500)/2 = \boxed{350}$. We also could have found the common difference by noting that there are four steps between the fourth term and the eighth term. So, if d is the common difference, we have $4d = 500 - 200 = 300$. Therefore, we find $d = 75$. The sixth term is two steps after the fourth, or $200 + 2d = \boxed{350}$.

21.26 The sixth term is exactly between the fourth and the eighth term, so it is the geometric mean of these two terms. Therefore, the sixth term is $\sqrt{(200)(800)} = \sqrt{160000} = \boxed{400}$. We also could have found the common ratio, r, by noting that we must multiply the fourth term by r four times to get to the eighth term. So, we have $200r^4 = 800$, or $r^4 = 4$. Because all the terms in the sequence are positive, we have $r = \sqrt{2}$. We get the sixth term of the sequence by multiplying the fourth term by r twice, to give $200r^2 = \boxed{400}$.

21.27 *Solution 1: Find the first term and common difference.* Let the first term be a and the common difference be d. Because the second term is -7, we have $a + d = -7$. Because the fifth term is 16, we have $a + 4d = 16$. Subtracting $a + d = -7$ from $a + 4d = 16$ gives $3d = 23$, so $d = 23/3$. Therefore, $a = -7 - d = -44/3$. So, the fourteenth term is $a + 13d = -44/3 + 13(23/3) = \boxed{85}$.

Solution 2: Use our understanding of arithmetic sequences. To get from the second term to the fifth, we take three steps. Those three steps take us from -7 to 16, so together, the three steps add 23. To get from the fifth term to the fourteenth term, we take three steps three times. Each of these blocks of three steps adds 23, so the fourteenth term is $16 + 3(23) = \boxed{85}$.

21.28 *Solution 1: Find the first term and common ratio.* Let the first term be a and the common ratio be r. Because the second term is -2, we have $ar = -2$. Because the fifth term is 16, we have $ar^4 = 16$. Dividing this by $ar = -2$, we have $r^3 = -8$. Therefore, $r = -2$. So, $a = -2/r = 1$. Therefore, the fourteenth term is $ar^{13} = (1)(-2)^{13} = \boxed{-8192}$.

Solution 2: Use our understanding of geometric sequences. To get from the second term to the fifth term, we multiply by the common ratio, r, three times. Therefore, multiplying -2 by r^3 gives us 16. So, $r^3 = -8$. Rather than finding r, we note that to get to the fourteenth term from the fifth term, we multiply by r nine times, which is the same as multiplying by r^3 three times. So, the fourteenth term is $16(-8)^3 = \boxed{-8192}$.

21.29

(a) The common difference is $5 - 2 = 3$. Let there be n terms. The last term is 101 and the first term is 2, so we have $2 + 3(n - 1) = 101$, so $n = 34$. Therefore, our sum is $34(2 + 101)/2 = 17(103) = \boxed{1751}$.

(b) The common difference is $-1/3$. Let there be n terms. The first term is 8 and the last term is $4/3$, so we must have $8 + (-1/3)(n - 1) = 4/3$. Solving this equation gives $n = 21$, so our sum is $21(8 + 4/3)/2 = \boxed{98}$.

21.30

(a) This infinite geometric series has first term 12 and common ratio $(-3)/12 = -1/4$, so its sum is $12/[1 - (-1/4)] = \boxed{48/5}$.

(b) Let $S = 6 + 18 + 54 + \cdots + 1458$. The common ratio is $18/6 = 3$, so we multiply by 3 to find $3S = 18 + 54 + \cdots + 4374$. Subtracting our equation for S from this, we have $2S = 4374 - 6 = 4368$, so $S = \boxed{2184}$.

21.31

(a) The sum of the first n integers is $n(n+1)/2$, so the sum of the first 50 integers is $50(51)/2 = 2550/2 = \boxed{1275}$.

(b) Because the sum of the first k integers is $k(k + 1)/2$, we must have $k(k + 1)/2 = 990$. Therefore, $k^2 + k = 1980$, so $k^2 + k - 1980 = 0$. Factoring then gives us $(k + 45)(k - 44) = 0$. Since k must be positive, we have $k = \boxed{44}$. Checking our answer, we see that $44(44 + 1)/2 = 22(45) = 990$.

21.32 *Solution 1: Find the first and last term.* Let a be the first term and d be the common difference. Because the second term is 4, we have $a + d = 4$. Because the ninth term is -7, we have $a + 8d = -7$. Subtracting $a + d = 4$ from this gives $7d = -11$, so $d = -11/7$. Therefore, $a = 4 - d = 39/7$. Because the first term is $39/7$ and the common difference is $-11/7$, the tenth term is $a + 9d = -60/7$. Finally, we see that the sum of the first 10 terms of the series is

$$10 \cdot \frac{\frac{39}{7} + \left(-\frac{60}{7}\right)}{2} = 10 \cdot \frac{-3}{2} = \boxed{-15}.$$

Solution 2: Use a little cleverness. To find our sum, we multiply the number of terms, 10, by the average of the first and last terms. The first term is one step before the second term and the last term (the tenth term) is one step after the ninth term. So, the sum of the first and tenth terms equals the sum of the second and ninth terms. Therefore, the average of the first and last terms of our series equals the average of the second and ninth. So, the sum equals $10[4 + (-7)]/2 = \boxed{-15}$.

21.33 If we pair the terms $1-2, 3-4, 5-6$, and so on, we see that the expression equals $-1-1-1-\cdots-1$, where there are a total of fifty -1's. So, the expression equals $\boxed{-50}$.

21.34 Let the first term be a. Because the sum of the series is 45, we have $45 = a/[1 - (-1/2)] = a/(3/2) = 2a/3$. Therefore, $a = \boxed{135/2}$.

21.35 For $0.\overline{72}$, we have

$$\begin{aligned} 0.\overline{72} &= \frac{7}{10} + \frac{2}{10^2} + \frac{7}{10^3} + \frac{2}{10^4} + \cdots \\ &= \frac{72}{100} + \frac{72}{100^2} + \frac{72}{100^3} + \cdots \\ &= \frac{\frac{72}{100}}{1 - \frac{1}{100}} = \frac{72}{99} = \boxed{\frac{8}{11}}. \end{aligned}$$

To find $0.63\overline{36}$, we first let $x = 0.63\overline{36}$. Multiplying both sides by 100 gives $100x = 63.\overline{36}$. Subtracting our expression for x from this gives

$$99x = 63.\overline{36} - 0.63\overline{36} = 63.3636\ldots - 0.6336\ldots = 63.36 - 0.63 = 62.73.$$

Therefore,

$$x = \frac{62.73}{99} = \frac{6273}{9900} = \boxed{\frac{697}{1100}}.$$

21.36 An infinite sequence that consists of the same number over and over is both an arithmetic and a geometric sequence. There are no other infinite sequences that are both arithmetic and geometric. To see why, consider the geometric sequence $a, ar, ar^2, \ldots$. If this sequence is also arithmetic, we must have $ar^2 - ar = ar - a$. Dividing by a (assuming $a \neq 0$, otherwise our sequence is all 0s), we have $r^2 - r = r - 1$, or $r^2 - 2r + 1 = 0$. Factoring gives $(r - 1)^2 = 0$, so $r = 1$. Therefore, our sequence is $\boxed{a, a, a, \ldots}$.

21.37 Multiplying all parts of the inequality chain by π gives us $-5\pi \leq x \leq 10\pi$. Since π is approximately 3.14, this inequality is approximately $-15.7 \leq x \leq 31.4$. The integers that satisfy this inequality are $-15, -14, -13, \ldots, 29, 30, 31$. To add these, we first notice that the sum of the numbers from -15 to 15 is zero, because each nonzero number cancels with its opposite. Then, we are left with $16 + 17 + 18 + \cdots + 31$. This arithmetic series has 16 terms (the first 31 positive integers, excluding the first 15 positive integers), first term 16, and last term 31, so its sum is $16(16 + 31)/2 = 8(47) = \boxed{376}$.

21.38 *Solution 1: Find C and D.* The sum of the first 100 positive even integers is

$$2 + 4 + 6 + \cdots + 198 + 200 = 2(1 + 2 + 3 + \cdots + 100) = 2\left(\frac{100 \cdot 101}{2}\right) = 10100.$$

The sum of the first 100 positive odd numbers is $100^2 = 10000$. Therefore, $(C + D)/(C - D) = 20100/100 = \boxed{201}$.

Solution 2: Find C + D and C − D. Since C is the sum of the first 100 even positive integers and D is the sum of the first 100 odd positive integers, $C + D$ is the sum of the first 200 positive integers, which equals $200(201)/2 = 20100$. Similarly, to compute $C - D$, we add the first 100 even integers, then subtract the first 100 odd integers. A little reorganizing shows that

$$C - D = 2 - 1 + 4 - 3 + 6 - 5 + \cdots + 198 - 197 + 200 - 199 = 1 + 1 + 1 + \cdots + 1.$$

There are 100 1's in this sum, one for each pair of corresponding numbers in C and D. Therefore, we have $C - D = 100$, so $(C + D)/(C - D) = \boxed{201}$.

21.39 Let the middle integer be a. Then, the sum of the seven integers is $7a$. We are given that this equals 273, so $7a = 273$. Therefore, $a = 39$. This is the middle integer, so the largest is $39 + 3(2) = \boxed{45}$.

21.40 Let a be the first term. To get to the nth term, we multiply by r exactly $n - 1$ times, so $b = a \cdot r^{n-1}$. Therefore, $\boxed{a = b/r^{n-1}}$.

21.41 The ball initially falls 100 feet. On its first bounce, it rises $100(3/4)$ feet, then falls that distance, traveling a total of $2 \cdot 100(3/4)$ feet. On its second bounce, it rises $100(3/4) \cdot (3/4) = 100(3/4)^2$ feet, then falls $100(3/4)^2$ feet, traveling a total of $2 \cdot 100(3/4)^2$ feet. And so on. Adding the initial drop and each subsequent bounce, the ball travels a total of

$$100 + 2 \cdot 100 \cdot \frac{3}{4} + 2 \cdot 100 \left(\frac{3}{4}\right)^2 + 2 \cdot 100 \left(\frac{3}{4}\right)^3 + 2 \cdot 100 \left(\frac{3}{4}\right)^4 + \cdots$$

feet. After the first term, this series is an infinite geometric series with first term $2 \cdot 100(3/4)$ and common ratio $3/4$. Therefore, the total distance the ball travels is

$$100 + \frac{2 \cdot 100 \cdot \frac{3}{4}}{1 - \frac{3}{4}} = 100 + \frac{150}{\frac{1}{4}} = \boxed{700 \text{ feet}}.$$

Challenge Problems

21.42 $\boxed{\text{Yes}}$. If the series is geometric, then we must have $a/1 = b/a$, since the ratio between consecutive terms must always be the same. Therefore, we have $b = a^2$. Since the sequence is arithmetic, the difference between consecutive terms must always be the same, so $a - 1 = b - a$, or $b = 2a - 1$. Equating these two expressions for b gives $a^2 = 2a - 1$, or $a^2 - 2a + 1 = 0$. Factoring gives $(a - 1)^2 = 0$, so $a = 1$. Therefore, $b = a^2 = 1$, as well.

21.43 All the numbers in the expression are perfect squares. We can write the expression as

$$30^2 - 29^2 + 28^2 - 27^2 + \cdots + 4^2 - 3^2 + 2^2 - 1^2.$$

Because $(n + 1)^2 - n^2 = (n + 1 + n)(n + 1 - n) = (n + 1) + n$, the difference between two consecutive squares equals the sum of the numbers that are squared. So, our sum is

$$(30^2 - 29^2) + (28^2 - 27^2) + \cdots + (4^2 - 3^2) + (2^2 - 1^2) = 59 + 55 + \cdots + 7 + 3.$$

This sum is an arithmetic series with first term 59, last term 3, and common difference -4. There were 30 terms in our original expression, so there are 15 terms in this arithmetic series. Therefore, its sum is $15(59 + 3)/2 = 15(31) = \boxed{465}$.

21.44

(a) We multiply the equation by $n(n + 2)$ to give $1 = A(n + 2) + Bn$. This equation must hold for all n, so we find A and B by strategically choosing n to eliminate A or B. When $n = 0$, we have

$1 = A(0 + 2) + B(0) = 2A$, so $A = 1/2$. When $n = -2$, we have $1 = A(-2 + 2) + B(-2) = -2B$, so $B = -1/2$. Therefore, we have

$$\frac{1}{n(n + 2)} = \frac{1/2}{n} - \frac{1/2}{n + 2}.$$

(b) Using the partial fraction decomposition we found in part (a), we have

$$\frac{1}{1 \cdot 3} + \frac{1}{2 \cdot 4} + \frac{1}{3 \cdot 5} + \cdots + \frac{1}{98 \cdot 100} = \frac{1/2}{1} - \frac{1/2}{3} + \frac{1/2}{2} - \frac{1/2}{4} + \frac{1/2}{3} - \frac{1/2}{5} + \cdots + \frac{1/2}{97} - \frac{1/2}{99} + \frac{1/2}{98} - \frac{1/2}{100}.$$

Most of the terms cancel, leaving only

$$\frac{1/2}{1} + \frac{1/2}{2} - \frac{1/2}{99} - \frac{1/2}{100} = \frac{1}{2}\left(1 + \frac{1}{2} - \frac{1}{99} - \frac{1}{100}\right) = \boxed{\frac{14651}{19800}}.$$

21.45 The last number of each row equals the number of integers that have been written in the triangle up to that point. There is 1 number in the first row, 2 in the second, 3 in the third, and so on, so there are $1 + 2 + 3 + \cdots + n$ numbers in the first n rows. So, after 16 rows, Sasha has used $1 + 2 + 3 + \cdots + 16 = 16(17)/2 = 136$ numbers, and 136 is the last number in the 16th row. So, the first number in the 17th row is 137, and the last number is $137 + 16 = 153$. (Make sure you see why it is not $137 + 17$.) The sum of these two numbers is $137 + 153 = \boxed{290}$.

21.46

(a) $\displaystyle\sum_{i=1}^{10}(2i - 5) = 2 \cdot 1 - 5 + 2 \cdot 2 - 5 + \cdots + 2 \cdot 10 - 5 = -3 - 1 + 1 + 3 + 5 + 7 + 9 + 11 + 13 + 15 = \boxed{60}$.

Notice that this sum is an arithmetic series.

(b) The expression $\displaystyle\sum_{i=1}^{72} 5$ means we add 5 exactly 72 times. This gives us a total of $5(72) = \boxed{360}$.

(c) $\displaystyle\sum_{i=1}^{7} 3^i = 3^1 + 3^2 + 3^3 + \cdots + 3^7$. This is a geometric series with first term 3, common ratio 3, and 7 terms. So, its sum is $3(3^7 - 1)/(3 - 1) = \boxed{3279}$.

(d) We have

$$\sum_{i=0}^{\infty} \frac{3^i + 5^i}{8^i} = \frac{3^0 + 5^0}{8^0} + \frac{3^1 + 5^1}{8^1} + \frac{3^2 + 5^2}{8^2} + \frac{3^3 + 5^3}{8^3} + \cdots.$$

We group the terms with 3's together and the terms with 5's together to see that

$$\sum_{i=0}^{\infty} \frac{3^i + 5^i}{8^i} = \left(\frac{3^0}{8^0} + \frac{3^1}{8^1} + \frac{3^2}{8^2} + \cdots\right) + \left(\frac{5^0}{8^0} + \frac{5^1}{8^1} + \frac{5^2}{8^2} + \cdots\right).$$

The expressions in parentheses are each infinite geometric series. They both have first term 1. The first series has common ratio 3/8 and the second has common ratio 5/8. Therefore, we have

$$\sum_{i=0}^{\infty} \frac{3^i + 5^i}{8^i} = \frac{1}{1 - \frac{3}{8}} + \frac{1}{1 - \frac{5}{8}} = \frac{8}{5} + \frac{8}{3} = \boxed{\frac{64}{15}}.$$

21.47 Because the sequence is an arithmetic series, the common difference is $(x - y) - (x + y) = -2y$. Therefore, we add $-2y$ to get from the second term to the third, so $x - y - 2y = xy$. Solving for y in terms of x gives us $y = \frac{x}{x+3}$. So, our sequence now is

$$\frac{x^2 + 4x}{x + 3}, \frac{x^2 + 2x}{x + 3}, \frac{x^2}{x + 3}, x + 3.$$

Because this is an arithmetic sequence, the difference between the fourth and third terms must equal the difference between the second and first, so we must have

$$x + 3 - \frac{x^2}{x + 3} = \frac{x^2 + 2x}{x + 3} - \frac{x^2 + 4x}{x + 3} \quad \Rightarrow \quad x + 3 = \frac{x^2 - 2x}{x + 3} \quad \Rightarrow \quad (x + 3)^2 = x^2 - 2x$$

$$\Rightarrow \quad x^2 + 6x + 9 = x^2 - 2x \quad \Rightarrow \quad 8x = -9 \quad \Rightarrow \quad x = -9/8.$$

Therefore, we have $y = (-9/8)/(-9/8 + 3) = -3/5$, and our sequence is

$$-\frac{69}{40}, -\frac{21}{40}, \frac{27}{40}, \frac{75}{40}.$$

So, the common difference is $75/40 - 27/40 = 48/40$, which means the next term is $75/40 + 48/40 = \boxed{123/40}$.

21.48 Each number k appears k times. So, in the numbers up to and including the last k, we have a total of $1 + 2 + 3 + \cdots + k$ numbers. This gives us a total of $k(k + 1)/2$ numbers. We want the 1993rd number, so we seek the smallest value of k such that $k(k + 1)/2 \geq 1993$. From here, we can use a little trial and error. We have $k(k + 1)/2 = 1953$ when $k = 62$ and $k(k + 1)/2 = 2016$ when $k = 63$. Therefore, the 1954th through 2016th numbers in the list are all 63. So, the 1993rd term is $\boxed{63}$.

21.49

(a) *Solution 1: Algebra.* The sum of the integers is the average of the first and last number times the number of integers. Let the sum of the first and last integer be s and the number of integers be n, so we have $sn/2 = 210$, or $sn = 420$. Both s and n must be positive integers. The largest n can be is 420, which makes $s = 1$. However, if the sum of the first and last integers is 1, then one of the integers is not positive.

We can find the first integer in terms of s and n, so we can easily test when it's positive. Let the first integer be a. Then, the last integer is $a + n - 1$, so we have $s = a + a + n - 1$, so $a = [s - (n - 1)]/2$. Returning to $sn = 420$, we see that $s = 420/n$. At this point, we could use trial-and-error to find the largest n that leaves a positive, or we could substitute:

$$a = \frac{s - (n - 1)}{2} = \frac{420}{2n} - \frac{n - 1}{2} = \frac{-n^2 + n + 420}{2n}.$$

We know that n must be a divisor of 420, since $s = 420/n$. We also know that $-n^2 + n + 420$ must be positive. Fortunately, we can factor this quadratic as $-(n - 21)(n + 20)$. Therefore, if $n \geq 21$, then the quadratic is nonpositive. Since $n = 20$ makes our quadratic positive and 20 is a divisor of 420, there must be at most 20 terms. The 20 consecutive integers form an arithmetic sequence whose terms sum to 210. This sum is 20 times the average term of the sequence, so the average of all the terms in the sequence is $210/20 = 10.5$. This tells us that the 20 consecutive positive integers that add to 210 are $1, 2, 3, \ldots, \boxed{20}$.

Solution 2: Educated guess. If 210 can be written as the sum of consecutive positive integers starting from 1, then this will be the largest group of positive consecutive integers that add to 210. Suppose there are k such integers. Then, we must have $k(k + 1)/2 = 210$, or $k^2 + k - 420 = 0$. So, we have $(k + 21)(k - 20) = 0$. Since k must be even, we have $k = 20$, and 210 is the sum of the first 20 positive integers. The largest of these integers is $\boxed{20}$.

(b) *Solution 1: Algebra.* From above, we had $s = 420/n$. If we have no restrictions on the consecutive integers, except that they are integers, then the largest n can be is $\boxed{420}$.

Solution 2: Reason it out. We can let the largest of our numbers be 210, then choose enough numbers smaller than 210 so that each number and its negative are in the list. Since the integers must be consecutive, they must be the numbers from -209 through 210. There are $\boxed{420}$ such numbers. We cannot have any numbers larger than 210, since then the sum would be either larger than 210 (every negative cancels with its opposite, but our largest number is not canceled) or the sum would be nonpositive (if the largest number gets canceled by its negative, then all the positive integers are canceled).

21.50 Each term in the sequence is itself a geometric series. Each term is of the form

$$1 + 2 + 2^2 + 2^3 + \cdots + 2^k.$$

Using our formula for finite geometric series, we have

$$1 + 2 + 2^2 + 2^3 + \cdots + 2^k = \frac{1(2^{k+1} - 1)}{2 - 1} = 2^{k+1} - 1.$$

So, the first n terms of the sequence are

$$2 - 1, 2^2 - 1, 2^3 - 1, \ldots, 2^n - 1.$$

So, the sum of these is the result of adding -1 exactly n times to the geometric series $2 + 2^2 + 2^3 + \cdots + 2^n$. Therefore, the sum equals

$$\frac{2(2^n - 1)}{2 - 1} + n(-1) = \boxed{2^{n+1} - 2 - n}.$$

21.51 Our product is

$$\frac{1 \cdot 5}{3^2} \cdot \frac{2 \cdot 6}{4^2} \cdot \frac{3 \cdot 7}{5^2} \cdots \frac{12 \cdot 16}{14^2} \cdot \frac{13 \cdot 17}{15^2}.$$

All the numbers from 5 through 13 appear twice in the numerator, so all the denominators from 5^2 through 13^2 cancel all the 5's and 13's in the numerators. There's one 3, one 4, one 14, and one 15 in the numerator that each cancel with a corresponding factor in the denominator. So, all we have left is

$$\frac{1 \cdot 2 \cdot 16 \cdot 17}{3 \cdot 4 \cdot 14 \cdot 15} = \boxed{\frac{68}{315}}.$$

21.52 Because x, y, z is a geometric sequence with common ratio r, we have $y = xr$ and $z = yr = xr^2$. So, our arithmetic sequence is $x, 2xr, 3xr^2$. Because this is an arithmetic sequence, we have $2xr - x = 3xr^2 - 2xr$. Because $x \neq y$, we know that $x \neq 0$. So, we can divide our equation by x to find $2r - 1 = 3r^2 - 2r$, or $3r^2 - 4r + 1 = 0$. Factoring gives $(3r - 1)(r - 1) = 0$. Because $x \neq y$, we can't have $r = 1$. So, we have $r = \boxed{1/3}$.

21.53

(a) The sum $1 + 3 + 5 + \cdots + (2k - 1)$ is an arithmetic series with first term 1, last term $2k - 1$, and k terms. So, it has sum $k(1 + 2k - 1)/2 = k^2$.

(b) If we write each square as the sum of odd numbers as described in part (a), then we have

$$S = (1) + (1 + 3) + (1 + 3 + 5) + \cdots + (1 + 3 + 5 + \cdots + 2n - 1).$$

The odd number 1 appears in all n of these sums. Adding these gives a total of $(1)(n)$. The odd number 3 appears in all but the first sum (the one corresponding to 1^2), so it appears $n - 1$ times. Adding these gives a total of $(3)(n - 1)$. The odd number 5 appears in all but the first two sums, so it appears $n - 2$ times. Adding these gives a total of $(5)(n - 2)$. We can continue in this way all the way up to $2n - 1$, which appears exactly once. Combining all these gives us

$$S = (1)(n) + (3)(n - 1) + (5)(n - 2) + \cdots (2n - 1)(1).$$

(c) If we multiply $S = 1^2 + 2^2 + 3^2 + \cdots + n^2$ by 2, we have

$$2S = 2 \cdot 1^2 + 2 \cdot 2^2 + 2 \cdot 3^2 + \cdots + 2 \cdot n^2 = (2)(1) + (4)(2) + (6)(3) + \cdots + (2n)(n).$$

(d) The numbers $n, n - 1, n - 2, \ldots, 1$ appear in that order in the first series, and in the opposite order among the terms in the second series. So, we reverse the first series to help us add:

S	$=$	$(2n-1)(1)$	$+$	$(2n-3)(2)$	$+$	$(2n-5)(3)$	$+$	$\cdots$	$+$	$(3)(n-1)$	$+$	$(1)(n)$
$2S$	$=$	$(2)(1)$	$+$	$(4)(2)$	$+$	$(6)(3)$	$+$	$\cdots$	$+$	$(2n-2)(n-1)$	$+$	$(2n)(n)$
$3S$	$=$	$(2n+1)(1)$	$+$	$(2n+1)(2)$	$+$	$(2n+1)(3)$	$+$	$\cdots$	$+$	$(2n+1)(n-1)$	$+$	$(2n+1)(n)$

So, we have

$$3S = (2n + 1)(1 + 2 + 3 + \cdots + n) = (2n + 1) \cdot \frac{n(n + 1)}{2}.$$

Dividing by 3 gives $S = n(n + 1)(2n + 1)/6$, as desired.

21.54 From the quadratic formula, we have $x = \frac{2 \pm \sqrt{26}}{11}$. Because $\sqrt{26}$ is just slightly more than 5, we see that both of these solutions are between -1 and 1. Therefore, a and b are both between -1 and 1. So, the sum of the infinite geometric series $1 + a + a^2 + a^3 + \cdots$ equals $\frac{1}{1-a}$. Similarly, we have $1 + b + b^2 + b^3 + \cdots = \frac{1}{1-b}$, so our desired product equals

$$\frac{1}{1 - a} \cdot \frac{1}{1 - b} = \frac{1}{(1 - a)(1 - b)} = \frac{1}{1 - a - b + ab} = \frac{1}{1 - (a + b) + ab}.$$

We can use the coefficients of our quadratic to find the sum of the roots, $a + b$, and the product of the roots, ab. Specifically, we have $a + b = -(-4)/11 = 4/11$ and $ab = -2/11$, so we have

$$\frac{1}{1 - (a + b) + ab} = \frac{1}{1 - \frac{4}{11} - \frac{2}{11}} = \boxed{\frac{11}{5}}.$$

(Notice that when we let $x = 1$ in our quadratic, we get $11x^2 - 4x - 2 = 5$. How could we have used this fact to evaluate $1 - (a + b) + ab$ quickly?)

21.55 The expression $1 + x + x^2 + x^3$ is a geometric series with first term 1 and common ratio x, so we have

$$1 + x + x^2 + x^3 = \frac{1(x^4 - 1)}{x - 1} = \frac{x^4 - 1}{x - 1}.$$

Because $1 + x + x^2 + x^3 = 0$, we know that x is not 1, so dividing by $x - 1$ is OK. Because $1 + x + x^2 + x^3 = (x^4 - 1)/(x - 1)$, our given equation is now $(x^4 - 1)/(x - 1) = 0$. Multiplying both sides by $x - 1$ gives $x^4 - 1 = 0$. We can factor the left side as a difference of squares to get $(x^2 - 1)(x^2 + 1) = 0$. So, our possible values of x are 1, -1, i, and $-i$. We discard $x = 1$, since we already noted that x cannot be 1.

Evaluating $x^4 + 2x^3 + 2x + 1$ for each of the other possible values of x gives us $\boxed{-2 \text{ and } 2}$ as the possible values of $x^4 + 2x^3 + 2x + 1$.

21.56 Because the interest rate is 6% compounded monthly, the present value of my first payment of $\$x$ is $\$x/(1 + 0.06/12)$. My second payment of $\$x$ is 2 months from now, so the present value of this payment is $\$x/(1 + 0.06/12)^2$. Similarly, the present value of the payment n months from now is $\$x/(1 + 0.06/12)^n$. The sum of the present values of all of these payments must equal the amount of money I borrow today, so we have

$$\frac{x}{1 + \frac{0.06}{12}} + \frac{x}{\left(1 + \frac{0.06}{12}\right)^2} + \frac{x}{\left(1 + \frac{0.06}{12}\right)^3} + \cdots + \frac{x}{\left(1 + \frac{0.06}{12}\right)^{360}} = 200000.$$

The series on the left side is a geometric series with common ratio $\frac{1}{1+0.06/12}$, first term $\frac{x}{1+0.06/12}$, and 360 terms. So, its sum is

$$\frac{\frac{x}{1+0.06/12}\left[\left(\frac{1}{1+0.06/12}\right)^{360} - 1\right]}{\frac{1}{1+0.06/12} - 1} \approx 166.7916x.$$

Solving $166.7916x = 200000$ gives us $x = \boxed{1199.10}$ as my payment amount.

21.57 The numerators of this series form an arithmetic sequence and the denominators form a geometric one. We're not sure how to deal with this series, but we know that

$$\frac{1}{2^1} + \frac{1}{2^2} + \frac{1}{2^3} + \cdots = \frac{\frac{1}{2}}{1 - \frac{1}{2}} = 1.$$

So, letting the series in the problem be S, we have

$$S = \frac{1}{2^1} + \frac{2}{2^2} + \frac{3}{2^3} + \frac{4}{2^4} + \cdots,$$
$$1 = \frac{1}{2^1} + \frac{1}{2^2} + \frac{1}{2^3} + \cdots.$$

The denominators line up nicely, so we subtract, giving

$$S - 1 = \frac{1}{2^2} + \frac{2}{2^3} + \frac{3}{2^4} + \cdots.$$

The sum on the right is $1/2$ of our original series, so we have $S - 1 = S/2$, so $S = \boxed{2}$.

We also could have tackled this problem by viewing the sum as an infinite sum of infinite series:

$$\frac{1}{2^1} + \frac{1}{2^2} + \frac{1}{2^3} + \frac{1}{2^4} + \cdots = 1,$$

$$\frac{1}{2^2} + \frac{1}{2^3} + \frac{1}{2^4} + \cdots = \frac{1}{2},$$

$$\frac{1}{2^3} + \frac{1}{2^4} + \cdots = \frac{1}{2^2},$$

$$\frac{1}{2^4} + \cdots = \frac{1}{2^3},$$

$$\vdots$$

Adding all these equations gives us

$$\frac{1}{2^1} + \frac{2}{2^2} + \frac{3}{2^3} + \frac{4}{2^4} + \cdots = 1 + \frac{1}{2^1} + \frac{1}{2^2} + \frac{1}{2^3} + \cdots = 2.$$

CHAPTER 22

Special Manipulations

Exercises for Section 22.1

22.1.1

(a) Squaring both sides of $a + \frac{1}{a} = 3$ gives $a^2 + 2a \cdot \frac{1}{a} + \frac{1}{a^2} = 9$, so $a^2 + \frac{1}{a^2} = 9 - 2 = \boxed{7}$.

(b) Squaring both sides of $a^2 + \frac{1}{a^2} = 7$ gives $(a^2)^2 + 2a^2 \cdot \frac{1}{a^2} + \left(\frac{1}{a^2}\right)^2 = 49$. Therefore, $a^4 + \frac{1}{a^4} = 49 - 2 = \boxed{47}$.

(c) *Solution 1: Multiply $a + \frac{1}{a}$ by $a^2 + \frac{1}{a^2}$.* We need cubes, so we multiply $a + \frac{1}{a}$ by $a^2 + \frac{1}{a^2}$, since we know what these both equal:

$$\left(a + \frac{1}{a}\right)\left(a^2 + \frac{1}{a^2}\right) = 3 \cdot 7 \quad \Rightarrow \quad a^3 + \frac{1}{a} + a + \frac{1}{a^3} = 21$$

$$\Rightarrow \quad a^3 + \frac{1}{a^3} = 21 - \left(a + \frac{1}{a}\right) = 21 - 3 = \boxed{18}.$$

Solution 2: Cube the original equation. We need cubes, so we cube $a + \frac{1}{a}$. Because $(x + y)^3 = x^3 + 3x^2y + 3xy^2 + y^3$, we have

$$\left(a + \frac{1}{a}\right)^3 = 3^3 \quad \Rightarrow \quad a^3 + 3a^2 \cdot \frac{1}{a} + 3a \cdot \frac{1}{a^2} + \frac{1}{a^3} = 27$$

$$\Rightarrow \quad a^3 + 3a + \frac{3}{a} + \frac{1}{a^3} = 27 \quad \Rightarrow \quad a^3 + \frac{1}{a^3} = 27 - 3\left(a + \frac{1}{a}\right) = 27 - 9 = \boxed{18}.$$

22.1.2 Suppose my numbers are x and y. We are told that $x + y = 14$ and $xy = 46$. We seek $x^2 + y^2$, so we square our equation for $x + y$, giving $x^2 + 2xy + y^2 = 196$. Since $xy = 46$, we have $x^2 + y^2 = 196 - 2xy = \boxed{104}$.

22.1.3 We let $x = \sqrt{7 - \sqrt{13}} - \sqrt{7 + \sqrt{13}}$. We try to get rid of the radicals by squaring both sides. This gives us

$$x^2 = \left(\sqrt{7 - \sqrt{13}}\right)^2 - 2\sqrt{7 - \sqrt{13}}\sqrt{7 + \sqrt{13}} + \left(\sqrt{7 + \sqrt{13}}\right)^2$$

$$= 7 - \sqrt{13} - 2\sqrt{(7 - \sqrt{13})(7 + \sqrt{13})} + 7 + \sqrt{13}$$

$$= 14 - 2\sqrt{49 - 13} = 14 - 12 = 2.$$

Because $x^2 = 2$, we must have either $x = \sqrt{2}$ or $x = -\sqrt{2}$. Because $7 + \sqrt{13}$ is greater than $7 - \sqrt{13}$, we know that the number $\sqrt{7 - \sqrt{13}} - \sqrt{7 + \sqrt{13}}$ is negative. Therefore, our answer is $\boxed{-\sqrt{2}}$.

22.1.4 We let $x = \sqrt[3]{2 + \sqrt{5}} + \sqrt[3]{2 - \sqrt{5}}$. We want to get rid of the cube roots, so we cube both sides. Because $(a + b)^3 = a^3 + 3a^2b + 3ab^2 + b^3$, we have

$$x^3 = \left(\sqrt[3]{2 + \sqrt{5}}\right)^3 + 3\left(\sqrt[3]{2 + \sqrt{5}}\right)^2 \sqrt[3]{2 - \sqrt{5}} + 3\sqrt[3]{2 + \sqrt{5}}\left(\sqrt[3]{2 - \sqrt{5}}\right)^2 + \left(\sqrt[3]{2 + \sqrt{5}}\right)^3$$

$$= 2 + \sqrt{5} + 3\sqrt[3]{(2 + \sqrt{5})(2 + \sqrt{5})(2 - \sqrt{5})} + 3\sqrt[3]{(2 + \sqrt{5})(2 - \sqrt{5})(2 - \sqrt{5})} + 2 - \sqrt{5}$$

$$= 4 + 3\sqrt[3]{(2 + \sqrt{5})(4 - 5)} + 3\sqrt[3]{(4 - 5)(2 - \sqrt{5})}$$

$$= 4 - 3\left(\sqrt[3]{2 + \sqrt{5}} + \sqrt[3]{2 - \sqrt{5}}\right).$$

The expression in parentheses is the number we seek, so it equals x. Therefore, our equation is $x^3 = 4 - 3x$, or $x^3 + 3x = 4$. Clearly, $x = 1$ is one solution to this equation. For any value of x greater than 1, the expression $x^3 + 3x$ is greater than $1 + 3 = 4$. For any value of x less than 1, we have $x^3 < 1$ and $3x < 3$, so $x^3 + 3x < 4$. So, the only possible real value of x is $x = \boxed{1}$.

Exercises for Section 22.2

22.2.1 Letting $x = \sqrt{12 + \sqrt{12 + \sqrt{12 + \sqrt{12 + \cdots}}}}$, we have $x = \sqrt{12 + x}$. Squaring both sides gives $x^2 = 12 + x$, so $x^2 - x - 12 = 0$. Factoring the left side gives $(x - 4)(x + 3) = 0$. Therefore, $x = 4$ or $x = -3$. Clearly x must be positive, so we have $x = \boxed{4}$.

22.2.2 Letting

$$x = 3 + \cfrac{1}{3 + \cfrac{1}{3 + \cfrac{1}{3 + \cdots}}},$$

we have $x = 3 + \frac{1}{x}$. Multiplying by x gives $x^2 = 3x + 1$, or $x^2 - 3x - 1 = 0$. Applying the quadratic formula gives $x = \frac{3 \pm \sqrt{13}}{2}$. Clearly x must be positive, so $x = \boxed{\dfrac{3 + \sqrt{13}}{2}}$.

22.2.3 Because $x + \cfrac{1}{x + \cfrac{1}{x + \cdots}}$ equals $2x$, the denominator of $\cfrac{1}{x + \cfrac{1}{x + \cdots}}$ equals $2x$. Therefore, we have $x + \frac{1}{2x} = 2x$. Subtracting x from both sides gives $\frac{1}{2x} = x$, so $2x^2 = 1$. Therefore, $x^2 = 1/2$, so $x = \pm\sqrt{1/2} = \boxed{\pm\sqrt{2}/2}$. Notice that both solutions work in this case because our self-similar expression is itself in terms of a variable, and is not expressed in terms of constants in such a way that the self-similar expression must clearly be positive (as we have in the other continued fraction problems in the text).

22.2.4 The expression in the denominator does not exactly match the whole given expression. Therefore, we can't just let the whole expression equal x, then replace the denominator with x. However, if we look just at the expression in the denominator, we have

$$1 + \cfrac{2}{1 + \cfrac{2}{1 + \cfrac{2}{1 + \cfrac{2}{1 + \cdots}}}}.$$

If we let x equal this expression, we have

$$x = 1 + \cfrac{2}{1 + \cfrac{2}{1 + \cfrac{2}{1 + \cfrac{2}{1 + \cdots}}}} = 1 + \frac{2}{x}.$$

Multiplying $x = 1 + \frac{2}{x}$ by x gives $x^2 = x + 2$, or $x^2 - x - 2 = 0$. Therefore, $(x-2)(x+1) = 0$, so $x = 2$ or $x = -1$. Since x must clearly be positive, we have $x = 2$. Returning to the expression in the problem, we have

$$1 + \cfrac{1}{1 + \cfrac{2}{1 + \cfrac{2}{1 + \cfrac{2}{1 + \cdots}}}} = 1 + \frac{1}{2} = \boxed{\frac{3}{2}}.$$

Exercises for Section 22.3

22.3.1 Each variable appears on the left side in four of the five equations. So, when we add all five equations, we have $4a + 4b + 4c + 4d + 4e = 18$. Dividing by 4 gives us $a + b + c + d + e = 9/2$. Subtracting $a + b + c + d = 9$ from this gives $e = -9/2$. Subtracting $a + b + c + e = -3$ from this gives $d = 15/2$. Continuing in this way with each of the other three equations gives $(a, b, c, d, e) = \boxed{\left(\frac{43}{2}, -\frac{21}{2}, -\frac{19}{2}, \frac{15}{2}, -\frac{9}{2}\right)}$.

22.3.2 Simplifying the expression we seek gives us $2a + 2b + 2c + 2d = 2(a + b + c + d) = 2(1111) = \boxed{2222}$.

22.3.3 Multiplying all three equations gives us

$$\frac{x}{y} \cdot \frac{y}{z} \cdot \frac{z}{w} = 3 \cdot 8 \cdot \frac{1}{2} \quad \Rightarrow \quad \frac{x}{w} = 12.$$

Taking the reciprocal of both sides of this equation gives $w/x = \boxed{1/12}$.

22.3.4 We can describe the problem with a system of five equations:

$$A + B + C + D + E = 60,$$
$$A + B = 24,$$
$$B + C = 15,$$
$$C + D = 18,$$
$$D + E = 30.$$

The symmetry of the last four equations gets us looking for a quick way to find A. We see that $B + C$ and $D + E$ together have every variable except A. If we use these with the first equation, we have

$$A + (B + C) + (D + E) = 60 \quad \Rightarrow \quad A + 15 + 30 = 60 \quad \Rightarrow \quad A = \boxed{15}.$$

Review Problems

22.8 The r and $\frac{4}{r}$ in the expression we seek are the squares of the terms on the left side of the equation we are given. So, we square the equation:

$$\left(\sqrt{r} + \frac{2}{\sqrt{r}} \right)^2 = 6^2 \quad \Rightarrow \quad r + 4 + \frac{4}{r} = 36 \quad \Rightarrow \quad r + \frac{4}{r} = \boxed{32}.$$

22.9 Letting $x = \sqrt{12 - \sqrt{12 - \sqrt{12 - \cdots}}}$, we have $x = \sqrt{12 - x}$. Therefore, $x^2 = 12 - x$, so $x^2 + x - 12 = 0$, or $(x + 4)(x - 3) = 0$. Clearly, x must be positive, so $x = \boxed{3}$.

22.10 Letting x equal the continued fraction, we see that x appears again in the denominator of the large fraction. So, we have $x = 3 + \frac{10}{x}$. Multiplying this equation by x gives $x^2 = 3x + 10$, or $x^2 - 3x - 10 = 0$. Factoring gives $(x - 5)(x + 2) = 0$. Since x is clearly positive, we have $x = \boxed{5}$.

22.11 The symmetry of the left sides of the equations suggests adding them all, which gives $2p + 2q + 2r + 2s = 52$, so $p + q + r + s = 26$. Subtracting $p + q + r - s = 32$ from this equation gives $2s = -6$, so $s = -3$. In much the same way, we subtract each of the other three given equations in turn from $p + q + r + s = 26$ to find $(p, q, r, s) = \boxed{\left(\frac{5}{2}, 20, \frac{13}{2}, -3 \right)}$.

22.12 Let $x = \sqrt{14 - 5\sqrt{3}} + \sqrt{14 + 5\sqrt{3}}$. Squaring this equation gives

$$x^2 = 14 - 5\sqrt{3} + 2\sqrt{14 - 5\sqrt{3}}\sqrt{14 + 5\sqrt{3}} + 14 + 5\sqrt{3} = 28 + 2\sqrt{(14 - 5\sqrt{3})(14 + 5\sqrt{3})}.$$

Since $(14 - 5\sqrt{3})(14 + 5\sqrt{3}) = 14^2 - (5\sqrt{3})^2 = 121$, we have

$$x^2 = 28 + 2\sqrt{121} = 28 + 22 = 50.$$

Since x must be positive, we have $x = \sqrt{50} = \boxed{5\sqrt{2}}$.

22.13 Suppose we label the numbers a, b, c, d, e, and f so that when the spouse of the person who chose a adds the five other numbers to $3a$, he or she gets 43. Similarly, when the spouse of the person who chose b adds the five other numbers to $3b$, he or she gets 51. We assign the other labels accordingly, so that the information in the problem gives us

$$3a + b + c + d + e + f = 43,$$
$$a + 3b + c + d + e + f = 51,$$
$$a + b + 3c + d + e + f = 61,$$
$$a + b + c + 3d + e + f = 32,$$
$$a + b + c + d + 3e + f = 81,$$
$$a + b + c + d + e + 3f = 52.$$

The symmetry suggests adding all six equations, which gives

$$8a + 8b + 8c + 8d + 8e + 8f = 320.$$

Therefore, $a + b + c + d + e + f = 40$. Subtracting this from $3a + b + c + d + e + f = 43$ gives $2a = 3$, so $a = 3/2$. Similarly, we subtract $a + b + c + d + e + f = 40$ from each equation in our system to determine that $(a, b, c, d, e, f) = \left(\frac{3}{2}, \frac{11}{2}, \frac{21}{2}, -4, \frac{41}{2}, 6\right)$. Therefore, the largest number chosen is $\boxed{41/2}$. (We can also write this answer as 20.5.)

22.14 From $a + 1 = a + b + c + d + 5$, we have $b + c + d + 4 = 0$. Similarly, $b + 2 = a + b + c + d + 5$ gives us $a + c + d + 3 = 0$, $c + 3 = a + b + c + d + 5$ gives $a + b + d + 2 = 0$, and $d + 4 = a + b + c + d + 5$ gives $a + b + c + 1 = 0$, so we have the system

$$b + c + d + 4 = 0,$$
$$a + c + d + 3 = 0,$$
$$a + b + d + 2 = 0,$$
$$a + b + c + 1 = 0.$$

The symmetry of the left sides suggests adding them all. We do so, and we find $3(a + b + c + d) + 10 = 0$, so $a + b + c + d = \boxed{-10/3}$.

22.15 We have symmetry on the left sides of the equations, since each variable is squared in one equation, cubed in another, and raised to the fourth power in yet another. Adding the three equations won't get us anywhere, but multiplying them will give us the same total power of each variable in the product:

$$(a^4 b^3 c^2)(a^3 b^2 c^4)(a^2 b^4 c^3) = (32)(8)(2) \quad \Rightarrow \quad (abc)^9 = 512.$$

Taking the ninth root of both sides gives $abc = \boxed{2}$.

22.16 To find the average of A, B, and C, we must find the sum of A, B, and C. While our two given equations do not have symmetric expressions, the $1001B$ and $1001C$ show us that adding them will make the coefficients of B and C the same. A quick look at the A terms show that adding the equations gives us $1001A$ on the left as well. So, adding the equations gives $1001(A + B + C) = 9009$. Therefore, $A + B + C = 9$, so the average of A, B, and C is $(A + B + C)/3 = \boxed{3}$.

22.17 Let x and y be the legs of the triangle. We are given $x + y = 18$ and $xy/2 = 37$. We wish to find $\sqrt{x^2 + y^2}$. We can find an expression for $x^2 + y^2$ by squaring $x + y = 18$, which gives

$$(x + y)^2 = 324 \quad \Rightarrow \quad x^2 + 2xy + y^2 = 324 \quad \Rightarrow \quad x^2 + y^2 = 324 - 2xy.$$

From $xy/2 = 37$, we have $xy = 74$. Therefore, $x^2 + y^2 = 324 - 2(74) = 176$. So, the length of the hypotenuse is $\sqrt{x^2 + y^2} = \sqrt{176} = \boxed{4\sqrt{11}}$.

Challenge Problems

22.18 The denominator of the continued fraction does not equal the whole continued fraction, because it begins with 2 rather than 1. So, we see that the denominator is 1 greater than the whole given expression. Therefore, if we let the expression equal x, we have

$$x = 1 + \cfrac{8}{2 + \cfrac{8}{2 + \cfrac{8}{2 + \cfrac{8}{2 + \cdots}}}} \quad \Rightarrow \quad x = 1 + \frac{8}{x + 1}.$$

Subtracting 1 from both sides gives $x - 1 = \frac{8}{x+1}$, then multiplying by $x + 1$ gives $x^2 - 1 = 8$, so $x^2 = 9$. Since the continued fraction is clearly positive, we have $x = \boxed{3}$.

22.19 We factor xy out of the first two terms of the second equation to give

$$xy(x + y) + x + y = 63.$$

Since we know that $xy = 6$, we have $6(x + y) + x + y = 63$, so $7(x + y) = 63$. Therefore, we have $x + y = 9$. We want $x^2 + y^2$, so we square $x + y = 9$ to find

$$(x + y)^2 = 9^2 \quad \Rightarrow \quad x^2 + 2xy + y^2 = 81 \quad \Rightarrow \quad x^2 + y^2 = 81 - 2xy = 81 - 2 \cdot 6 = \boxed{69}.$$

22.20

(a) Let r and s be the roots of the quadratic. The sum of the roots of $ax^2 + bx + c = 0$ is $-b/a$, so we have $r + s = \boxed{-4}$.

(b) The product of the roots of $ax^2 + bx + c = 0$ is c/a, so $rs = 1$. We want $r^2 + s^2$, so we square $r + s = -4$ to get $r^2 + 2rs + s^2 = 16$. Therefore, $r^2 + s^2 = 16 - 2rs = \boxed{14}$.

(c) We can either cube $r + s$ or multiply it by $r^2 + s^2$. The latter approach gives:

$$(r + s)(r^2 + s^2) = (-4)(14) \quad \Rightarrow \quad r^3 + rs^2 + r^2 s + s^3 = -56$$
$$\Rightarrow \quad r^3 + s^3 = -56 - r^2 s - rs^2 = -56 - rs(r + s) = \boxed{-52}.$$

22.21 The symmetry of the left sides suggests adding the equations, which gives

$$(n + 1)x + (n + 1)y + (n + 1)z = 3.$$

Factoring $n + 1$ out of the left side gives $(n + 1)(x + y + z) = 3$. Therefore, if $n = \boxed{-1}$, this equation is $0 = 3$, which has no solution.

22.22   Let x, y, and z be the dimensions of the box. The information about the areas of the sides of the box gives us

$$xy = 24,$$
$$yz = 32,$$
$$xz = 36.$$

The volume of the box is xyz. Since we seek xyz, the symmetry of the left sides of the equations in our system suggests multiplying the three equations, which gives $x^2 y^2 z^2 = (24)(32)(36)$. Since xyz must be positive, we have

$$xyz = \sqrt{(24)(32)(36)} = 2 \cdot 4 \cdot 6 \sqrt{(6)(2)(1)} = \boxed{96\sqrt{3}}.$$

22.23   *Solution 1: Factor $a^3 - b^3$.* We know that $a - b$ is a factor of $a^3 - b^3$, so we factor $a^3 - b^3$, which gives us

$$(a - b)(a^2 + ab + b^2) = 120 \quad \Rightarrow \quad 3(a^2 + ab + b^2) = 120 \quad \Rightarrow \quad a^2 + ab + b^2 = 40.$$

We don't know much about $a^2 + b^2$ or ab, but we can get an expression involving both by squaring $a - b = 3$. This gives us $a^2 - 2ab + b^2 = 9$. We seek ab, and subtracting $a^2 - 2ab + b^2 = 9$ from $a^2 + ab + b^2 = 40$ eliminates the $a^2 + b^2$, leaving $3ab = 31$. Therefore, $ab = \boxed{31/3}$.

Solution 2: Cube $a - b$. Since $(x + y)^3 = x^3 + 3x^2 y + 3xy^2 + y^3$, we have

$$(a - b)^3 = 3^3 \quad \Rightarrow \quad a^3 - 3a^2 b + 3ab^2 - b^3 = 27 \quad \Rightarrow \quad a^3 - b^3 - 3ab(a - b) = 27.$$

We are given $a^3 - b^3 = 120$ and $a - b = 3$, so this equation becomes

$$120 - 3ab(3) = 27 \quad \Rightarrow \quad ab = \boxed{31/3}.$$

22.24   We let $x + \frac{1}{x} = r$. We know $x^2 + \frac{1}{x^2}$, so we square our equation for r, which gives

$$x^2 + 2 + \frac{1}{x^2} = r^2 \quad \Rightarrow \quad r^2 = 2 + \left(x^2 + \frac{1}{x^2}\right) = 2 + 9 = 11.$$

Therefore, $r = \boxed{\pm\sqrt{11}}$.

22.25

(a) Because $(a + b)^3 = a^3 + 3a^2 b + 3ab^2 + b^3$, we have

$$\left(x + \frac{1}{x}\right)^3 = x^3 + 3x^2 \cdot \frac{1}{x} + 3x \cdot \frac{1}{x^2} + \frac{1}{x^3} = x^3 + 3x + \frac{3}{x} + \frac{1}{x^3}$$
$$= x^3 + \frac{1}{x^3} + 3\left(x + \frac{1}{x}\right).$$

Letting $A = x + \frac{1}{x}$ and $C = x^3 + \frac{1}{x^3}$, this equation is $A^3 = C + 3A$, or $C = A^3 - 3A$.

(b) We have $B = x^2 + \frac{1}{x^2}$, so squaring B will give us x^4 and $\frac{1}{x^4}$ terms:

$$B^2 = x^4 + 2 + \frac{1}{x^4} = D + 2.$$

Since $B = A^2 - 2$, we have $B^2 = A^4 - 4A^2 + 4$. Substituting this into $B^2 = D + 2$, we have $D = \boxed{A^4 - 4A^2 + 2}$.

(c) We have B, C, and D in terms of A, so if we can express E in terms of B, C, and D, we can write it in terms of A. We can get x^5 and $\frac{1}{x^5}$ by multiplying $x^4 + \frac{1}{x^4}$ by $x + \frac{1}{x}$:

$$DA = \left(x^4 + \frac{1}{x^4}\right)\left(x + \frac{1}{x}\right) = x^5 + \frac{1}{x^5} + x^3 + \frac{1}{x^3} = E + C.$$

Therefore, $E = DA - C = A^5 - 4A^3 + 2A - A^3 + 3A = \boxed{A^5 - 5A^3 + 5A}$.

(d) We can continue in the same vein as the last part by multiplying $x^{n-1} + \frac{1}{x^{n-1}}$ by $x + \frac{1}{x}$:

$$\left(x^{n-1} + \frac{1}{x^{n-1}}\right)\left(x + \frac{1}{x}\right) = x^n + \frac{1}{x^n} + x^{n-2} + \frac{1}{x^{n-2}}.$$

Therefore,

$$x^n + \frac{1}{x^n} = \left(x^{n-1} + \frac{1}{x^{n-1}}\right)\left(x + \frac{1}{x}\right) - x^{n-2} - \frac{1}{x^{n-2}}.$$

So, if we can write $x^{n-2} + \frac{1}{x^{n-2}}$ and $x^{n-1} + \frac{1}{x^{n-1}}$ in terms of $x + \frac{1}{x}$, then we have an expression for $x^n + \frac{1}{x^n}$ in terms of $x + \frac{1}{x}$. We already have expressions for $x^n + \frac{1}{x^n}$ for $n = 1, 2, 3, 4,$ and 5. (In the previous part, we used the expressions for $n = 3$ and $n = 4$ to make one for $n = 5$.) Since we have expressions for $n = 4$ and $n = 5$, we can make one for $n = 6$. Then we will have expressions for $n = 5$ and $n = 6$, so we can make one for $n = 7$, and so on. This approach is called induction, and we will study it more later in the Art of Problem Solving series.

22.26 The first denominator is not the same as our whole expression, but the second is! So, if we let our whole expression equal x, then we have

$$x = 3 + \cfrac{1}{2 + \cfrac{1}{3 + \cfrac{1}{2 + \cdots}}} = 3 + \cfrac{1}{2 + \frac{1}{x}}.$$

So, we have

$$x = 3 + \frac{1}{\frac{2x}{x} + \frac{1}{x}} = 3 + \frac{1}{\frac{2x+1}{x}} = 3 + \frac{x}{2x + 1}.$$

Multiplying both sides of $x = 3 + \frac{x}{2x+1}$ by $2x + 1$ gives $2x^2 + x = 3(2x + 1) + x = 7x + 3$, so $2x^2 - 6x - 3 = 0$. The quadratic formula then gives $x = \frac{6 \pm 2\sqrt{15}}{4} = \frac{3 \pm \sqrt{15}}{2}$. Since our continued fraction is clearly positive, we have $x = \boxed{\dfrac{3 + \sqrt{15}}{2}}$.

22.27 That 3 in the numerator of the fraction is pretty annoying, but the whole denominator is a continued fraction we know how to evaluate. Letting x equal this denominator, we have

$$x = 2 + \cfrac{1}{2 + \cfrac{1}{2 + \cfrac{1}{2 + \cdots}}} = 2 + \frac{1}{x}.$$

Therefore, $x = 2 + \frac{1}{x}$, so $x^2 = 2x + 1$, or $x^2 - 2x - 1 = 0$. The quadratic formula gives us $x = \frac{2 \pm \sqrt{8}}{2} = 1 \pm \sqrt{2}$. Since x must be positive, we have $x = 1 + \sqrt{2}$. Substituting this for the denominator of the original continued fraction, we have

$$2 + \cfrac{3}{2 + \cfrac{1}{2 + \cfrac{1}{2 + \cfrac{1}{2 + \cdots}}}} = 2 + \frac{3}{1 + \sqrt{2}} = 2 + \frac{3}{1 + \sqrt{2}} \cdot \frac{1 - \sqrt{2}}{1 - \sqrt{2}}$$

$$= 2 + \frac{3 - 3\sqrt{2}}{-1} = 2 - 3 + 3\sqrt{2} = \boxed{-1 + 3\sqrt{2}}.$$

22.28 If we let $x = \sqrt{a + \sqrt{b}} + \sqrt{a - \sqrt{b}}$, we can square both sides to have

$$x^2 = a + \sqrt{b} + 2\sqrt{a + \sqrt{b}}\sqrt{a - \sqrt{b}} + a - \sqrt{b} = 2a + 2\sqrt{(a + \sqrt{b})(a - \sqrt{b})} = 2a + 2\sqrt{a^2 - b}.$$

If x is the square root of an integer, then its square must be an integer. Therefore, $\sqrt{a^2 - b}$ must be an integer. So, $a^2 - b$ must be a perfect square, which is choice $\boxed{(B)}$.

22.29

(a) Let x equal the whole expression. We've already seen a couple values of b that give integer values of x. For example, if $b = 12$, we have

$$x = \sqrt{12 + x} \quad \Rightarrow \quad x^2 = 12 + x \quad \Rightarrow \quad (x - 4)(x + 3) = 0.$$

Since x must be positive, we have $x = 4$. We might continue to try trial and error, but we could also look at our solution method above to deduce other values of b that work. Specifically, we have

$$x = \sqrt{b + x} \quad \Rightarrow \quad x^2 = b + x \quad \Rightarrow \quad b = x^2 - x = x(x - 1).$$

Each of these steps is reversible, since x and $b + x$ are positive. So, if our expression is an integer, we see that we can write b as the product of two consecutive integers. Moreover, for any positive integer x greater than 1, if we let $b = x(x - 1)$, then the continued radical expression in the problem equals x, an integer. We can now quickly find values of b for which the expression is an integer. For $x = 2$, we have $b = 2$. For $x = 3, 4$, and 5, we find $b = 6, 12$, and 20, respectively.

(b) This was answered in the previous part.

(c) We just saw that $\sqrt{n + \sqrt{n + \sqrt{n + \cdots}}}$ is an integer only if n is the product of two consecutive integers. Since $4{,}000{,}000 = (2000)^2$, we know that $1999 \cdot 2000 < 4000000 < 2000 \cdot 2001$, so the largest acceptable value of n is $1999 \cdot 2000 = \boxed{3998000}$.

(d) $\boxed{\text{Yes}}$. Let $x = \sqrt{a + \sqrt{a + \sqrt{a + \sqrt{a + \cdots}}}}$. Then,

$$x = \sqrt{a + x} \quad\Rightarrow\quad x^2 = a + x \quad\Rightarrow\quad x^2 - x - a = 0.$$

So, since x is positive, we have $x = \frac{1 + \sqrt{1+4a}}{2}$. Now, let $y = \sqrt{a - \sqrt{a - \sqrt{a - \sqrt{a - \cdots}}}}$. Then,

$$y = \sqrt{a - y} \quad\Rightarrow\quad y^2 = a - y \quad\Rightarrow\quad y^2 + y - a = 0.$$

So, since y is positive, we have $y = \frac{-1 + \sqrt{1+4a}}{2} = x - 1$. Therefore, if x is an integer, so is y. (Extra challenge: How is our value of x in this problem related to our solutions to parts (a) and (b)?)

22.30 First, we simplify the left side. We have

$$x = \sqrt{a + \sqrt{a + \sqrt{a + \cdots}}} \quad\Rightarrow\quad x = \sqrt{a + x} \quad\Rightarrow\quad x^2 = a + x \quad\Rightarrow\quad x^2 - x - a = 0.$$

Therefore, $x = \frac{1 \pm \sqrt{1+4a}}{2}$. Since x must be positive, we have

$$x = \sqrt{a + \sqrt{a + \sqrt{a + \cdots}}} = \frac{1 + \sqrt{1 + 4a}}{2}.$$

Next, we evaluate the continued fraction:

$$x = 1 + \cfrac{a}{1 + \cfrac{a}{1 + \cfrac{a}{1 + \cdots}}} \quad\Rightarrow\quad x = 1 + \frac{a}{x} \quad\Rightarrow\quad x^2 = x + a \quad\Rightarrow\quad x^2 - x - a = 0.$$

That equation looks familiar. The quadratic formula gives $x = \frac{1 \pm \sqrt{1+4a}}{2}$. Since x must be positive, we have

$$x = 1 + \cfrac{a}{1 + \cfrac{a}{1 + \cfrac{a}{1 + \cdots}}} = \frac{1 + \sqrt{1 + 4a}}{2}.$$

This equals our expression for $\sqrt{a + \sqrt{a + \sqrt{a + \cdots}}}$, so we have

$$\sqrt{a + \sqrt{a + \sqrt{a + \cdots}}} = 1 + \cfrac{a}{1 + \cfrac{a}{1 + \cfrac{a}{1 + \cdots}}}.$$

22.31 We want xyz. The symmetric form of the equations suggests multiplying them or adding them. Let's try multiplying them first, since we know that will give us an xyz term. The product of the left sides of the equations is

$$\left(x + \frac{1}{y}\right)\left(y + \frac{1}{z}\right)\left(z + \frac{1}{x}\right) = \left(xy + \frac{x}{z} + 1 + \frac{1}{yz}\right)\left(z + \frac{1}{x}\right)$$

$$= xyz + y + x + \frac{1}{z} + z + \frac{1}{x} + \frac{1}{y} + \frac{1}{xyz}$$

$$= xyz + x + y + z + \frac{1}{x} + \frac{1}{y} + \frac{1}{z} + \frac{1}{xyz}.$$

So, the product of the three equations gives us

$$xyz + x + y + z + \frac{1}{x} + \frac{1}{y} + \frac{1}{z} + \frac{1}{xyz} = \frac{28}{3}.$$

Now, if we can evaluate $x + y + z + \frac{1}{x} + \frac{1}{y} + \frac{1}{z}$, we will have an equation we can solve for xyz. This sum is just the sum of the left sides of the given equations, so we add the three equations to give

$$x + y + z + \frac{1}{x} + \frac{1}{y} + \frac{1}{z} = \frac{22}{3}.$$

Therefore, we have

$$xyz + \frac{1}{xyz} = \frac{28}{3} - \left(x + y + z + \frac{1}{x} + \frac{1}{y} + \frac{1}{z} + \frac{1}{xyz} \right) = 2.$$

Letting $P = xyz$, we have $P + \frac{1}{P} = 2$, so $P^2 + 1 = 2P$, or $P^2 - 2P + 1 = 0$. Therefore, $(P - 1)^2 = 0$, so $P = \boxed{1}$.

22.32   Seeing that the expression starting from the second cube root is exactly the same as the original expression, we have

$$x = \sqrt[3]{3 - 2\sqrt[3]{3 - 2\sqrt[3]{3 - 2\sqrt[3]{3 - \cdots}}}} = \sqrt[3]{3 - 2x}.$$

Cubing both sides of $x = \sqrt[3]{3 - 2x}$ gives $x^3 = 3 - 2x$, or $x^3 + 2x = 3$. Clearly, $x = 1$ is a solution to this equation. To see that this is the only solution, we note that $x^3 + 2x > 1$ when $x > 1$ and $x^3 + 2x < 1$ when $x < 1$. Therefore, $\sqrt[3]{3 - 2\sqrt[3]{3 - 2\sqrt[3]{3 - 2\sqrt[3]{3 - \cdots}}}} = \boxed{1}$.

22.33   We let $x = \sqrt[3]{18 + 5\sqrt{13}} + \sqrt[3]{18 - 5\sqrt{13}}$. Once again, the fact that the numbers inside the radicals are conjugates gives us some confidence that raising this equation to a power will simplify it. However, those cube roots tell us that squaring won't do. We'll have to cube both sides:

$$x^3 = \left(\sqrt[3]{18 + 5\sqrt{13}} + \sqrt[3]{18 - 5\sqrt{13}} \right)^3.$$

To expand that right side, we have to cube a binomial. We expand $(a + b)^3$ to use as a model:

$$(a + b)^3 = (a + b)(a + b)^2 = (a + b)(a^2 + 2ab + b^2) = a^3 + 3a^2b + 3ab^2 + b^3.$$

With this as a guide, we cube our nasty expression:

$$\begin{aligned} x^3 &= \left(\sqrt[3]{18 + 5\sqrt{13}} \right)^3 + 3 \left(\sqrt[3]{18 + 5\sqrt{13}} \right)^2 \left(\sqrt[3]{18 - 5\sqrt{13}} \right) \\ &\quad + 3 \left(\sqrt[3]{18 + 5\sqrt{13}} \right) \left(\sqrt[3]{18 - 5\sqrt{13}} \right)^2 + \left(\sqrt[3]{18 - 5\sqrt{13}} \right)^3. \end{aligned}$$

The first and last terms are easy to deal with. It's the other two that give us problems. The conjugates give us an idea, though. Instead of evaluating

$$3 \left(\sqrt[3]{18 + 5\sqrt{13}} \right)^2 \left(\sqrt[3]{18 - 5\sqrt{13}} \right)$$

as

$$3\sqrt[3]{\left(18+5\sqrt{13}\right)^2}\left(\sqrt[3]{18-5\sqrt{13}}\right),$$

we take advantage of the conjugates:

$$3\left(\sqrt[3]{18+5\sqrt{13}}\right)^2\left(\sqrt[3]{18-5\sqrt{13}}\right)=3\sqrt[3]{\left(18+5\sqrt{13}\right)\left(18+5\sqrt{13}\right)\left(18-5\sqrt{13}\right)}$$
$$=3\sqrt[3]{\left(18+5\sqrt{13}\right)\left[\left(18+5\sqrt{13}\right)\left(18-5\sqrt{13}\right)\right]}$$
$$=3\sqrt[3]{\left(18+5\sqrt{13}\right)\left[324-325\right]}$$
$$=-3\sqrt[3]{18+5\sqrt{13}}$$

Similarly, we have

$$3\left(\sqrt[3]{18+5\sqrt{13}}\right)\left(\sqrt[3]{18-5\sqrt{13}}\right)^2=-3\sqrt[3]{18-5\sqrt{13}}.$$

Now, our equation is

$$x^3=(18+5\sqrt{13})-3\sqrt[3]{18+5\sqrt{13}}-3\sqrt[3]{18-5\sqrt{13}}+(18-5\sqrt{13})$$
$$=36-3\sqrt[3]{18+5\sqrt{13}}-3\sqrt[3]{18-5\sqrt{13}}$$

Oh no! We didn't get rid of all the radicals. Before we cube again, we notice that the radicals in our expression look familiar. We recall that

$$x=\sqrt[3]{18+5\sqrt{13}}+\sqrt[3]{18-5\sqrt{13}},$$

so we can write the radicals in our equation for x^3 simply in terms of x:

$$x^3=36-3\left(\sqrt[3]{18+5\sqrt{13}}+\sqrt[3]{18-5\sqrt{13}}\right)=36-3x.$$

Our equation now is just $x^3=36-3x$, or $x^3+3x=36$. Clearly, $x=3$ is one solution to this equation. If $x>3$, then we have $x^3+3x>3^3+9=36$, so there are no solutions to $x^3+3x=36$ for which $x>3$. Similarly, if $x<3$, then $x^3+3x<36$, so there are no solutions to the equation that are less than 3. Therefore, we have

$$\sqrt[3]{18+5\sqrt{13}}+\sqrt[3]{18-5\sqrt{13}}=\boxed{3}.$$

That's a pretty long way to say "three!"

22.34 Letting the number of dollars spent by Ashley, Betty, Carlos, Dick, and Elgin be A, B, C, D, and E, respectively, we have the system

$$A+B+C+D+E=56,$$
$$|A-B|=19,$$
$$|B-C|=7,$$
$$|C-D|=5,$$
$$|D-E|=4,$$
$$|E-A|=11.$$

Dealing with those absolute value signs is tough. One way to handle them is to notice that $|A - B| = 19$ means that $A - B = 19$ or $A - B = -19$. For now, we write this as $A - B = \pm 19$, since doing this with all the absolute value equations makes it easier to combine equations by adding them. (The symmetry of the expressions inside the absolute value signs makes us think of adding the equations.) So, we can write our absolute value equations as

$$A - B = \pm 19,$$
$$B - C = \pm 7,$$
$$C - D = \pm 5,$$
$$D - E = \pm 4,$$
$$E - A = \pm 11.$$

Now, we note that adding all the left sides gives us 0, so adding all these equations gives

$$0 = \pm 19 \pm 7 \pm 5 \pm 4 \pm 11.$$

So, now we must choose $+$ or $-$ for each $\pm$ to make this equation true. The only ways that work are

$$0 = 19 - 7 - 5 + 4 - 11 = -19 + 7 + 5 - 4 + 11.$$

(We see this by noting that $19 + 7 + 5 + 4 + 11 = 46$, so we are breaking the five numbers into two sets that add to 23.) Taking the first case, we have the system

$$A + B + C + D + E = 56, \ A - B = 19, \ B - C = -7, \ C - D = -5, \ D - E = 4, \ E - A = -11.$$

Solving for each of A, B, C, and D in terms of E, we have:

$$A = E + 11, \ B = A - 19 = E - 8, \ D = E + 4, \ C = D - 5 = E - 1.$$

Substituting these in $A + B + C + D + E = 56$ gives

$$E + 11 + E - 8 + E - 1 + E + 4 + E = 56 \quad \Rightarrow \quad 5E = 50 \quad \Rightarrow \quad E = \boxed{10}.$$

If we try the other case, we have

$$A + B + C + D + E = 56, \ A - B = -19, \ B - C = 7, \ C - D = 5, \ D - E = -4, \ E - A = 11.$$

Solving this system for E as before yields a non-integer for E, so $E = 10$ is the only possible answer.

22.35 We have

$$\sqrt{104\sqrt{6} + 468\sqrt{10} + 144\sqrt{15} + 2006} = a\sqrt{2} + b\sqrt{3} + c\sqrt{5}.$$

We square both sides to get rid of some radicals. Squaring the left is easy. On the right, we have

$$(a\sqrt{2} + b\sqrt{3} + c\sqrt{5})^2$$
$$= (a\sqrt{2} + b\sqrt{3} + c\sqrt{5})(a\sqrt{2} + b\sqrt{3} + c\sqrt{5})$$
$$= (a\sqrt{2} + b\sqrt{3} + c\sqrt{5})(a\sqrt{2}) + (a\sqrt{2} + b\sqrt{3} + c\sqrt{5})(b\sqrt{3}) + (a\sqrt{2} + b\sqrt{3} + c\sqrt{5})(c\sqrt{5})$$
$$= 2a^2 + ab\sqrt{6} + ac\sqrt{10} + ab\sqrt{6} + 3b^2 + bc\sqrt{15} + ac\sqrt{10} + bc\sqrt{15} + 5c^2$$
$$= 2a^2 + 3b^2 + 5c^2 + 2ab\sqrt{6} + 2ac\sqrt{10} + 2bc\sqrt{15}.$$

Setting this equal to $2006 + 104\sqrt{6} + 468\sqrt{10} + 144\sqrt{15}$, we have $2ab = 104$, $2ac = 468$, and $2bc = 144$ from the coefficients of $\sqrt{6}$, $\sqrt{10}$, and $\sqrt{15}$, respectively. Dividing each of these equations by 2 gives $ab = 52$, $ac = 234$, and $bc = 72$. We want abc, so we take the product of all three of these equations, to get

$$(ab)(ac)(bc) = (52)(234)(72) = (2^2 \cdot 13)(2 \cdot 3^2 \cdot 13)(2^3 \cdot 3^2) = 2^6 \cdot 3^4 \cdot 13^2.$$

Therefore, $(abc)^2 = (2^3 \cdot 3^2 \cdot 13)^2$, so $abc = 8 \cdot 9 \cdot 13 = (72)(13) = \boxed{936}$.

22.36 The expression we seek is part of the equation we already have, so we isolate that expression in the equation by moving all the terms with even denominators from the left to the right:

$$\frac{1}{1^2} + \frac{1}{3^2} + \frac{1}{5^2} + \cdots = \frac{\pi^2}{6} - \left(\frac{1}{2^2} + \frac{1}{4^2} + \frac{1}{6^2} + \cdots\right).$$

If we can find the sum of the terms with even denominators, then we can find the sum of the terms with odd denominators. The even denominator terms have a common factor, $\frac{1}{4}$, so we factor that out, hoping to be able to deal with the resulting sum:

$$\begin{aligned}
\frac{1}{1^2} + \frac{1}{3^2} + \frac{1}{5^2} + \cdots &= \frac{\pi^2}{6} - \left(\frac{1}{2^2} + \frac{1}{4^2} + \frac{1}{6^2} + \cdots\right) \\
&= \frac{\pi^2}{6} - \left(\frac{1}{2^2 \cdot 1^2} + \frac{1}{2^2 \cdot 2^2} + \frac{1}{2^2 \cdot 3^2} + \cdots\right) \\
&= \frac{\pi^2}{6} - \frac{1}{4}\left(\frac{1}{1^2} + \frac{1}{2^2} + \frac{1}{3^2} + \cdots\right)
\end{aligned}$$

Success! We are given that the sum in parentheses on the right hand side equals $\pi^2/6$, so we now have

$$\frac{1}{1^2} + \frac{1}{3^2} + \frac{1}{5^2} + \cdots = \frac{\pi^2}{6} - \frac{1}{4}\left(\frac{\pi^2}{6}\right) = \frac{3}{4}\left(\frac{\pi^2}{6}\right) = \boxed{\frac{\pi^2}{8}}.$$

www.artofproblemsolving.com

The Art of Problem Solving (AoPS) is:

- ## Books

  For over 25 years, *the Art of Problem Solving* books have been used by students as a resource for the American Mathematics Competitions and other national and local math events.

  > *Every school should have this in their math library.*
  > – Paul Zeitz, past coach of the U.S. International Mathematical Olympiad team

  Visit our site to learn about our textbooks, which form a full math curriculum for high-performing students in grades 6-12.

- ## Classes

  The Art of Problem Solving offers online classes on topics such as number theory, counting, geometry, algebra, and more at beginning, intermediate, and Olympiad levels.

  > *All the children were very engaged. It's the best use of technology I have ever seen.*
  > – Mary Fay-Zenk, coach of National Champion California MATHCOUNTS teams

- ## Forum

  As of April 2019, the Art of Problem Solving Forum has over 395,000 members who have posted over 8,100,000 messages on our discussion board. Members can also join any of our free "Math Jams".

  > *I'd just like to thank the coordinators of this site for taking the time to set it up... I think this is a great site, and I bet just about anyone else here would say the same...*
  > – AoPS Community Member

- ## Resources

  We have links to summer programs, book resources, problem sources, national and local competitions, and a LaTeX tutorial.

  > *I'd like to commend you on your wonderful site. It's informative, welcoming, and supportive of the math community. I wish it had been around when I was growing up.*
  > – AoPS Community Member

- ## ...and more!

Membership is **FREE**! Come join the Art of Problem Solving community today!